Pestman / Albe

Wiebe R. Pestman
Ivo B. Alberink

Mathematical Statistics

Problems and Detailed Solutions

Walter de Gruyter
Berlin · New York 1998

Authors
Wiebe R. Pestman
Ivo B. Alberink
University of Nijmegen
NL-6525 ED Nijmegen
The Netherlands

1991 Mathematics Subject Classification: 62-01
Keywords: Estimation theory, hypothesis testing, regression analysis, non-parametrics, stochastic analysis, vectorial (multivariate) statistics

Printed on acid-free paper which falls within the guidelines of the ANSI to ensure permanence and durability.

Library of Congress – Cataloging-in-Publication Data

Pestman, Wiebe R., 1954–
Mathematical statistics : problems and detailed solutions / Wiebe R. Pestman, Ivo B. Alberink.
p. cm. – (De Gruyter textbook)
ISBN 3-11-015359-9 (hardcover : alk. paper). – ISBN 3-11-015358-0 (pbk. : alk. paper)
1. Mathematical statistics–Problems, exercises, etc.
I. Alberink, Ivo B., 1971. II. Title. III. Series.
QA276.2.P47 1998
519.5'076–dc21 98-20084
CIP

Die Deutsche Bibliothek – Cataloging-in-Publication Data

Pestman, Wiebe R.:
Mathematical statistics - problems and detailed solutions / Wiebe R. Pestman ; Ivo B. Alberink. - Berlin ; New York : de Gruyter, 1998
(De Gruyter textbook)
Erg. zu: Pestman, Wiebe R.: Mathematical statistics - an introduction
ISBN 3-11-015358-0 brosch.
ISBN 3-11-015359-9 Gb.

Printed in Germany.
Typesetting using the authors' TEX files: I. Zimmermann, Freiburg.
Printing and Binding: WB-Druck GmbH & Co., Rieden/Allgäu.

Preface

The book that is now in front of you is a presentation of detailed solutions to various exercises (some 260) in mathematical statistics. It is a companion volume and solutions manual for the textbook *Mathematical Statistics – An Introduction* (Walter de Gruyter, Berlin – New York, 1998) by the second author.

Though an exercise book, we tried to keep it as self-contained as was humanly possible. Each chapter starts with a summary of the corresponding chapter in the textbook, followed by formulations of and solutions to the corresponding exercises. In this way the book is very suitable for self-study.

If you find any mistakes or typing errors, please let us know. Of course suggestions on nice, useful or interesting additional problems are welcome too.

Thanks go to Alessandro di Bucchianico and Mark van de Wiel for computing most of the statistical tables.

Nijmegen, March 1998

Ivo B. Alberink
Wiebe R. Pestman

Table of contents

Chapter 1. Probability theory

Chapter 2. Statistics and their probability distributions, estimation theory

Chapter 3. Hypothesis testing

Chapter 4. Simple regression analysis

Chapter 5. Normal analysis of variance

Chapter 6. Non-parametric methods

Chapter 7. Stochastic analysis and its applications in statistics

Chapter 8. Vectorial variables

Chapter 1
Probability Theory

1.1 Summary of Chapter I

1.1.1 Probability spaces (I.1)

We are dealing with experiments in which randomness is playing a role. The set of all possible outcomes of the experiment is denoted by Ω, the *sample space.* By definition an *event* is a subset of the sample space. A σ*-algebra* of subsets of Ω is understood to be a collection $\mathfrak{A}$ of subsets, enjoying the following properties:

(i) $\Omega \in \mathfrak{A}$,

(ii) If $A \in \mathfrak{A}$ then $A^c \in \mathfrak{A}$,

(iii) If $A_1, A_2, \ldots \in \mathfrak{A}$ then $\bigcup_{i=1}^{\infty} A_i \in \mathfrak{A}$.

A *measure* on such a σ-algebra is a map $\mu : \mathfrak{A} \to [0, +\infty]$ that has the following properties:

(i) For every countable family $A_1, A_2, \ldots$ of mutually disjoint elements of $\mathfrak{A}$ one has

$$\mu\left(\bigcup_{i=1}^{\infty} A_i\right) = \sum_{i=1}^{\infty} \mu(A_i),$$

(ii) $\mu(\emptyset) = 0$.

The smallest σ-algebra on $\mathbb{R}^n$ containing all open sets will be denoted by $\mathfrak{B}^n$. The elements of $\mathfrak{B}^n$ are called *Borel sets* and a measure on $\mathfrak{B}^n$ is said to be a *Borel measure.* A fundamental result in measure theory is the existence of a unique Borel measure λ on $\mathfrak{B}^n$ such that $\lambda([0,1]^n) = 1$ and $\lambda(A + \mathbf{a}) = \lambda(A)$ for all $A \in \mathfrak{B}^n$, $\mathbf{a} \in \mathbb{R}^n$: this measure is called the *Lebesgue measure on* $\mathbb{R}^n$.

A function $f : \mathbb{R}^n \to \mathbb{R}^m$ is said to be a *Borel function* if for all open sets $O \subset \mathbb{R}^m$ the set $f^{-1}(O)$ is a Borel set in $\mathbb{R}^n$. Every continuous function is a Borel function.

The rest of §I.1 is a summary of the elements of measure theory. Important concepts are that of an *integral* and that of a Borel measure having a *density with respect to the Lebesgue measure.*

A measure $\mathbb{P}$ on $(\Omega, \mathfrak{A})$ is said to be a *probability measure* if $\mathbb{P}(\Omega) = 1$. A *probability space* is understood to be an ordered triplet $(\Omega, \mathfrak{A}, \mathbb{P})$, where Ω is a sample space, $\mathfrak{A}$ a σ-algebra of subsets of Ω and $\mathbb{P}$ a probability measure on Ω (we ought to say 'on $\mathfrak{A}$').

1.1.2 Stochastic variables (I.2)

Given a probability space $(\Omega, \mathfrak{A}, \mathbb{P})$, a function $\mathbf{X} : \Omega \to \mathbb{R}^n$ is said to be $\mathfrak{A}$-*measurable* if $\mathbf{X}^{-1}(A) \in \mathfrak{A}$ for all $A \in \mathfrak{B}^n$. In statistics and probability theory such a function is usually called a *stochastic n-vector.* In the case where $n = 1$ we shall be speaking about a *stochastic variable.*

To every stochastic n-vector there is in a canonical way an associated Borel measure $\mathbb{P}_{\mathbf{X}}$ on $\mathbb{R}^n$; it is defined by $\mathbb{P}_{\mathbf{X}}(A) := \mathbb{P}\left[\mathbf{X}^{-1}(A)\right]$ for every A that is Borel in $\mathbb{R}^n$. $\mathbb{P}_{\mathbf{X}}$ is called the *probability distribution of* $\mathbf{X}$.

If the range of $\mathbf{X}$ is countable, then $\mathbf{X}$ is said to be *discretely distributed.* Examples of discrete probability distributions are the binomial, the geometrical and the Poisson distribution. A stochastic n-vector $\mathbf{X}$ is said to enjoy an *absolutely continuous distribution* if there exists a Borel function $f : \mathbb{R}^n \to [0, +\infty)$ such that $\mathbb{P}_{\mathbf{X}} = f\lambda$, that is to say $\mathbb{P}_{\mathbf{X}}(A) = \int 1_A f \, d\lambda$ for all A that are Borel in $\mathfrak{B}^n$. Here the function f is called the *probability density of* $\mathbf{X}$.

In measure theory $\mathbb{P}_{\mathbf{X}}$ is also called the *image of* $\mathbb{P}$ *under* $\mathbf{X}$. If $X_1, \dots, X_n$ are real-valued stochastic variables, then we can form the stochastic n-vector $\mathbf{X}$ by setting

$$\mathbf{X}(\omega) := (X_1(\omega), \dots, X_n(\omega)) \qquad (\omega \in \Omega).$$

We shall frequently denote $\mathbb{P}_{X_1,\dots,X_n}$ instead of $\mathbb{P}_{\mathbf{X}}$.

1.1.3 Product measures and statistical independence (I.3)

A probability measure $\mathbb{P}$ on $\mathbb{R}^n$ is called the *product measure* of the probability measures $\mathbb{P}_1, \dots, \mathbb{P}_n$ on $\mathbb{R}$ if for all $A_1, \dots, A_n \in \mathfrak{B}$ one has:

$$\mathbb{P}(A_1 \times \dots \times A_n) = \mathbb{P}_1(A_1) \cdots \mathbb{P}_n(A_n).$$

We indicate this by writing $\mathbb{P} = \mathbb{P}_1 \otimes \dots \otimes \mathbb{P}_n$.

The stochastic variables $X_1, X_2, \dots, X_n$ are said to be *statistically independent* if

$$\mathbb{P}_{(X_1,\dots,X_n)} = \mathbb{P}_{X_1} \otimes \dots \otimes \mathbb{P}_{X_n}.$$

Events $B_1, \dots, B_n$ are called independent if $1_{B_1}, \dots, 1_{B_n}$ are independent variables. Statistical independence of stochastic vectors is defined in exactly the same way.

The *tensor product* of a couple of functions $f_1, \dots, f_n : \mathbb{R} \to \mathbb{R}$ is the function $f : \mathbb{R}^n \to \mathbb{R}$ defined by $f(x_1, \dots, x_n) := f(x_1) \cdots f(x_n)$; we denote $f = f_1 \otimes \cdots \otimes f_n$. If $X_1, \dots, X_n$ are stochastic variables having densities $f_1, \dots, f_n$, then the system $X_1, \dots, X_n$ is statistically independent if and only if the stochastic n-vector $(X_1, \dots, X_n)$ has as its density the function $f = f_1 \otimes \cdots \otimes f_n$.

1.1.4 Functions of stochastic vectors (I.4)

If $\mathbf{X}$ is a stochastic m-vector and $f : \mathbb{R}^m \to \mathbb{R}^p$ a Borel function, then we can form the composition $f(\mathbf{X})$, being a stochastic p-vector. If $\mathbf{X}$ and $\mathbf{Y}$ are statistically independent stochastic vectors and f, g are Borel functions, then the stochastic vectors $f(\mathbf{X})$ and $g(\mathbf{Y})$ are automatically statistically independent.

If $\varphi : \mathbb{R}^n \to \mathbb{R}$ and $f : \mathbb{R}^m \to \mathbb{R}^n$ are Borel, then for every stochastic m-vector $\mathbf{X}$ we have (Theorem I.4.3)

$$\int \varphi \, d\mathbb{P}_{f(\mathbf{X})} = \int \varphi \circ f \, d\mathbb{P}_{\mathbf{X}}.$$

1.1.5 Expectation value, variance and covariance of stochastic variables (I.5)

Given a stochastic m-vector $\mathbf{X}$ and a Borel function $g : \mathbb{R}^m \to \mathbb{R}$ and provided that $\int |g| \, d\mathbb{P}_{\mathbf{X}} < +\infty$, the *expectation value* (or *mean*) of $g(\mathbf{X})$ is defined by

$$\mathbb{E}\,[g(\mathbf{X})] := \int g \, d\mathbb{P}_{\mathbf{X}}.$$

So for every stochastic variable X we have $\mathbb{E}(X) = \int x \, d\mathbb{P}_X(x)$, provided $\int |x| \, d\mathbb{P}_X(x) < +\infty$.

If $\mathbf{X}$ is discretely distributed with range $W = \{\mathbf{a}_1, \mathbf{a}_2, \dots\}$, then we have

$$\mathbb{E}\,[g(\mathbf{X})] = \sum_{\mathbf{a} \in W} g(\mathbf{a}) \, \mathbb{P}(\mathbf{X} = \mathbf{a}),$$

provided that $\sum_{\mathbf{a} \in W} |g(\mathbf{a})| \, \mathbb{P}(\mathbf{X} = \mathbf{a}) < +\infty$. If $\mathbf{X}$ enjoys an absolutely continuous distribution with density function f, then

$$\mathbb{E}\,[g(\mathbf{X})] = \int g(\mathbf{x}) f(\mathbf{x}) \, d\mathbf{x},$$

provided that $\int |g(\mathbf{x})| f(\mathbf{x}) \, d\mathbf{x} < +\infty$.

The action of taking the expectation value of stochastic variables is a linear operator, that is to say:

$$\mathbb{E}(aX) = a\,\mathbb{E}(X) \quad \text{and} \quad \mathbb{E}(X+Y) = \mathbb{E}(X) + \mathbb{E}(Y)$$

whenever $a \in \mathbb{R}$ and $\mathbb{E}(X)$ and $\mathbb{E}(Y)$ exist. If X and Y are statistically independent we have moreover that $\mathbb{E}(XY) = \mathbb{E}(X)\,\mathbb{E}(Y)$.

If $\mathbb{E}(X^2)$ exists, then automatically also $\mathbb{E}(X)$ exists; we then define the *variance* of X by

$$\operatorname{var}(X) := \mathbb{E}(X^2) - \mathbb{E}(X)^2.$$

We have $\operatorname{var}(X) = \mathbb{E}\left[(X - \mathbb{E}(X))^2\right]$; it follows that always $\operatorname{var}(X) \geq 0$. If $\operatorname{var}(X) = 0$ then the variable X will show with probability 1 the constant value $\mu = \mathbb{E}(X)$ as its outcome.

The *covariance* of a pair of stochastic variables X and Y is defined by

$$\operatorname{cov}(X, Y) := \mathbb{E}(XY) - \mathbb{E}(X)\mathbb{E}(Y),$$

provided $\mathbb{E}(X^2)$ and $\mathbb{E}(Y^2)$ exist. It is a direct consequence of this definition that $\operatorname{cov}(X, X) = \operatorname{var}(X)$ and $\operatorname{cov}(X, Y) = 0$ if X and Y are independent. Furthermore, writing $\mu_X := \mathbb{E}(X)$ and $\mu_Y := \mathbb{E}(Y)$, we have $\operatorname{cov}(X, Y) = \mathbb{E}\left[(X - \mu_X)(Y - \mu_Y)\right]$. The map $(X, Y) \mapsto \operatorname{cov}(X, Y)$ defines a semi inner product.

If $X_1, \ldots, X_n$ are stochastic variables satisfying $\operatorname{cov}(X_i, X_j) = 0$ for all $i \neq j$, then

$$\operatorname{var}(X_1 + \cdots + X_n) = \sum_{i=1}^{n} \operatorname{var}(X_i).$$

Covariance is invariant under translations, that is to say, for all $a, b \in \mathbb{R}$ we have:

$$\operatorname{cov}(X + a, Y + b) = \operatorname{cov}(X, Y).$$

The *standard deviation* of a stochastic variable X is defined by

$$\sigma_X := \sqrt{\operatorname{var}(X)},$$

provided $\mathbb{E}(X^2)$ exists. The *correlation coefficient* of two variables X and Y is given by

$$\rho(X, Y) := \frac{\operatorname{cov}(X, Y)}{\sigma_X\, \sigma_Y},$$

provided this expression exists. If so, then we have $-1 \leq \rho(X, Y) \leq 1$. The values ± 1 are taken on if and only if the variables X and Y are linked by a relation of type $aX + bY = c$, where a, b, c are constants.

1.1.6 Independent normally distributed stochastic variables (I.6)

A stochastic variable X is said to be *normally distributed with parameters μ and σ^2* (where $\sigma > 0$) if it enjoys an absolutely continuous distribution with density

$$f(x) = \frac{1}{\sqrt{2\pi}\,\sigma} \exp\left[-\frac{1}{2}\left(\frac{x - \mu}{\sigma}\right)^2\right].$$

We often indicate this by saying that X is $N(\mu, \sigma^2)$-distributed. If so, then $\mu_X = \mathbb{E}(X) = \mu$ and $\sigma_X^2 = \text{var}(X) = \sigma^2$. If X is $N(\mu, \sigma^2)$-distributed, then $pX + q$ (where $p \neq 0$) is $N(p\mu + q,\ p^2\sigma^2)$-distributed (Theorem I.6.1). In particular: the variable $(X - \mu)/\sigma$ enjoys a $N(0, 1)$-distribution, also called the *standard normal distribution.* In general we say that $(X - \mu_X)/\sigma_X$ is the *standardized* of a variable X.

From now on we assume that $X_1, \dots, X_n$ are statistically independent variables, all of them with a common $N(0, \sigma^2)$-distribution. Then $\mathbb{P}_{X_1,\dots,X_n}$ has a density given by

$$f(\mathbf{x}) = \frac{1}{\sigma^n\,(2\pi)^{n/2}} \exp\left(-\frac{1}{2}\frac{\langle\mathbf{x}, \mathbf{x}\rangle}{\sigma^2}\right) \qquad (\mathbf{x} \in \mathbb{R}^n).$$

A probability distribution $\mathbb{P}_\mathbf{Y}$ of a stochastic n-vector $\mathbf{Y}$ is called *rotation invariant* if for every orthogonal linear operator $\mathbf{Q} : \mathbb{R}^n \to \mathbb{R}^n$ we have that $\mathbb{P}_\mathbf{Y} = \mathbb{P}_{\mathbf{QY}}$. (Equivalently: $\mathbb{P}_\mathbf{Y}(A) = \mathbb{P}_\mathbf{Y}(\mathbf{Q}A)$ for all $A \in \mathfrak{B}^n$ and all orthogonal linear operators $\mathbf{Q}$). Under these conditions $\mathbb{P}_{X_1,\dots,X_n}$ is rotation invariant.

Next, suppose that $X_1, \dots, X_n$ is a statistically independent system of variables and suppose that X_i is $N(\mu_i, {\sigma_i}^2)$-distributed for $i = 1, 2, \dots, n$. Then (Theorem I.6.6) the sum $S := X_1 + \cdots + X_n$ is $N(\mu_1 + \cdots + \mu_n,\ \sigma_1^2 + \cdots + \sigma_n^2)$-distributed. Let

$$\mathfrak{V} := \left\{ \sum_{i=1}^{n} c_i X_i \ : \ c_i \in \mathbb{R} \ \ (\text{all } i) \right\}$$

be the linear span of the variables $X_1, \dots, X_n$. A very useful criterion for statistical independence is Theorem I.6.7. It states: if $M_1, \dots, M_p, N_1, \dots, N_q \in \mathfrak{V}$ and if for all possible i and j we have $\text{cov}(M_i, N_j) = 0$, then the stochastic vectors $(M_1, \dots, M_p)$ and $(N_1, \dots, N_q)$ are statistically independent.

In particular, if $M, N \in \mathfrak{V}$, then:

$$\text{cov}(M, N) = 0 \qquad \Longleftrightarrow \qquad M \text{ and } N \text{ are statistically independent.}$$

1.1.7 Distribution functions and probability distributions (I.7)

The *distribution function* $F_X : \mathbb{R} \to [0, 1]$ belonging to a stochastic variable X is defined by

$$F_X(x) := \mathbb{P}\,[X \leq x] = \mathbb{P}_X\,[(-\infty, x]].$$

Such functions always have the following properties:

(i) F_X is increasing and right-continuous;

(ii) $\lim_{x\to-\infty} F_X(x) = 0$ and $\lim_{x\to+\infty} F_X(x) = 1$.

Conversely, every function F satisfying these conditions is of the form $F = F_X$ where X is some stochastic variable. If X enjoys an absolutely continuous distribution with probability density f, then for all $x \in \mathbb{R}$ we have

$$F_X(x) = \int_{-\infty}^{x} f(t)\,dt.$$

If f is continuous in x_0, then $f(x_0) = F'_X(x_0)$.

For a stochastic n-vector $\mathbf{X} = (X_1, \ldots, X_n)$ the so-called *joint distribution function* $F_{\mathbf{X}} : \mathbb{R}^n \to [0, 1]$ is defined by

$$F_{\mathbf{X}}(x_1, \ldots, x_n) := \mathbb{P}\left(X_1 \leq x_1, \ldots, X_n \leq x_n\right).$$

If $F_{\mathbf{X}}$ has continuous partial derivatives up to order n then $\mathbf{X}$ has a probability density $f_{\mathbf{X}}$ given by

$$f_{\mathbf{X}}(\mathbf{x}) = \frac{\partial^n}{\partial x_1 \ldots \partial x_n} F_{\mathbf{X}}(\mathbf{x}) \qquad (\mathbf{x} \in \mathbb{R}^n).$$

The components $X_1, \ldots, X_n$ of $\mathbf{X}$ form a statistically independent system if and only if

$$F_{\mathbf{X}}(x_1, \ldots, x_n) = F_{X_1}(x_1) \cdots F_{X_n}(x_n) \quad \text{for all } \mathbf{x} \in \mathbb{R}^n.$$

1.1.8 Moments, moment generating functions and characteristic functions (I.8)

For every stochastic variable X the nth *moment*, provided it exists, is defined by $\mu_n := \mathbb{E}(X^n)$.

The function $t \mapsto \mathbb{E}(e^{tX}) =: M_X(t)$ is said to be the *moment generating function of* X. Its domain consists of all t for which the expression $\mathbb{E}(e^{tX})$ makes sense. If for some $\varepsilon > 0$ the expression $\mathbb{E}(e^{\varepsilon|X|})$ exists, then all moments of X exist and (Theorem I.8.1) M_X is on the interval $(-\varepsilon, \varepsilon)$ given by

$$M_X(t) = \sum_{n=0}^{\infty} \frac{t^n\,\mathbb{E}(X^n)}{n!}.$$

We then have:

$$\mathbb{E}(X^n) = \left[M_X^{(n)}(t)\right]_{t=0}.$$

Furthermore, for two variables X and Y the following four statements are under mild conditions equivalent:

(i) $\mathbb{P}_X = \mathbb{P}_Y$,

(ii) $F_X = F_Y$,

(iii) $M_X = M_Y$ on some interval $(-\varepsilon, +\varepsilon)$,

(iv) $\mathbb{E}(X^n) = \mathbb{E}(Y^n)$ for all $n \in \mathbb{N}$.

We say that a sequence of stochastic variables $X_1, X_2, \ldots$ *converges in distribution* (or: *converges weakly*) to X if

$$\lim_{n\to\infty} F_{X_n}(x) = F_X(x)$$

for all points x in which F_X is continuous. By Theorem I.8.4 this kind of convergence occurs if

$$\lim_{n\to\infty} M_{X_n}(t) = M_X(t)$$

for all t in some interval $(-\infty, \xi]$.

The *characteristic function* χ (or χ_X) belonging to a stochastic variable X is defined by

$$\chi(t) := \mathbb{E}(e^{itX}).$$

This expression exists for *all* t in $\mathbb{R}$ and the function $t \mapsto \chi(t)$ is always continuous on $\mathbb{R}$. Lévy's theorem states that a sequence $X_1, X_2, \ldots$ converges in distribution to X if and only if

$$\lim_{n\to\infty} \chi_{X_n}(t) = \chi_X(t) \quad \text{for all } t \text{ in } \mathbb{R}.$$

A probability distribution $\mathbb{P}_X$ is completely characterized by χ_X. More precisely: $\chi_X = \chi_Y$ is equivalent to $\mathbb{P}_X = \mathbb{P}_Y$.

If the stochastic variables X and Y are statistically independent then

$$M_{X+Y}(t) = M_X(t)M_Y(t) \quad \text{and} \quad \chi_{X+Y}(t) = \chi_X(t)\,\chi_Y(t).$$

As to the effect of scale transformations on characteristic and moment generating functions we have the following. If $Y = pX + q$, then

1. $M_Y(t) = e^{qt}\, M_X(pt)$ for all $t \in \mathbb{R}$ where these expressions make sense,

2. $\chi_Y(t) = e^{iqt}\, \chi_X(pt)$ for all $t \in \mathbb{R}$.

If X is $N(\mu, \sigma^2)$-distributed, then:

1. $M_X(t) = e^{\mu t + \frac{1}{2}\sigma^2 t^2}$ for all $t \in \mathbb{R}$,

2. $\chi_X(t) = e^{i\mu t - \frac{1}{2}\sigma^2 t^2}$ for all $t \in \mathbb{R}$.

1.1.9 The central limit theorem (I.9)

In this section the *central limit theorem* is proved. Its content is the following. If $X_1, X_2, \dots$ is a statistically independent sequence of identically distributed stochastic variables with expectation value μ and variance σ^2, then the sequence

$$\sqrt{n}\ \{\tfrac{1}{n} S_n - \mu\} = \sqrt{n}\ \{\tfrac{1}{n}(X_1 + \dots + X_n) - \mu\}$$

converges in distribution to the $N(0, \sigma^2)$-distribution. In other words, $\frac{1}{n} S_n$ is for large n approximately $N(\mu, \sigma^2/n)$-distributed.

1.1.10 Transformation of probability densities (I.10)

Given a variable X (or two statistically independent variables X and Y) how can we determine the probability densities of for example X/Y, $\sqrt{X}$ or X^2 ? In I.10 techniques are discussed how to solve these problems.

1.2 Exercises to Chapter I

Exercise 1

Suppose $\mathfrak{A}$ is a σ-algebra. Prove that for arbitrary $A, B \in \mathfrak{A}$ also

$$A \cup B \in \mathfrak{A}, \ A \cap B \in \mathfrak{A} \ \text{ and } \ A \backslash B \in \mathfrak{A}.$$

Proof. If $A, B \in \mathfrak{A}$, then certainly $A^c, B^c, A \cup B = A \cup B \cup \emptyset \cup \emptyset \cdots \in \mathfrak{A}$. Consequently $\mathfrak{A}$ is closed under the operation of taking unions. Because of this also $A^c \cup B^c \in \mathfrak{A}$ and in turn it follows that $A \cap B = (A^c \cup B^c)^c \in \mathfrak{A}$, so that $\mathfrak{A}$ is also closed under the operation of taking intersections. In this way we see that also $A \backslash B = A \cap B^c \in \mathfrak{A}$. □

Exercise 2

Prove that for $A_1, A_2, \dots \in \mathfrak{A}$, also $\bigcap_{i=1}^{\infty} A_i \in \mathfrak{A}$.

Proof. By the laws of De Morgan we have (see Exercise 1)

$$\bigcap_{i=1}^{\infty} A_i = \left(\bigcup_{i=1}^{\infty} A_i^c \right)^c \in \mathfrak{A}.$$

□

Exercise 3

Let $\mathfrak{F} \subset \mathcal{P}(\Omega)$ and set

$$\mathfrak{A} := \bigcap_{\mathfrak{C} \in \mathcal{Z}} \mathfrak{C}$$

where $\mathcal{Z}$ is the collection of all σ-algebras on Ω containing $\mathfrak{F}$.

(i) Prove that $\mathfrak{A}$ is a σ-algebra.

Proof.

(i) For all $\mathfrak{C} \in \mathcal{Z}$ we have $\Omega \in \mathfrak{C}$. Consequently $\Omega \in \mathfrak{A}$.

(ii) If $A \in \mathfrak{A}$ then $A \in \mathfrak{C}$ for all $\mathfrak{C}$ in $\mathcal{Z}$. Hence also $A^c \in \mathfrak{C}$ for all $\mathfrak{C}$ in $\mathcal{Z}$, which implies that $A^c \in \mathfrak{A}$.

(iii) If $A_1, A_2, \ldots \in \mathfrak{A}$ then also $A_1, A_2, \ldots \in \mathfrak{C}$ for all $\mathfrak{C}$ in $\mathcal{Z}$. It follows that $\bigcup_i A_i \in \mathfrak{C}$ for all $\mathfrak{C}$ in $\mathcal{Z}$, hence $\bigcup_i A_i \in \mathfrak{A}$. □

(ii) Prove that for each σ-algebra $\mathfrak{B}$ which contains $\mathfrak{F}$ one has $\mathfrak{A} \subset \mathfrak{B}$.

Proof. Suppose $A \in \mathfrak{A}$. Then (by construction of $\mathfrak{A}$) $A \in \mathfrak{B}$. Therefore $\mathfrak{A} \subset \mathfrak{B}$. □

Exercise 4

Is the collection of all open sets in $\mathbb{R}$ a σ-algebra ?

Solution. This collection is not a σ-algebra. To see this, note that $(-\infty, 0)$ is an element of this collection, whereas $(-\infty, 0)^c = [0, +\infty)$ is not. □

Exercise 5

Suppose $f : \mathbb{R}^m \to \mathbb{R}^n$ is a Borel function. Define

$$\mathfrak{A} := \{A \subset \mathbb{R}^n : f^{-1}(A) \in \mathfrak{B}^m\}.$$

(i) Prove that $\mathfrak{A}$ is a σ-algebra containing all open sets in $\mathbb{R}^n$.

Proof.

(i) $f^{-1}(\mathbb{R}^n) = \mathbb{R}^m \in \mathfrak{B}^m$, therefore $\mathbb{R}^n \in \mathfrak{A}$.

(ii) Suppose $A \in \mathfrak{A}$, that is $f^{-1}(A) \in \mathfrak{B}^m$. Then $f^{-1}(A^c) = \left(f^{-1}(A)\right)^c \in \mathfrak{B}^m$ and consequently $A^c \in \mathfrak{A}$.

(iii) If $A_1, A_2, \ldots \in \mathfrak{A}$, then $f^{-1}(A_i) \in \mathfrak{B}^m$ for all i. Therefore we have $f^{-1}(\bigcup_i A_i) = \bigcup_i f^{-1}(A_i) \in \mathfrak{B}^m$; hence $\bigcup_i A_i \in \mathfrak{A}$.

This shows that $\mathfrak{A}$ is a σ-algebra. If O is open in $\mathbb{R}^n$, then (Definition I.1.4) $f^{-1}(O) \in \mathfrak{B}^m$, which is the same as saying that $O \in \mathfrak{A}$. It follows that $\mathfrak{A}$ contains all open sets in $\mathbb{R}^n$. □

(ii) Prove that $f^{-1}(A)$ is Borel in $\mathbb{R}^m$ for all Borel sets A in $\mathbb{R}^n$.

Proof. Take for $\mathfrak{F}$ the collection of open sets in $\mathbb{R}^n$ and apply Exercise 3. □

Exercise 6

Prove that the composition of two Borel functions is a Borel function.

Proof. Suppose that $f : \mathbb{R}^p \to \mathbb{R}^q$ and $g : \mathbb{R}^q \to \mathbb{R}^n$ are Borel functions. For every open set $O \in \mathbb{R}^n$ we then have (applying Exercise 5(ii)):

$$(g \circ f)^{-1}(O) = f^{-1}(g^{-1}(O)) \in \mathfrak{B}^p. \qquad \square$$

Exercise 7

If f and g are real-valued Borel functions then so are fg and $f+g$. Prove this.

Proof. Define $F_1, F_2 : \mathbb{R}^2 \to \mathbb{R}$ by

$$F_1(x,y) := xy \quad \text{and} \quad F_2(x,y) := x+y.$$

These functions are continuous, so they are surely Borel. The function $G : \mathbb{R} \to \mathbb{R}^2$ defined by $G(x) := (f(x), g(x))$ is also a Borel function (see Appendix B). Applying Exercise 6 we see that $F_1 \circ G$ and $F_2 \circ G$ are Borel functions. However, $F_1 \circ G$ is the function $x \mapsto f(x)g(x)$ and $F_2 \circ G$ is the function $x \mapsto f(x) + g(x)$. $\square$

Exercise 8

Let $\mathfrak{A}$ be a σ-algebra and let μ be a measure on $\mathfrak{A}$. If $A, B \in \mathfrak{A}$ and $A \subset B$, then $\mu(A) \le \mu(B)$. Prove this.

Proof. By Exercise 1 we have $B \backslash A \in \mathfrak{A}$. Moreover we have

$$B = A \cup (B\backslash A) \cup \emptyset \cup \emptyset \cup \cdots .$$

Because these sets are mutually disjoint, we may write

$$\mu(B) = \mu(A) + \mu(B\backslash A) + 0 + 0 + \cdots = \mu(A) + \mu(B\backslash A) \ge \mu(A). \qquad \square$$

Exercise 9

Prove that if μ is a measure on $\mathfrak{A}$ and $A_1, A_2, \ldots \in \mathfrak{A}$, then

$$\mu\left(\bigcup_{i=1}^{\infty} A_i\right) \le \sum_{i=1}^{\infty} \mu(A_i).$$

Proof. For $k = 1, 2, \ldots$ set $B_k := A_k \setminus [\bigcup_{i=1}^{k-1} A_i]$. Now the B_k are mutually disjoint, and we have $\bigcup_{k=1}^{\infty} B_k = \bigcup_{k=1}^{\infty} A_k$ and $B_k \subset A_k$ for all k. By Exercise 8 we may write

$$\mu\left(\bigcup_{k=1}^{\infty} A_k\right) = \mu\left(\bigcup_{k=1}^{\infty} B_k\right) = \sum_{k=1}^{\infty} \mu(B_k) \le \sum_{k=1}^{\infty} \mu(A_k). \qquad \square$$

Exercise 10

Suppose that μ is a measure on $\mathfrak{A}$ and that $A_1, A_2, \ldots \in \mathfrak{A}$ where $A_1 \supset A_2 \supset \cdots$ and $\mu(A_1) < +\infty$. Prove that

$$\lim_{N\to\infty} \mu(A_N) = \mu\left(\bigcap_{k=1}^{\infty} A_k\right).$$

Proof. Prelude. If $D_1, D_2 \in \mathfrak{A}$ satisfy $D_1 \subset D_2$ and $\mu(D_2) < +\infty$, then $\mu(D_2 \setminus D_1) + \mu(D_1) = \mu(D_2)$. Therefore $\mu(D_2 \setminus D_1) = \mu(D_2) - \mu(D_1)$ (Exercise 8 guarantees that we will not be confronted with the undefined expression $(+\infty) - (+\infty)$ here).

For all $k \in \mathbb{N}^*$ we write $B_k := A_1 \setminus A_k$; then $B_1 \subset B_2 \subset \cdots$. Let $C_1 := B_1 (= \emptyset)$ and $C_k := B_k \setminus B_{k-1}$ for all $k = 2, 3, \ldots$. Then the C_k are mutually disjoint and $\bigcup_{k=1}^{N} C_k = \bigcup_{k=1}^{N} B_k$ for all $N = 2, 3, \ldots$. Consequently

$$\begin{aligned}
\mu(A_1) - \mu\left(\bigcap_{k=1}^{\infty} A_k\right) &= \mu\left(A_1 \setminus \bigcap_{k=1}^{\infty} A_k\right) = \mu\left(\bigcup_{k=1}^{\infty} (A_1 \setminus A_k)\right) \\
&= \mu\left(\bigcup_{k=1}^{\infty} B_k\right) = \mu\left(\bigcup_{k=1}^{\infty} C_k\right) \\
&= \sum_{k=1}^{\infty} \mu(C_k) = \sum_{k=2}^{\infty} \mu(B_k \setminus B_{k-1}) \\
&= \sum_{k=2}^{\infty} \mu(A_{k-1} \setminus A_k) = \sum_{k=2}^{\infty} (\mu(A_{k-1}) - \mu(A_k)).
\end{aligned}$$

The last member presents a 'telescopic series': for all N we have

$$\sum_{k=2}^{N} (\mu(A_{k-1}) - \mu(A_k)) = \mu(A_1) - \mu(A_N).$$

The limit $\lim_{N\to\infty} \mu(A_N)$ exists (the sequence in question is decreasing and ≥ 0), therefore as a consequence of the foregoing

$$\mu(A_1) - \mu\left(\bigcap_{k=1}^{\infty} A_k\right) = \mu(A_1) - \lim_{N\to\infty} \mu(A_N).$$

Because $\mu(A_1) < +\infty$, this implies that

$$\lim_{N\to\infty} \mu(A_N) = \mu\left(\bigcap_{k=1}^{\infty} A_k\right).$$

(The condition that $\mu(A_1) < +\infty$ cannot be dropped. To see this, take for μ the Lebesgue measure on $\mathbb{R}^2$ and set $A_k := (0, \frac{1}{k}) \times \mathbb{R}$ for $k = 1, 2, \ldots$. Then we have a decreasing set of Borel sets for which $\lim_{N\to\infty} \mu(A_N) = +\infty$ whereas $\mu(\bigcap_{k=1}^{\infty} A_k) = \mu(\emptyset) = 0$.) □

Exercise 11

Suppose $\mathbf{X}$ is a stochastic n-vector. Prove that the map $\mathbb{P}_{\mathbf{X}} : A \mapsto \mathbb{P}(\mathbf{X}^{-1}(A))$ (all $A \in \mathfrak{B}^n$) defines a Borel measure on $\mathbb{R}^n$.

Proof. Let $(\Omega, \mathfrak{A}, \mathbb{P})$ be the underlying probability space. We know that the map $\mathbf{X} : \Omega \to \mathbb{R}^n$ is $\mathfrak{A}$-measurable. To prove that $\mathbb{P}_{\mathbf{X}}$ defines a Borel measure, let $A_1, A_2, \ldots$ be a sequence of mutually disjoint elements in $\mathfrak{B}^n$. Then

$$\begin{aligned}\mathbb{P}_{\mathbf{X}}\left(\bigcup_{i=1}^{\infty} A_i\right) &= \mathbb{P}\left[\mathbf{X}^{-1}\left(\bigcup_{i=1}^{\infty} A_i\right)\right] = \mathbb{P}\left[\bigcup_{i=1}^{\infty}(\mathbf{X}^{-1}(A_i))\right] \\ &= \sum_{i=1}^{\infty} \mathbb{P}[\mathbf{X}^{-1}(A_i)] = \sum_{i=1}^{\infty} \mathbb{P}_{\mathbf{X}}(A_i).\end{aligned}$$

Here the third equality holds because for $k \neq l$ we have

$$\mathbf{X}^{-1}(A_k) \cap \mathbf{X}^{-1}(A_l) = \emptyset.$$

□

Exercise 12

We perform an experiment in which we throw a fair die: X is the squared outcome. What does $\mathbb{P}_X$ look like?

Solution. We take $\Omega = \{1, \ldots, 6\}$, $\mathfrak{A} = \mathcal{P}(\Omega)$ and $\mathbb{P}$ on $\mathfrak{A}$ defined by $\mathbb{P}(A) := \#A/n$. There are now two probability spaces in sight, namely $(\Omega, \mathfrak{A}, \mathbb{P})$ and $(\mathbb{R}, \mathfrak{B}, \mathbb{P}_X)$.

For $k \in \{1, 2, 3, 4, 5, 6\}$ we have $\mathbb{P}_X(\{k^2\}) = \mathbb{P}(\{k\})$. Hence for all $A \in \mathfrak{B}$:

$$\begin{aligned}\mathbb{P}_X(A) &= \mathbb{P}_X(A \cap \{1, 4, 9, 16, 25, 36\}) \\ &= \mathbb{P}_X((A \cap \{1\}) \cup \cdots \cup (A \cap \{36\})) \\ &= \sum_{k=1}^{6} \mathbb{P}_X(A \cap \{k^2\}) = \sum_{k=1}^{6} \mathbb{P}_X(\{k^2\})\, \delta_{k^2}(A) \\ &= \sum_{k=1}^{6} \mathbb{P}(\{k\})\, \delta_{k^2}(A).\end{aligned}$$

In this way we learn:

$$\mathbb{P}_X = \tfrac{1}{6}\,(\delta_1 + \delta_4 + \cdots + \delta_{36}).$$

□

Exercise 13

Given is a probability space $(\Omega, \mathfrak{A}, \mathbb{P})$ and a map $X : \Omega \to \mathbb{R}^n$ for which $X^{-1}(O) \in \mathfrak{A}$ for all open O in $\mathbb{R}^n$. Prove that X is $\mathfrak{A}$-measurable.

Proof. Every open set in $\mathbb{R}^n$ is an element of $\{A \subset \mathbb{R}^n : X^{-1}(A) \in \mathfrak{B}^m\}$. Now replace in Exercise 5 the function f by X and $(\mathbb{R}^m, \mathfrak{B}^m)$ by $(\Omega, \mathfrak{A})$. It then follows that $X^{-1}(A) \in \mathfrak{A}$ for all $A \in \mathfrak{B}^n$. □

Exercise 14

Let $(\Omega, \mathfrak{A}, \mathbb{P})$ be a probability space and $A \in \mathfrak{A}$. It is easily seen that $X = 1_A$ is a stochastic variable. Express $\mathbb{P}_X$ in terms of δ_0 and δ_1.

Solution. Quite analogous to Exercise 12: for every $B \in \mathfrak{B}$ we have

$$\mathbb{P}_X(B) = \cdots = \mathbb{P}(A^c)\,\delta_0(B) + \mathbb{P}(A)\,\delta_1(B).$$

That is to say:

$$\mathbb{P}_X = \mathbb{P}(A^c)\,\delta_0 + \mathbb{P}(A)\,\delta_1.$$

□

Exercise 15

Given is a probability space $(\Omega, \mathfrak{A}, \mathbb{P})$ and events $B_1, B_2 \in \mathfrak{A}$. We set $X_1 := 1_{B_1}$ and $X_2 := 1_{B_2}$. Prove that the following statements are equivalent:

(i) X_1 and X_2 are statistically independent,

(ii) $\mathbb{P}(B_1 \cap B_2) = \mathbb{P}(B_1)\,\mathbb{P}(B_2)$.

Proof. First we prove that (i) $\Rightarrow$ (ii). If X_1 and X_2 are independent, then for all $A_1, A_2 \in \mathfrak{B}$ we have $\mathbb{P}(X_1 \in A_1 \text{ and } X_2 \in A_2) = \mathbb{P}(X_1 \in A_1)\,\mathbb{P}(X_2 \in A_2)$. Now just take $A_1 = A_2 = \{1\}$.

Next we turn to (ii) $\Rightarrow$ (i). If $\mathbb{P}(B_1 \cap B_2) = \mathbb{P}(B_1)\mathbb{P}(B_2)$, do we then have

$$\mathbb{P}(X_1 \in A_1 \text{ and } X_2 \in A_2) = \mathbb{P}(X_1 \in A_1)\,\mathbb{P}(X_2 \in A_2) \qquad (*)$$

for all $A_1, A_2 \in \mathfrak{B}$?

Because the range of X_1, X_2 consists of the points 0 and 1 we may replace A_1, A_2 by $A_1 \cap \{0,1\}$, $A_2 \cap \{0,1\}$. In this way we have to distinguish four cases:

- $A_1 = \emptyset$ or $A_2 = \emptyset$. Then $(*)$ is surely valid.
- $A_1 = \{0,1\}$ or $A_2 = \{0,1\}$. Then too $(*)$ is trivial.
- $A_1 = \{0\}$. Now there are two possibilities: $A_2 = \{0\}$ or $A_2 = \{1\}$.
  In the case $A_2 = \{1\}$ we have to check whether $\mathbb{P}(B_1^c \cap B_2) = \mathbb{P}(B_1^c)\mathbb{P}(B_2)$. This is true, for we have

  $$\mathbb{P}(B_1^c \cap B_2) = \mathbb{P}(B_2) - \mathbb{P}(B_1 \cap B_2) = \mathbb{P}(B_2) - \mathbb{P}(B_1)\mathbb{P}(B_2) = \mathbb{P}(B_2)\mathbb{P}(B_1^c).$$

  In the case where $A_2 = \{0\}$ we have to check whether $\mathbb{P}(B_1^c \cap B_2^c) = \mathbb{P}(B_1^c)\mathbb{P}(B_2^c)$. This is also true:

  $$\mathbb{P}(B_1^c \cap B_2^c) = \mathbb{P}(B_1^c) - \mathbb{P}(B_1^c \cap B_2) = \mathbb{P}(B_1^c) - \mathbb{P}(B_1^c)\mathbb{P}(B_2) = \mathbb{P}(B_1^c)\mathbb{P}(B_2^c).$$

- The case where $A_1 = \{1\}$ can be handled in the same way.

□

Exercise 16

Given is a probability space $(\Omega, \mathfrak{A}, \mathbb{P})$ and events $B_1, B_2, B_3 \in \mathfrak{A}$. We set $X_1 := 1_{B_1}, X_2 := 1_{B_2}$ and $X_3 := 1_{B_3}$. Consider the following two statements:

(i) X_1, X_2, X_3 is a statistically independent system,

(ii) $\mathbb{P}(B_i \cap B_j) = \mathbb{P}(B_i)\mathbb{P}(B_j)$ for all $i \neq j$ and

$$\mathbb{P}(B_1 \cap B_2 \cap B_3) = \mathbb{P}(B_1)\mathbb{P}(B_2)\mathbb{P}(B_3).$$

Prove that (i) and (ii) are equivalent.

Proof. Concerning (i) $\Rightarrow$ (ii): Independence of X_1, X_2 and X_3 implies that

$$\mathbb{P}_{X_1,X_2,X_3}(A_1 \times A_2 \times A_3) = \mathbb{P}_{X_1}(A_1)\mathbb{P}_{X_2}(A_2)\mathbb{P}_{X_3}(A_3)$$

for all $A_1, A_2, A_3 \in \mathfrak{B}$. Taking $A_3 = \{0,1\}$, $A_1 = A_2 = \{1\}$, we obtain $\mathbb{P}(B_1 \cap B_2) = \mathbb{P}(B_1)\mathbb{P}(B_2)$. Taking $A_1 = A_2 = A_3 = \{1\}$ we get $\mathbb{P}(B_1 \cap B_2 \cap B_3) = \mathbb{P}(B_1)\mathbb{P}(B_2)\mathbb{P}(B_3)$. Etcetera.

Next we consider (ii) $\Rightarrow$ (i). Suppose the given equalities for B_1, B_2 and B_3 hold. To prove that

$$\mathbb{P}_{X_1,X_2,X_3}(A_1 \times A_2 \times A_3) = \mathbb{P}_{X_1}(A_1)\mathbb{P}_{X_2}(A_2)\mathbb{P}_{X_3}(A_3)$$

holds for all $A_1, A_2, A_3 \in \mathfrak{B}$ it is sufficient to consider subsets A_1, A_2, A_3 of $\{0,1\}$. We exclude the trivial cases where A_i equals $\emptyset$ or $\{0,1\}$ for some i. Now eight cases are left. We work out one of them, namely the case in which $A_1 = A_2 = A_3 = \{0\}$. In that case we have to prove that

$$\mathbb{P}(B_1^c \cap B_2^c \cap B_3^c) = \mathbb{P}(B_1^c)\,\mathbb{P}(B_2^c)\,\mathbb{P}(B_3^c).$$

This can be seen as follows:

$$\begin{aligned}
\mathbb{P}(B_1^c \cap B_2^c \cap B_3^c) &= \mathbb{P}(B_1^c \cap B_2^c) - \mathbb{P}(B_1^c \cap B_2^c \cap B_3) \\
&= \mathbb{P}(B_1^c)\,\mathbb{P}(B_2^c) - (\mathbb{P}(B_1^c \cap B_3) - \mathbb{P}(B_1^c \cap B_2 \cap B_3)) \\
&= \mathbb{P}(B_1^c)\,\mathbb{P}(B_2^c) - \mathbb{P}(B_1^c)\,\mathbb{P}(B_3) \\
&\quad + (\mathbb{P}(B_2 \cap B_3) - \mathbb{P}(B_1 \cap B_2 \cap B_3)) \\
&= \mathbb{P}(B_1^c)\,\mathbb{P}(B_2^c) - \mathbb{P}(B_1^c)\,\mathbb{P}(B_3) \\
&\quad + \mathbb{P}(B_2)\,\mathbb{P}(B_3) - \mathbb{P}(B_1)\mathbb{P}(B_2)\mathbb{P}(B_3) \\
&= \mathbb{P}(B_1^c)\,\mathbb{P}(B_2^c) - \mathbb{P}(B_1^c)\,\mathbb{P}(B_3) + \mathbb{P}(B_2)\,\mathbb{P}(B_3)\{\mathbb{P}(B_1^c)\} \\
&= \mathbb{P}(B_1^c)\{\mathbb{P}(B_2^c) - \mathbb{P}(B_3) + \mathbb{P}(B_2)\,\mathbb{P}(B_3)\} \\
&= \mathbb{P}(B_1^c)\{\mathbb{P}(B_2^c) - \mathbb{P}(B_3)\{1 - \mathbb{P}(B_2)\}\} \\
&= \mathbb{P}(B_1^c)\,\mathbb{P}(B_2^c)\,\mathbb{P}(B_3^c).
\end{aligned}$$

□

Exercise 17

If X is a discretely distributed stochastic variable which has as its range $\{a_1, a_2, a_3, \ldots\}$, then

$$\mathbb{P}_X = \sum_{i=1}^{\infty} \mathbb{P}(X = a_i)\ \delta_{a_i}.$$

Prove this.

Proof. We have to prove that for an arbitrary $A \in \mathfrak{B}$ one has

$$\mathbb{P}_X(A) = \sum_{i=1}^{\infty} \mathbb{P}(X = a_i)\delta_{a_i}(A).$$

Without loss of generality we assume that $A \subset \{a_1, a_2, a_3, \ldots\}$. Now the proof is the same as that of Exercises 12 and 14. □

Exercise 18

For each $n \in \mathbb{N}$ and for every stochastic variable X we have

$$\int |x|^n \, d\,\mathbb{P}_X(x) \ \leq \ \int |x|^{n+1} \, d\,\mathbb{P}_X(x) + 1.$$

Prove this.

Proof. Define $B := \{x \in \mathbb{R} : |x| \leq 1\} = [-1, +1]$. Then

$$\int_B |x|^n \, d\,\mathbb{P}_X(x) \ \leq \ \int_B 1 \, d\,\mathbb{P}_X(x) \ \leq \ 1$$

and

$$\int_{B^c} |x|^n \, d\,\mathbb{P}_X(x) \ \leq \ \int_{B^c} |x|^{n+1} \, d\,\mathbb{P}_X(x) \ \leq \ \int |x|^{n+1} \, d\,\mathbb{P}_X(x).$$

Consequently

$$\begin{aligned} \int |x|^n \, d\,\mathbb{P}_X(x) \ &= \ \int_B |x|^n \, d\,\mathbb{P}_X(x) + \int_{B^c} |x|^n \, d\,\mathbb{P}_X(x) \\ &\leq \ 1 + \int |x|^{n+1} \, d\,\mathbb{P}_X(x). \end{aligned}$$

So the existence of $\mathbb{E}(X^{n+1})$ guarantees the existence of $\mathbb{E}(X^n)$. In particular the existence of $\mathbb{E}(X^2)$ guarantees the existence of $\mathbb{E}(X)$ and therefore the existence of $\mathrm{var}(X) = \mathbb{E}(X^2) - \mathbb{E}(X)^2$. □

Exercise 19

Suppose that the stochastic variable X has a probability density given by

$$f_X(x) = \frac{1}{\beta - \alpha} 1_{[\alpha,\beta]}(x).$$

Prove that

$$\mathrm{var}(X) = \tfrac{1}{12}(\beta - \alpha)^2.$$

Proof. As we have $(\beta - \alpha)(\beta^2 + \alpha\beta + \alpha^2) = \beta^3 - \alpha^3$,

$$\mathbb{E}(X^2) = \int x^2 f_X(x) dx = \int_\alpha^\beta \frac{1}{\beta - \alpha} x^2 \, dx = \frac{1}{\beta - \alpha} \left[\tfrac{1}{3}x^3\right]_\alpha^\beta = \tfrac{1}{3}(\beta^2 + \alpha\beta + \alpha^2).$$

Also,

$$\mathbb{E}(X) = \int x f_X(x) dx = \frac{1}{\beta - \alpha} \int_\alpha^\beta x \, dx = \frac{1}{\beta - \alpha} \left[\tfrac{1}{2}x^2\right]_\alpha^\beta = \tfrac{1}{2}(\beta + \alpha).$$

Hence

$$\begin{aligned} \mathrm{var}\ (\mathrm{X}) = \mathbb{E}(X^2) - \mathbb{E}(X)^2 &= \tfrac{1}{3}(\beta^2 + \alpha\beta + \alpha^2) - \tfrac{1}{4}(\beta + \alpha)^2 \\ &= \tfrac{1}{12}(\beta^2 - 2\alpha\beta + \alpha^2) = \tfrac{1}{12}(\beta - \alpha)^2. \end{aligned}$$

□

Exercise 20

Suppose X has the following distribution function:

$$F_X(x) = \begin{cases} 1 - e^{-x} & \text{if } x \geq 0, \\ 0 & \text{elsewhere.} \end{cases}$$

Determine the left-hand and right-hand derivative of F_X in the point $x = 0$.

Solution. We have

$$\lim_{x \uparrow 0} \frac{F_X(x) - F_X(0)}{x - 0} = \lim_{x \uparrow 0} \frac{0 - 0}{x} = 0.$$

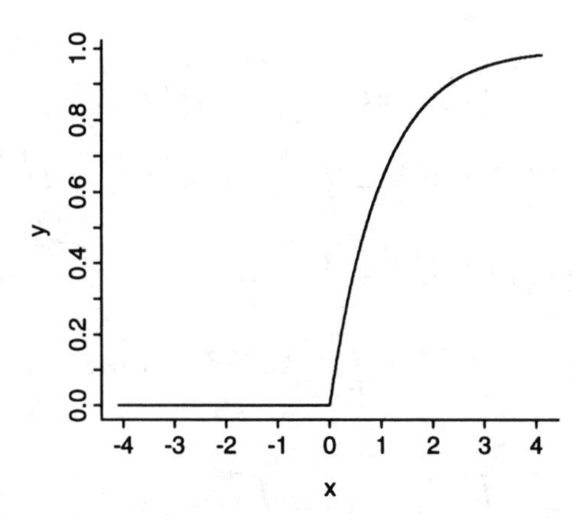

That is to say, the left-hand derivative in $x = 0$ equals zero. From the right side we expect the derivative of $1 - e^{-x}$ to appear, that is $e^{-0} = 1$. To verify this we notice that for $x > 0$

$$\begin{aligned}|1 - e^{-x} - x| &= \left|\int_0^x (e^{-t} - 1)\,dt\right| = \int_0^x (1 - e^{-t})\,dt \\ &\leq x \sup_{t\in[0,x]} (1 - e^{-t}) = x\,(1 - e^{-x}).\end{aligned}$$

Consequently, for $x > 0$ we have

$$\left|\frac{F_X(x) - F_X(0)}{x} - 1\right| = \left|\frac{1 - e^{-x} - 0}{x} - 1\right| \leq 1 - e^{-x}.$$

This shows that

$$\lim_{x\downarrow 0} \frac{F_X(x) - F_X(0)}{x - 0} = 1,$$

so the right-hand derivative of F_X in $x = 0$ equals 1. □

Exercise 21

Given are two statistically independent stochastic variables X and Y, having probability densities

$$f_X(x) = \begin{cases} \frac{1}{2}x^2 e^{-x} & \text{if } x \geq 0 \\ 0 & \text{elsewhere} \end{cases} \quad \text{and} \quad f_Y(y) = \begin{cases} e^{-y} & \text{if } y \geq 0 \\ 0 & \text{elsewhere} \end{cases}$$

The variable Z is defined by $Z := Y/X$.

(i) Determine $\mathbb{E}(X)$, $\mathbb{E}(Y)$ and $\mathbb{E}(Z)$.

Solution. Firstly we have

$$\begin{aligned}
\mathbb{E}(X) &= \int_0^{+\infty} \tfrac{1}{2}x^3 e^{-x}\,dx = \int_0^{+\infty} -\tfrac{1}{2}x^3\,de^{-x} \\
&= \left[-\tfrac{1}{2}x^3 e^{-x}\right]_0^{+\infty} - \int_0^{+\infty} -\tfrac{3}{2}x^2 e^{-x}\,dx = \int_0^{+\infty} -\tfrac{3}{2}x^2\,de^{-x} \\
&= \left[-\tfrac{3}{2}x^2 e^{-x}\right]_0^{+\infty} + \int_0^{+\infty} 3x e^{-x}\,dx = \int_0^{+\infty} -3x\,de^{-x} \\
&= \left[-3x e^{-x}\right]_0^{+\infty} + \int_0^{+\infty} 3e^{-x}\,dx = 3\left[-e^{-x}\right]_0^{+\infty} = 3.
\end{aligned}$$

Moreover

$$\begin{aligned}
\mathbb{E}(Y) &= \int_0^{+\infty} y e^{-y}\,dy = \int_0^{+\infty} -y\,de^{-y} \\
&= \left[-y e^{-y}\right]_0^{+\infty} + \int_0^{+\infty} e^{-y}\,dy = \left[-e^{-y}\right]_0^{+\infty} = 1.
\end{aligned}$$

In order to calculate $\mathbb{E}(Z)$ we use the explicit form of f_Z which has been derived in §I.10, Example 4, namely:

$$\begin{aligned}
\mathbb{E}(Z) &= \int_0^{+\infty} 3z(1+z)^{-4}\,dz = \int_0^{+\infty} -z\,d(1+z)^{-3} \\
&= \left[-z(1+z)^{-3}\right]_0^{+\infty} + \int_0^{+\infty} (1+z)^{-3}\,dz \\
&= \left[-\tfrac{1}{2}(1+z)^{-2}\right]_0^{+\infty} = \tfrac{1}{2}.
\end{aligned}$$

□

(ii) Although the variables X and Y are statistically independent, we do not have $\mathbb{E}(Y/X) = \mathbb{E}(Y)/\mathbb{E}(X)$, as $\frac{1}{2} \neq \frac{1}{3}$.

Exercise 22

Suppose X and Y are stochastic variables and suppose that the stochastic 2-vector (X, Y) has a probability density given by

$$f_{X,Y}(x,y) = \begin{cases} x+y & \text{on } [0,1]\times[0,1], \\ 0 & \text{elsewhere.} \end{cases}$$

Determine the numerical value of $\mathbb{E}(X)$ and $\mathbb{P}(X \le Y)$.

Solution. For $a \in [0,1]$ we have

$$\begin{aligned}
\mathbb{P}(X \le a) &= \int_0^a \int_0^1 f_{X,Y}(x,y)\,dy\,dx = \int_0^a \int_0^1 (x+y)\,dy\,dx \\
&= \int_0^a (x + \tfrac{1}{2})\,dx = \left[\tfrac{1}{2}x^2 + \tfrac{1}{2}x\right]_0^a = \tfrac{1}{2}a(1+a).
\end{aligned}$$

For $a > 1$ one has $\mathbb{P}(X \leq a) = 1$ and for $a < 0$ one has $\mathbb{P}(X \leq a) = 0$. This describes F_X. The probability density f_X of X can be obtained by differentiation of F_X. This leads to $f_X(x) = (\frac{1}{2} + x)1_{[0,1]}(x)$. It follows from this that

$$\mathbb{E}(X) = \int_0^1 (x + \tfrac{1}{2})x\, dx = \left[\tfrac{1}{4}x^2 + \tfrac{1}{3}x^3\right]_0^1 = \tfrac{1}{4} + \tfrac{1}{3} = \tfrac{7}{12}.$$

By symmetry of $f_{X,Y}$ we conclude that $\mathbb{P}(X \leq Y) = \frac{1}{2}$. This can also be obtained by explicit calculation of the double integral in

$$\mathbb{P}(X \leq Y) = \int\int_A f_{X,Y}(x,y)\, dx\, dy$$

where $A = \{(x,y) \in [0,1] \times [0,1] : x \leq y\}$. □

Exercise 23

Given is a stochastic 2-vector (X,Y) for which

$$f_{X,Y}(x,y) = \begin{cases} xe^{-xy} & \text{if } (x,y) \in [0,1] \times [0,+\infty), \\ 0 & \text{elsewhere.} \end{cases}$$

(i) Determine, as far as they exist, $\mathbb{E}(X)$ and $\mathbb{E}(Y)$.

Solution. If $a > 1$ then $\mathbb{P}(X \leq a) = 1$ and $\mathbb{P}(X \leq a) = 0$ for $a < 0$. For $a \in [0,1]$ we have

$$\mathbb{P}(X \leq a) = \int_0^a \int_0^{+\infty} xe^{-xy}\, dy\, dx = \int_0^a [-e^{-xy}]_0^{+\infty} dx = \int_0^a 1\, dx = a.$$

It follows that X is uniformly distributed on the interval $(0,1)$. Therefore (see Exercise 19) we have $\mathbb{E}(X) = \frac{1}{2}$. Furthermore

$$\mathbb{E}(Y) = \int_0^1 \int_0^{+\infty} y\,xe^{-xy}\, dy\, dx = \int_0^1 \alpha(x)\, dx,$$

where

$$\begin{aligned} \alpha(x) &= \int_0^{+\infty} xye^{-xy}\, dy &&= \frac{1}{x}\int_0^{+\infty} xye^{-xy}\, d(xy) \\ &= \frac{1}{x}\int_0^{+\infty} ue^{-u}\, du &&= \frac{1}{x}\left\{[-ue^{-u}]_0^{+\infty} + \int_0^{+\infty} e^{-u}\, du\right\} \\ &= -\frac{1}{x}\,[e^{-u}]_0^{+\infty} &&= \frac{1}{x}. \end{aligned}$$

Hence $\mathbb{E}(Y) = \int_0^1 \frac{1}{x}\, dx = +\infty$. □

(ii) Determine F_Z, where $Z = XY$.

Solution.

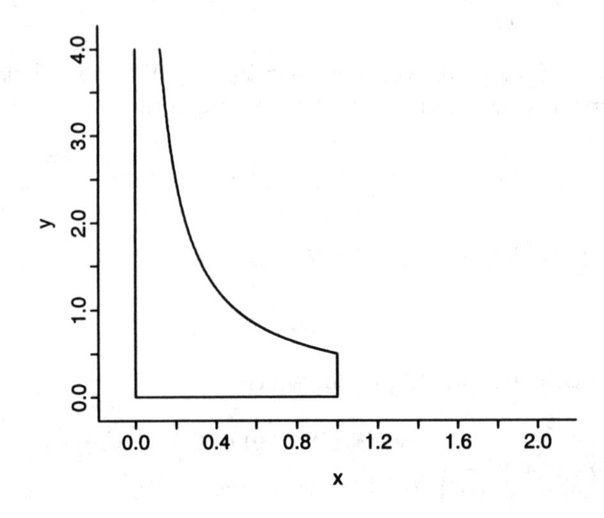

For $z \leq 0$ we have $\mathbb{P}(Z \leq z) = 0$. For $z > 0$ we can write

$$\mathbb{P}(Z \leq z) = \iint_A f_{X,Y}(x,y)\,dx\,dy$$

where the domain of integration is of type

$$A(z) = \{(x,y) \in [0,1] \times [0,+\infty) : xy \leq z\}$$

(in the picture we have indicated $A(\frac{1}{2})$). Here

$$\begin{aligned}\iint_A f_{X,Y}(x,y)\,dx\,dy &= \int_0^1 \int_0^{z/x} xe^{-xy}\,dy\,dx \\ &= \int_0^1 \left[-e^{-xy}\right]_0^{z/x} dx = \int_0^1 1 - e^{-z}\,dx \\ &= 1 - e^{-z}.\end{aligned}$$

In this way we learn that Z has an exponential distribution with parameter $\lambda = 1$; hence $\mathbb{E}(Z) = 1$. □

Exercise 24

Suppose X and Y are statistically independent stochastic variables, both with probability density

$$f(s) = e^{-s}\, 1_{[0,+\infty)}(s).$$

(i) Determine the probability density of the variable

$$Z := \sqrt{Y/X}.$$

Solution.

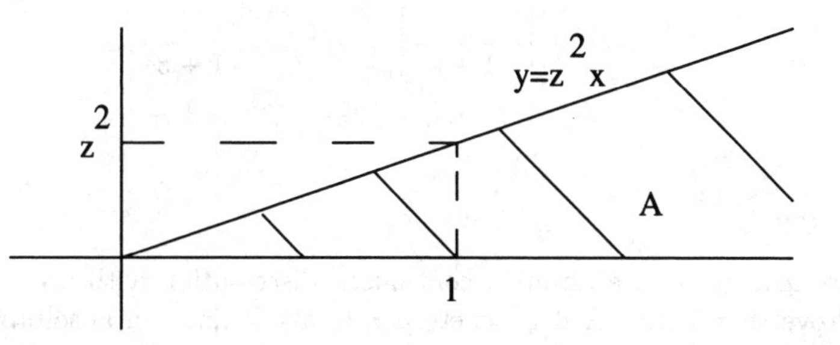

Of course we have $\mathbb{P}(Z \leq z) = 0$ if $z < 0$. Furthermore $\mathbb{P}(Z \leq z) = \mathbb{P}(Y \leq z^2 X)$ for $z \geq 0$. Equivalently: $\mathbb{P}(Z \leq z) = \mathbb{P}[(X,Y) \in A]$, where A is as in the figure above. Note that X and Y are 'surely' ≥ 0. Now we can write

$$\begin{aligned} F_Z(z) = \mathbb{P}(Z \leq z) &= \int_0^{+\infty} \int_0^{z^2 x} f_{X,Y}(x,y)\, dy\, dx \\ &= \int_0^{+\infty} e^{-x} \int_0^{z^2 x} e^{-y}\, dy\, dx \\ &= \int_0^{+\infty} e^{-x} [-e^{-y}]_0^{z^2 x}\, dx \\ &= \int_0^{+\infty} e^{-x}(1 - e^{-z^2 x})\, dx \\ &= \int_0^{+\infty} e^{-x}\, dx - \int_0^{+\infty} e^{-(1+z^2)x}\, dx \\ &= 1 + \frac{1}{1+z^2}\left[e^{-(1+z^2)x}\right]_0^{+\infty} \\ &= 1 - \frac{1}{1+z^2} = \frac{z^2}{1+z^2}. \end{aligned}$$

Differentiation of F_Z leads to

$$f_Z(z) = \frac{(1+z^2)\,2z - 2z\,z^2}{(1+z^2)^2} = \frac{2z}{(1+z^2)^2} \quad \text{if } z \geq 0$$

and $f_Z(z) = 0$ elsewhere. □

(ii) Determine $\mathbb{E}(Z)$.

Solution.

$$\begin{aligned}
\mathbb{E}(Z) &= \int_0^{+\infty} \frac{2z^2}{(1+z^2)^2}\,dz = \int_0^{+\infty} -z\,d(1+z^2)^{-1} \\
&= \left[-\frac{z}{1+z^2}\right]_0^{+\infty} + \int_0^{+\infty} \frac{1}{1+z^2}\,dz \\
&= -0+0+[\arctan z]_0^{+\infty} = \tfrac{1}{2}\pi.
\end{aligned}$$

□

Exercise 25

Suppose X enjoys an absolutely continuous distribution with probability density f_X. Prove that $Y := pX + q$, where $p \neq 0$, also enjoys an absolutely continuous distribution and that

$$f_Y(y) = \frac{1}{|p|} f_X\left(\frac{y-q}{p}\right).$$

Proof. If $p > 0$, then

$$\begin{aligned}
F_Y(y) = \mathbb{P}(Y \leq y) &= \mathbb{P}\left(X \leq \frac{y-q}{p}\right) = \int_{-\infty}^{(y-q)/p} f_X(x)\,dx \\
&= \int_{-\infty}^{y} f_X\left(\frac{t-q}{p}\right)\frac{1}{p}\,dt = \int_{-\infty}^{y} f_X\left(\frac{x-q}{p}\right)\frac{1}{|p|}\,dx.
\end{aligned}$$

This shows that Y enjoys an absolutely continuous distribution, its density as stated in the exercise. If $p < 0$, then

$$\begin{aligned}
F_Y(y) = \mathbb{P}\left(X \geq \frac{y-q}{p}\right) &= 1 - \int_{-\infty}^{(y-q)/p} f_X(x)\,dx \\
&= 1 - \int_{+\infty}^{y} f_X\left(\frac{t-q}{p}\right)\frac{1}{p}\,dt \\
&= 1 + \int_{y}^{+\infty} \frac{1}{p} f_X\left(\frac{t-q}{p}\right)dt \\
&= 1 - \int_{y}^{+\infty} \frac{1}{|p|} f_X\left(\frac{x-q}{p}\right)dx \\
&= \int_{-\infty}^{y} \frac{1}{|p|} f_X\left(\frac{x-q}{p}\right)dx.
\end{aligned}$$

The last equality in this chain is valid because

$$\int_{-\infty}^{+\infty} \frac{1}{|p|} f_X\left(\frac{x-q}{p}\right)dx = 1,$$

which can be verified by carrying out a substitution $t = (x-q)/p$. Summarizing, we see that also for $p < 0$ the variable Y enjoys an absolutely continuous distribution and its probability density is as stated. □

Exercise 26

(i) Let $X \geq 0$ be a stochastic variable with expectation value μ. Prove that $\mathbb{P}(X \geq \varepsilon) \leq \mu/\varepsilon$ for all $\varepsilon > 0$.

Proof. For all $\varepsilon > 0$ we have

$$\begin{aligned}
\varepsilon\,\mathbb{P}(X \geq \varepsilon) &= \varepsilon \int_{\varepsilon}^{+\infty} d\mathbb{P}_X(x) = \int_{\varepsilon}^{+\infty} \varepsilon\, d\mathbb{P}_X(x) \\
&\leq \int_{\varepsilon}^{+\infty} x\, d\mathbb{P}_X(x) \leq \int_{0}^{+\infty} x\, d\mathbb{P}_X(x) \\
&= \mathbb{E}(X) = \mu.
\end{aligned}$$

□

(ii) Let X be a stochastic variable with expectation value μ and variance σ^2. Prove that $\mathbb{P}((X-\mu)^2 \geq \varepsilon^2) \leq \sigma^2/\varepsilon^2$ for all $\varepsilon > 0$.

Proof. Consider the variable $(X-\mu)^2$. This variable has an expectation value equal to σ^2 and it satisfies the conditions posed on X in (i). Replacing in (i) X by $(X-\mu)^2$ and ε by ε^2, we get

$$\mathbb{P}((X-\mu)^2 \geq \varepsilon^2) \leq \sigma^2/\varepsilon^2.$$

□

Exercise 27

Suppose $\mathbf{X} = (X_1, \ldots, X_n)$ and $\mathbf{Y} = (Y_1, \ldots, Y_n)$ are identically distributed. If $f : \mathbb{R}^n \to \mathbb{R}$ is a Borel-function then $f(\mathbf{X})$ and $f(\mathbf{Y})$ are also identically distributed. Prove this.

Proof. For all $A \in \mathfrak{B}$ we have

$$\begin{aligned}
\mathbb{P}(f(\mathbf{X}) \in A) &= \mathbb{P}(\mathbf{X} \in f^{-1}(A)) = \mathbb{P}_{\mathbf{X}}(f^{-1}(A)) \\
&= \mathbb{P}_{\mathbf{Y}}(f^{-1}(A)) = \mathbb{P}(\mathbf{Y} \in f^{-1}(A)) = \mathbb{P}(f(\mathbf{Y}) \in A).
\end{aligned}$$

Hence $f(\mathbf{X})$ and $f(\mathbf{Y})$ are identically distributed. □

Exercise 28

(i) Suppose X_1 and X_2 are statistically independent $N(0,1)$-distributed variables. Set

$$A(c) := \{\mathbf{x} \in \mathbb{R}^2 : ax_1 + bx_2 \leq c\},$$

where $a, b > 0$. Prove that for all c we have

$$\mathbb{P}((X_1, X_2) \in A(c)) = \mathbb{P}\left[X_1 \leq \frac{c}{\sqrt{a^2+b^2}}\right].$$

Proof.

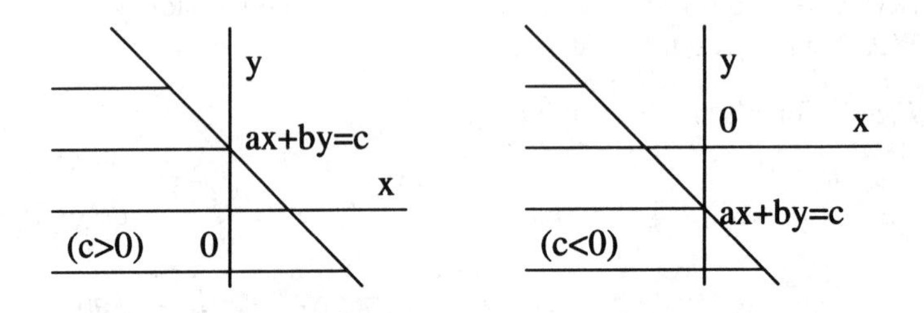

The probability distribution $\mathbb{P}_{X_1,X_2}$ of (X_1, X_2) is rotation invariant (Proposition I.6.2). That is to say, for every orthogonal linear operator $\mathbf{Q}$ one has

$$\mathbb{P}((X_1, X_2) \in A(c)) = \mathbb{P}\left((X_1, X_2) \in \mathbf{Q}A(c)\right).$$

There always is a rotation carrying over $A(c)$ into a set $B(q) = \{\mathbf{x} \in \mathbb{R}^2 : X_1 \leq q\}$ for some q.

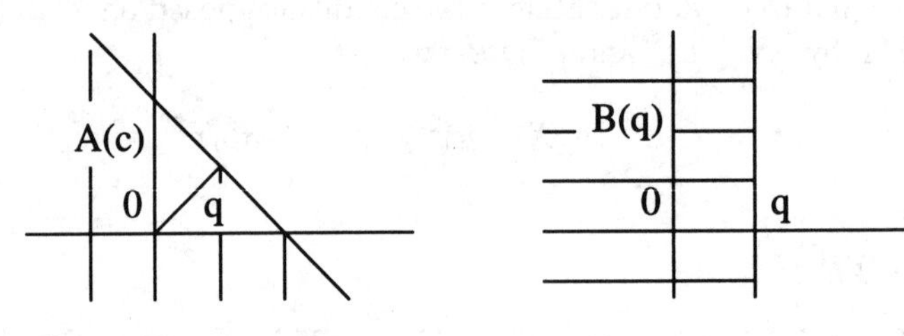

Consequently we have

$$\mathbb{P}((X_1, X_2) \in A(c)) = \mathbb{P}\left((X_1, X_2) \in B(c)\right) = \mathbb{P}(X_1 \leq q).$$

In the case indicated by the figure below we can express q in the following way in terms of a, b and c.

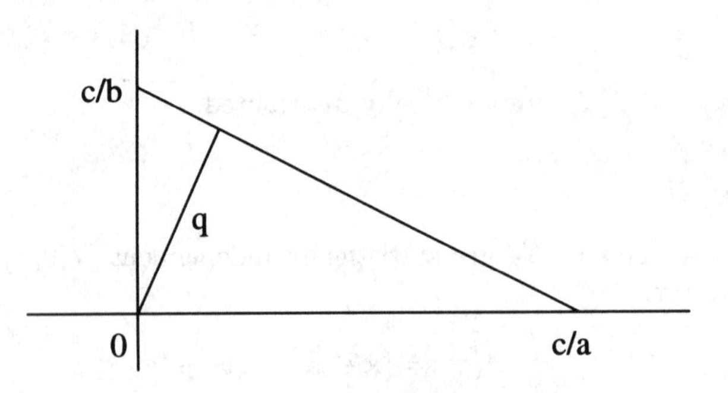

The area of the triangle is on the one hand given by $\frac{1}{2} \cdot c/a \cdot c/b$ and on the other hand by $\frac{1}{2}\, q \sqrt{(c/b)^2 + (c/a)^2}$. It follows that for $c \geq 0$

$$\begin{aligned} q &= \frac{c^2}{ab}\left(\sqrt{\frac{c^2}{b^2}+\frac{c^2}{a^2}}\right)^{-1} = \frac{c}{ab}\left(\sqrt{\frac{a^2+b^2}{a^2b^2}}\right)^{-1} \\ &= \frac{cab}{ab}\,\frac{1}{\sqrt{a^2+b^2}} = \frac{c}{\sqrt{a^2+b^2}}. \end{aligned}$$

Similar arguments apply if $c < 0$. □

(ii) Prove that $(aX_1 + bX_2)/\sqrt{a^2+b^2}$ enjoys a $N(0,1)$-distribution.

Proof.

$$\begin{aligned} \mathbb{P}\left[(aX_1+bX_2)/\sqrt{a^2+b^2} \leq z\right] &= \mathbb{P}\left[aX_1+bX_2 \leq z\sqrt{a^2+b^2}\right] \\ &= \mathbb{P}\left[(X_1,X_2) \in A(z\sqrt{a^2+b^2})\right] \\ &\overset{\text{(i)}}{=} \mathbb{P}\left[X_1 \leq z\right] \end{aligned}$$

This proves that $(aX_1 + bX_2)/\sqrt{a^2+b^2}$ is $N(0,1)$-distributed. □

(iii) Prove that $aX_1 + bX_2$ is $N(0, a^2+b^2)$-distributed.

Proof. Applying Theorem I.6.1 together with (ii) we learn that

$$\sqrt{a^2+b^2}\left(\frac{aX_1+bX_2}{\sqrt{a^2+b^2}}\right)$$

enjoys a $N(0, a^2+b^2)$-distribution. □

Exercise 29

Given is a statistically independent sequence $X_1, X_2, \ldots$ of variables, all of them enjoying a Bernoulli distribution with parameter $\theta = \frac{1}{2}$.

(i) Prove that (all $i = 1, 2, \ldots$) $M_{X_i}(t) = e^{t/2}\cosh(t/2)$.

Proof. We have $\mathbb{P}(X_i = 0) = \frac{1}{2}$ and $\mathbb{P}(X_i = 1) = \frac{1}{2}$, consequently

$$\begin{aligned} M_{X_i}(t) &= \int e^{tx}\, d\mathbb{P}_{X_i}(x) = \tfrac{1}{2}e^{t\cdot 0} + \tfrac{1}{2}e^{t\cdot 1} = \tfrac{1}{2}\left(e^t+1\right) \\ &= e^{t/2}\left(\frac{e^{t/2}+e^{-t/2}}{2}\right) = e^{t/2}\cosh(t/2). \end{aligned}$$

□

(ii) Let $S_n := \sum_{i=1}^n X_i$. Prove that $M_{S_n}(t) = (e^{t/2})^n(\cosh(t/2))^n$.

Proof. By a standard generalization of Proposition I.8.8 we can write

$$M_{S_n}(t) = \prod_{i=1}^n M_{X_i}(t) = (e^{t/2})^n(\cosh(t/2))^n.$$

□

(iii) Now let $Z_n := (S_n - \frac{1}{2}n)/\frac{1}{2}\sqrt{n} = (2/\sqrt{n})S_n - \sqrt{n}$. Prove that $M_{Z_n}(t) = (\cosh(t/\sqrt{n}))^n$.

Proof. We can express M_{Z_n} as follows (see Proposition I.8.6):

$$\begin{aligned} M_{Z_n}(t) = e^{-\sqrt{n}\,t} M_{S_n}(2t/\sqrt{n}) &= e^{-\sqrt{n}\,t} e^{\sqrt{n}\,t} (\cosh(t/\sqrt{n}))^n \\ &= (\cosh(t/\sqrt{n}))^n. \end{aligned}$$

□

(iv) Prove that $\cosh(t/\sqrt{n}) = 1 + t^2/(2n) + \alpha_n(t)$, where

$$\lim_{n\to\infty} n\alpha_n(t) = 0.$$

Proof. We exploit the Taylor expansion of e^x:

$$e^x = 1 + x + x^2/2! + \cdots .$$

It follows from this that

$$\begin{aligned} \cosh(s) &= \tfrac{1}{2}(e^s + e^{-s}) \\ &= \tfrac{1}{2}(1 + s + s^2/2 + s^3/3! + \cdots + 1 - s + s^2/2 - s^3/3! + \cdots) \\ &= 1 + s^2/2 + (s^4/4! + s^6/6! + \cdots) \;=:\; 1 + s^2/2 + r(s). \end{aligned}$$

For the function $s \mapsto r(s)$ defined by the above, we have

$$\lim_{s\to 0} \frac{r(s)}{s^2} = \lim_{s\to 0}\left(\frac{s^2}{4!} + \frac{s^4}{6!} + \cdots\right) = 0.$$

Taking $s = t/\sqrt{n}$ we get

$$\cosh(t/\sqrt{n}) = 1 + t^2/(2n) + r(t/\sqrt{n}) \quad \text{and} \quad \lim_{n\to\infty} \frac{r(t/\sqrt{n})}{t^2/n} = 0.$$

Setting $\alpha_n(t) := r(t/\sqrt{n})$ we have $\lim_{n\to\infty} n\alpha_n(t) = 0$ for all t. □

(v) Prove that Z_n converges in distribution to a $N(0,1)$-distribution.

Proof. For all $t \in \mathbb{R}$ we have

$$\lim_{n\to\infty} M_{Z_n}(t) = \lim_{n\to\infty} (\cosh(t/\sqrt{n}))^n = \lim_{n\to\infty} (1 + t^2/2 + r(t))^n.$$

According to Lemma I.9.1 the limit on the right side equals $\exp(\frac{1}{2}t^2)$. This is the moment generating function belonging to the standard normal distribution. Now we apply Lévy's theorem (Theorem I.8.4) to complete the proof.

□

Exercise 30

Consider the variables $X = 2$ and $X_n = 2 + 1/n$. Do we have $\lim_{n\to\infty} F_{X_n}(x) = F_X(x)$ for all $x \in \mathbb{R}$?

Solution. No, this is not true for $x = 2$. We have $F_{X_n}(2) = 0$ for all $n = 1, 2, \ldots$. So $\lim_{n\to\infty} F_{X_n}(2) = 0$, whereas $F_X(2) = 1$.

Does the sequence $\{X_n : n = 1, 2, \ldots\}$ converge in distribution to X?

Solution. Yes. We have

$$F_X(x) = 1_{[2,+\infty)}(x) \quad \text{and} \quad \lim_{n\to\infty} F_{X_n}(x) = 1_{(2,+\infty)}(x).$$

It follows that $\lim_{n\to\infty} F_{X_n}(x) = F_X(x)$ for all $x \neq 2$, that is, for all x in which F_X is continuous. □

Exercise 31

Tossing a coin unendingly, we define X_n as follows:

$$\begin{cases} X_n = 1 & \text{if we get a head in the } n^{\text{th}} \text{ tossing,} \\ X_n = 0 & \text{if we don't.} \end{cases}$$

We assume that $X_1, X_2, \ldots$ is a statistically independent sequence. Verify that for all fixed m we have $X_1, X_2, \ldots \to X_m$ in distribution.

Proof. Trivial: for *all* F_n we have $F_m = F_n$. So $\lim_{n\to\infty} F_n(x) = F_m(x)$ for all $x \in \mathbb{R}$. □

Exercise 32

Prove Exercise 29 using characteristic functions.

Proof. See Exercise 29 for notations and the general line.

For all $k = 1, 2, \ldots$ one has

$$\chi_{X_k}(t) = \mathbb{E}(e^{itX_k}) = \int e^{itx} d\mathbb{P}_{X_k}(x) = \tfrac{1}{2}e^0 + \tfrac{1}{2}e^{it} = \tfrac{1}{2} + \tfrac{1}{2}e^{it}.$$

Therefore, setting $S_n := X_1 + \cdots + X_n$,

$$\chi_{S_n}(t) = \prod_{i=1}^{n} \chi_{X_k}(t) = \left(\tfrac{1}{2}(1 + e^{it})\right)^n.$$

Next, setting $Z_n := (S_n - \frac{1}{2}n)/\frac{1}{2}\sqrt{n}$, using Proposition I.8.6 we obtain

$$\begin{aligned}
\chi_{Z_n}(t) &= e^{it\,(-\sqrt{n})}\,\chi_{S_n}(2t/\sqrt{n}) \\
&= e^{-i\sqrt{n}\,t}\,(\tfrac{1}{2}\,(1+e^{2it/\sqrt{n}}))^n \\
&= \left(\tfrac{1}{2}\,e^{-it/\sqrt{n}}(1+e^{2it/\sqrt{n}})\right)^n \\
&= \left(\tfrac{1}{2}\,(e^{-it/\sqrt{n}}+e^{it/\sqrt{n}})\right)^n = (\cos(t/\sqrt{n}))^n \\
&= (1-t^2/(2n)+t^4/(3!n^2)-\cdots)^n = (1-t^2/(2n)+\alpha_n(t))^n,
\end{aligned}$$

where $\lim_{n\to\infty} n\alpha_n(t) = 0$. Therefore for all t we have

$$\lim_{n\to\infty} \chi_{Z_n}(t) = \exp(-\tfrac{1}{2}t^2).$$

Applying Lévy's theorem as well as Proposition I.8.7, we learn that Z_n converges in distribution to the $N(0,1)$-distribution. □

Exercise 33

Suppose X is a stochastic variable that enjoys an absolutely continuous distribution with probability density f_X. Suppose that $\varphi : \mathbb{R} \to \mathbb{R}$ is continuously differentiable and that $\varphi' > 0$. Prove that $\varphi(X)$ also enjoys an absolutely continuous distribution and find an expression for its probability density.

Proof. Applying well-known results in basic mathematical analysis we conclude that φ is strictly increasing and that the following limits (finite or infinite) exist:

$$\varphi(+\infty) := \lim_{x\to+\infty} \varphi(x) \quad \text{and} \quad \varphi(-\infty) := \lim_{x\to-\infty} \varphi(x).$$

Now there is an inverse function $\varphi^{-1} : (\varphi(-\infty), \varphi(+\infty)) \to \mathbb{R}$, which is also continuously differentiable. Its derivative can be obtained by applying the chain rule on the identity $\varphi(\varphi^{-1}(y)) = y$. This gives us $\varphi'(\varphi^{-1}(y))\,(\varphi^{-1})'(y) = 1$, so

$$(\varphi^{-1})'(y) = 1/\varphi'(\varphi^{-1}(y)).$$

Next, we can write

$$\begin{aligned}
\mathbb{P}[\varphi(X) \le z] = \mathbb{P}[X \le \varphi^{-1}(z)] &= \int_{-\infty}^{\varphi^{-1}(z)} f_X(x)\,dx \\
&= \int_{\varphi(-\infty)}^{z} f_X(\varphi^{-1}(t))\,d\varphi^{-1}(t) \\
&= \int_{\varphi(-\infty)}^{z} f_X(\varphi^{-1}(t))\,\frac{1}{\varphi'(\varphi^{-1}(t))}\,dt \\
&= \int_{-\infty}^{z} g(t)\,dt
\end{aligned}$$

with

$$g(t) = f_X(\varphi^{-1}(t)) \frac{1}{\varphi'(\varphi^{-1}(t))} \quad \text{for all } t \in (\varphi(-\infty), \varphi(+\infty))$$

and $g(t) = 0$ elsewhere. This shows that $\varphi(X)$ enjoys an absolutely continuous distribution and that g is its probability density. □

Exercise 34

We toss a coin until the first time we get a head. The stochastic variable X denotes the payment: it is 2^k rouble if the first head appears in the kth tossing. Determine $\mathbb{E}(X)$.

Solution. Here we have

$$\mathbb{E}(X) = \sum_{k=1}^{\infty} 2^k \, \mathbb{P}[X = 2^k] = \sum_{k=1}^{\infty} 2^k \left(\tfrac{1}{2}\right)^{k-1} \tfrac{1}{2} = \sum_{k=1}^{\infty} 1 = +\infty.$$

□

Exercise 35

(i) From basic mathematical analysis we know that $\int_0^{+\infty} e^{-x^2}\, dx = \frac{1}{2}\sqrt{\pi}$. Prove that for any $\sigma > 0$ a probability density is defined by the function

$$f : x \mapsto \frac{1}{\sigma\sqrt{2\pi}} \, e^{-\frac{1}{2}(x-\mu)^2/\sigma^2}.$$

Proof. We prove that f satisfies the three characterizing properties of a probability density (see Proposition I.2.2).

(i) The function f is continuous and therefore surely Borel.

(ii) Of course $f \geq 0$.

(iii) Finally, by the substitution $y := \sqrt{\frac{1}{2}}\,(x - \mu)/\sigma$ we see that

$$\begin{aligned} \int_{-\infty}^{+\infty} f(x)\, dx &= \int_{-\infty}^{+\infty} \frac{1}{\sigma\sqrt{2\pi}} \, e^{-\frac{1}{2}(x-\mu)^2/\sigma^2}\, dx \\ &= \int_{-\infty}^{+\infty} \frac{\sqrt{2}}{\sqrt{2\pi}} \, e^{-y^2}\, dy \\ &= \frac{2}{\sqrt{\pi}} \int_0^{+\infty} e^{-y^2}\, dy = 1. \end{aligned}$$

□

(ii) A stochastic n-vector $(X_1, \ldots, X_n)$ enjoys by definition a *central vectorial normal distribution* if its probability density is defined by

$$f(\mathbf{x}) := C\, e^{-\frac{1}{2}\langle \mathbf{Tx}, \mathbf{x}\rangle} \quad (\mathbf{x} \in \mathbb{R}^n).$$

Here C is a constant and $\mathbf{T}$ is a positive ($\langle \mathbf{Tx}, \mathbf{x}\rangle \geq 0$ for all $\mathbf{x}$ in $\mathbb{R}^n$), symmetric and invertible linear operator. Express C in terms of $\det(\mathbf{T}), n$ and π.

Proof. $\mathbf{T}$ is a symmetric linear operator. For this reason there exists (by elementary linear algebra) an orthogonal linear operator $\mathbf{Q}$ and a diagonal operator $\mathbf{\Lambda}$ such that $\mathbf{T} = \mathbf{Q\Lambda Q^*}$. The diagonal elements of the matrix $[\mathbf{\Lambda}]$ of $\mathbf{\Lambda}$ are the eigenvalues $\lambda_1, \dots, \lambda_n$ of $\mathbf{T}$. These eigenvalues are strictly positive. (Namely, if λ_i is the eigenvalue belonging to an eigenvector $\mathbf{v}_i$ then $\langle \mathbf{Tv}_i, \mathbf{v}_i\rangle \geq 0$. This implies $\lambda_i\langle \mathbf{v}_i, \mathbf{v}_i\rangle = \lambda_i \|\mathbf{v}_i\|^2 \geq 0$, or $\lambda_i > 0$. The case $\lambda_i = 0$ cannot occur because $\mathbf{T}$ is invertible.) Moreover, for the linear operator $\mathbf{Q}$ we have $|\det(D\mathbf{Q})_\mathbf{u}| = |\det \mathbf{Q}| = 1$ for all $\mathbf{u} \in \mathbb{R}^n$.

By the substitution rule for multiple integrals we can write

$$\begin{aligned}
1 &= \int_{\mathbb{R}^n} f(\mathbf{x})d\mathbf{x} = \int_{\mathbf{Q}^*(\mathbb{R}^n)} f(\mathbf{Qu})\,|\det(D\mathbf{Q})_\mathbf{u}|d\mathbf{u} \\
&= \int_{\mathbb{R}^n} C\exp\left[-\tfrac{1}{2}\langle \mathbf{TQu}, \mathbf{Qu}\rangle\right] d\mathbf{u} \\
&= \int_{\mathbb{R}^n} C\exp\left[-\tfrac{1}{2}\langle \mathbf{Q^*TQu}, \mathbf{u}\rangle\right] d\mathbf{u} \\
&= \int_{\mathbb{R}^n} C\exp\left[-\tfrac{1}{2}\langle \mathbf{\Lambda u}, \mathbf{u}\rangle\right] d\mathbf{u} \\
&= \int_{\mathbb{R}^n} C\exp\left(-\tfrac{1}{2}[\lambda_1 {u_1}^2 + \cdots + \lambda_n {u_n}^2]\right) d\mathbf{u} \\
&= C\prod_{i=1}^n \int_{-\infty}^{+\infty} \exp(-\tfrac{1}{2}\lambda_i u_i^2)\, du_i \qquad (*)
\end{aligned}$$

where u_i denotes $\langle \mathbf{u}, \mathbf{e}_i\rangle$. Because

$$\int_{-\infty}^{+\infty} \exp(-\tfrac{1}{2}\lambda_i x^2)\, dx = \int_{-\infty}^{+\infty} \exp(-\tfrac{1}{2}y^2)\frac{1}{\sqrt{\lambda_i}}\, dy = \sqrt{\frac{2\pi}{\lambda_i}}$$

we have

$$\prod_{i=1}^n \int_{-\infty}^{+\infty} \exp(-\tfrac{1}{2}\lambda_i u_i^2)\, du_i = \frac{(\sqrt{2\pi})^n}{\sqrt{\det(\mathbf{T})}}.$$

Substituting this in $(*)$ we learn that

$$C = \sqrt{\frac{\det(\mathbf{T})}{(2\pi)^n}}.$$

□

Exercise 36

(i) Suppose the stochastic 2-vector (X, Y) enjoys an absolutely continuous distribution with probability density $f_{X,Y}$. Prove that $Z := Y/X$ also enjoys an absolutely continuous distribution and that

$$f_Z(z) = \int_{-\infty}^{+\infty} |x|\, f_{X,Y}(x, xz)\, dx.$$

Proof.

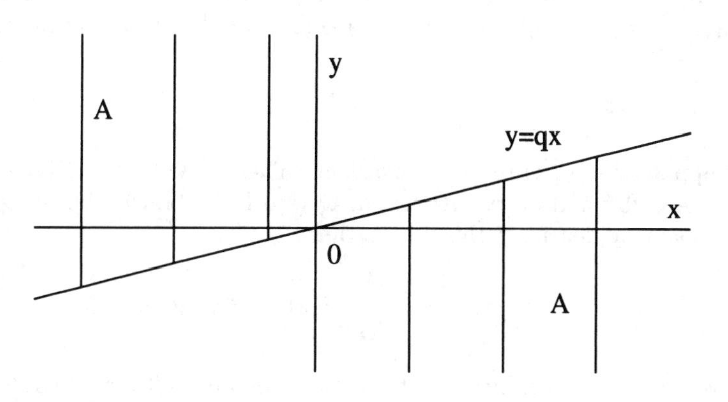

For any $q \in \mathbb{R}$ we have (see the figure)

$$\begin{aligned}
&\mathbb{P}[Z \leq q] \\
&= \mathbb{P}[Y/X \leq q] = \mathbb{P}[(X,Y) \in A] \\
&= \int_{-\infty}^{0}\int_{qx}^{+\infty} f_{X,Y}(x,y)\,dy\,dx + \int_{0}^{+\infty}\int_{-\infty}^{qx} f_{X,Y}(x,y)\,dy\,dx \\
&\overset{(*)}{=} \int_{-\infty}^{0}\int_{q}^{-\infty} f_{X,Y}(x,xz)\,x\,dz\,dx + \int_{0}^{+\infty}\int_{-\infty}^{q} f_{X,Y}(x,xz)\,x\;dz\,dx \\
&= \int_{-\infty}^{0}\int_{-\infty}^{q} |x|\,f_{X,Y}(x,xz)\,dz\,dx + \int_{0}^{+\infty}\int_{-\infty}^{q} |x|\,f_{X,Y}(x,xz)\;dz\,dx \\
&= \int_{-\infty}^{q}\int_{-\infty}^{+\infty} |x|\,f_{X,Y}(x,xz)\,dz\,dx = \int_{-\infty}^{q} f_Z(z)\,dz.
\end{aligned}$$

At $(*)$ we carried out the substitution $z := y/x$ in the 'inner' integral. □

(ii) Suppose X and Y are statistically independent and that both variables are $N(0,1)$-distributed. Prove that $Z = X/Y$ is Cauchy distributed with parameters 0 and 1.

Proof. We apply (i):

$$\begin{aligned}
f_Z(z) &= \int_{-\infty}^{+\infty} |x|\,f_{X,Y}(x,xz)\,dx = \int_{-\infty}^{+\infty} \frac{|x|}{2\pi} \exp\left[-\tfrac{1}{2}x^2(1+z^2)\right]\,dx \\
&= \frac{1}{\pi}\left[-\frac{1}{1+z^2}\exp\left(-\tfrac{1}{2}x^2(1+z^2)\right)\right]_0^{+\infty} = \frac{1}{\pi}\frac{1}{z^2+1}
\end{aligned}$$

which is what we had to prove. □

Exercise 37

The stochastic 2-vector $\mathbf{X} = (X_1, X_2)$ has an absolutely continuous distribution, its continuous density function given by $f : \mathbb{R}^2 \to [0,+\infty)$. Prove that the probability distribution $\mathbb{P}_{X_1,X_2}$ of (X_1, X_2) is rotation invariant if and only if $f(\mathbf{Q}(x_1, x_2)) = f(x_1, x_2)$ for all $(x_1, x_2) \in \mathbb{R}^2$ and all orthogonal maps $\mathbf{Q} : \mathbb{R}^2 \to \mathbb{R}^2$.

Proof.

($\Rightarrow$) Suppose $\mathbb{P}_{X_1,X_2}$ is rotation invariant. Then $\mathbb{P}(\mathbf{X} \in A) = \mathbb{P}(\mathbf{X} \in \mathbf{Q}(A))$ for all $A \in \mathfrak{B}^2$ and every orthogonal $\mathbf{Q}$ (see Definition I.6.1). In terms of the probability density f this means that

$$\int_A f(\mathbf{x})d\mathbf{x} = \int_{\mathbf{Q}A} f(\mathbf{x})d\mathbf{x} \quad \text{for all } A \in \mathfrak{B}^2.$$

We have, however, by the substitution rule for multiple integrals

$$\int_{\mathbf{Q}A} f(\mathbf{x})d\mathbf{x} = \int_A (f \circ \mathbf{Q})(\mathbf{x}) \, |\det (D\mathbf{Q})_{\mathbf{x}}| \, d\mathbf{x} = \int_A (f \circ \mathbf{Q})(\mathbf{x})d\mathbf{x}.$$

It follows that

$$\int_A f(\mathbf{x})d\mathbf{x} = \int_A (f \circ \mathbf{Q})(\mathbf{x})d\mathbf{x}$$

for all Borel sets $A \in \mathfrak{B}^2$. Using the continuity of f, it is easily deduced from this that $f = f \circ \mathbf{Q}$.

($\Leftarrow$) If $f = f \circ \mathbf{Q}$, then

$$\begin{aligned} \mathbb{P}(\mathbf{X} \in A) &= \int_A f(\mathbf{x})d\mathbf{x} = \int_A (f \circ \mathbf{Q})(\mathbf{x}) \, |\det (D\mathbf{Q})_{\mathbf{x}}| \, d\mathbf{x} \\ &= \int_{\mathbf{Q}A} f(\mathbf{x})d\mathbf{x} = \mathbb{P}(\mathbf{X} \in \mathbf{Q}(A)). \end{aligned}$$

This shows that $\mathbb{P}_{X_1,X_2}$ is rotation invariant. □

Exercise 38

Let $f_1, f_2 > 0$ be continuously differentiable probability densities on $\mathbb{R}$. Define $(f_1 \otimes f_2)(x_1, x_2) := f_1(x_1)f_2(x_2)$ for all $(x_1, x_2) \in \mathbb{R}^2$. Let X_1, X_2 be corresponding stochastic variables. Suppose that (X_1, X_2) is rotation invariant, that is to say (see Exercise 37), suppose that $(f_1 \otimes f_2)(x_1, x_2) = (f_1 \otimes f_2)(\mathbf{Q}(x_1, x_2))$ for every orthogonal $\mathbf{Q} : \mathbb{R}^2 \to \mathbb{R}^2$ and every $(x_1, x_2) \in \mathbb{R}^2$.

(i) Show that $f_i(x) = f_i(-x)$ for all $x \in \mathbb{R}$ $(i = 1, 2)$.

Solution. Consider the orthogonal linear operator $\mathbf{Q} : \mathbb{R}^2 \to \mathbb{R}^2$ corresponding to the (orthogonal) matrix $\begin{pmatrix} -1 & 0 \\ 0 & 1 \end{pmatrix}$.

The fact that $f_1 \otimes f_2 = (f_1 \otimes f_2) \circ \mathbf{Q}$ implies that

$$f_1(x_1)f_2(x_2) = f_1(-x_1)f_2(x_2) \quad \text{for all } (x_1, x_2) \in \mathbb{R}^2.$$

Because $f_2 > 0$, this leads to the conclusion that $f_1(x) = f_1(-x)$ for all $x \in \mathbb{R}$. A similar argument applies to f_2. □

(ii) Prove that $f_1 = f_2$.

Proof. Here, consider the orthogonal linear map $\mathbf{Q} : \mathbb{R}^2 \to \mathbb{R}^2$ belonging to the (orthogonal) matrix $\begin{pmatrix} 0 & 1 \\ 1 & 0 \end{pmatrix}$.

Now the fact that $f_1 \otimes f_2 = (f_1 \otimes f_2) \circ \mathbf{Q}$ is the same as saying that $f_1(x_1)f_2(x_2) = f_1(x_2)f_2(x_1)$ for all $x_1, x_2 \in \mathbb{R}$. Setting $x_2 := 0$ we obtain:

$$f_1(x) = \frac{f_1(0)}{f_2(0)} f_2(x) \quad \text{for all } x \in \mathbb{R}.$$

By the constraint $\int f_1(x)dx = \int f_2(x)dx = 1$ we necessarily have $f_1(0) = f_2(0)$: hence $f_1 = f_2$. □

(iii) Prove that there exist constants A and B such that

$$f(x) \; := \; f_1(x) = f_2(x) = A \exp^{\frac{1}{2}Bx^2}.$$

Proof. Define for all $t \in \mathbb{R}$ the orthogonal linear operator $\mathbf{Q}(t) : \mathbb{R}^2 \to \mathbb{R}^2$ by the (orthogonal) matrix

$$\begin{pmatrix} \cos t & -\sin t \\ \sin t & \cos t \end{pmatrix}.$$

Writing out the equality $f_1 \otimes f_2 = (f_1 \otimes f_2) \circ \mathbf{Q}$ we get

$$f(x_1 \cos t - x_2 \sin t)\, f(x_1 \sin t + x_2 \cos t) = f(x_1)f(x_2)$$

for all $t, x_1, x_2 \in \mathbb{R}$. Differentiation with respect to t tells us that for all $t, x_1, x_2 \in \mathbb{R}$ the following equation holds:

$$f(x_1 \cos t - x_2 \sin t) \cdot f'(x_1 \sin t + x_2 \cos t) \cdot (x_1 \cos t - x_2 \sin t) +$$

$$f(x_1 \sin t + x_2 \cos t) \cdot f'(x_1 \cos t - x_2 \sin t) \cdot$$

$$(-x_1 \sin t - x_2 \cos t) = 0\,.$$

Setting $t = 0$ we obtain $f(x_1)f'(x_2)x_1 = f(x_2)f'(x_1)x_2$; consequently for all $x_1, x_2 \neq 0$ we have

$$\frac{f'(x_1)}{x_1 f(x_1)} = \frac{f'(x_2)}{x_2 f(x_2)}.$$

It follows from this that the function $x \mapsto f'(x)/xf(x)$ is constant for $x \neq 0$. Hence $f'(x) = B\, xf(x)$ for some $B \in \mathbb{R}$ and all $x \neq 0$; by (i) it follows that $f'(0) = 0$. Solving the elementary differential equation which emerges in this way we get $f(x) = A\, e^{\frac{1}{2}Bx^2}$ for some A and some B. □

Exercise 39

Suppose X_1 and X_2 are statistically independent stochastic variables with positive, continuously differentiable probability densities. If the probability distribution of (X_1, X_2) is rotation invariant, then both variables are $N(0, \sigma^2)$-distributed. Prove this.

Proof. Under the imposed conditions on X_1 and X_2 the probability density of both variables is (exercise 38) of the form $f(x) = A\ e^{\frac{1}{2}Bx^2}$. By the constraint $1 = A \int_{-\infty}^{+\infty} e^{\frac{1}{2}Bx^2}\,dx$ the constant B is necessarily negative, therefore we may write $B = -1/\sigma^2$. In this notation we have

$$\int_{-\infty}^{+\infty} e^{\frac{1}{2}Bx^2}\,dx = \int_{-\infty}^{+\infty} e^{-\frac{1}{2}(x/\sigma)^2}\,dx = \int_{-\infty}^{+\infty} e^{-\frac{1}{2}y^2}\,d(\sigma y) = \sigma\sqrt{2\pi}.$$

Hence $A = 1/(\sigma\sqrt{2\pi})$ and it follows that

$$f(x) = \frac{1}{\sigma\sqrt{2\pi}}\ e^{-\frac{1}{2}(x^2/\sigma^2)}.$$

We see that both X_1 and X_2 are $N(0, \sigma^2)$-distributed. □

Exercise 40

Suppose $X_1, \dots, X_n$ is a statistically independent system of stochastic variables, all of them having a continuously differentiable strictly positive probability density. If the probability density of $(X_1, \dots, X_n)$ is rotation invariant then the X_i are all $N(0, \sigma^2)$-distributed. Prove this.

Proof.

Step 1. First generalize Exercise 37 to the case where n probability densities are involved. This is not more difficult than the case $n = 2$.

Step 2. Let $f_1, \dots, f_n$ be the probability densities of $X_1, \dots, X_n$. Then

$$(f_1 \otimes \cdots \otimes f_n)(x_1, \dots, x_n) = (f_1 \otimes \cdots \otimes f_n)(\mathbf{Q}(x_1, \dots, x_n)) \qquad (*)$$

Suppose $\mathbf{O} : \mathbb{R}^2 \to \mathbb{R}^2$ is an arbitrary orthogonal linear operator. Let $\mathbf{Q} : \mathbb{R}^n \to \mathbb{R}^n$ be the orthogonal linear operator defined by

$$\mathbf{Q}(x_1, \dots, x_n) := (\mathbf{O}(x_1, x_2), x_3, \dots, x_n).$$

Now by $(*)$ we learn that $f_1 \otimes f_2 = (f_1 \otimes f_2) \circ \mathbf{O}$ for every orthogonal $\mathbf{O} : \mathbb{R}^2 \to \mathbb{R}^2$. By Exercise 39 this implies that X_1 and X_2 both have the same $N(0, \sigma^2)$-distribution. Of course similar arguments can be applied to the other components of $(X_1, \dots, X_n)$. □

Exercise 41

Our goal is to prove the existence of two variables X and Y such that

(a) $\operatorname{cov}(X, Y) = 0$,

(b) X and Y are not statistically independent,

(c) X and Y both share the same $N(0,1)$-distribution.

We consider the probability space $(\mathbb{R}, \mathfrak{B}, \mathbb{P})$, where $\mathbb{P}$ is defined by

$$\mathbb{P}(A) := \int_A \frac{1}{\sqrt{2\pi}} e^{-\frac{1}{2}x^2}\, dx \quad \text{for all Borel sets } A$$

(that is, $\mathbb{P}$ is the probability measure belonging to the standard normal distribution). We define the stochastic variable $X : \Omega \to \mathbb{R}$ by $X : x \mapsto x$. Of course we now have $\mathbb{P}_X = \mathbb{P}$, so X enjoys a $N(0,1)$-distribution. For all $c \geq 0$ the stochastic variable Y_c is defined by

$$Y_c(x) := \begin{cases} x & \text{if } |x| \geq c, \\ -x & \text{if } |x| < c. \end{cases}$$

(i) Sketch a graph of Y_c.

Solution. A graph of Y_c, for $c > 0$, looks like the picture below:

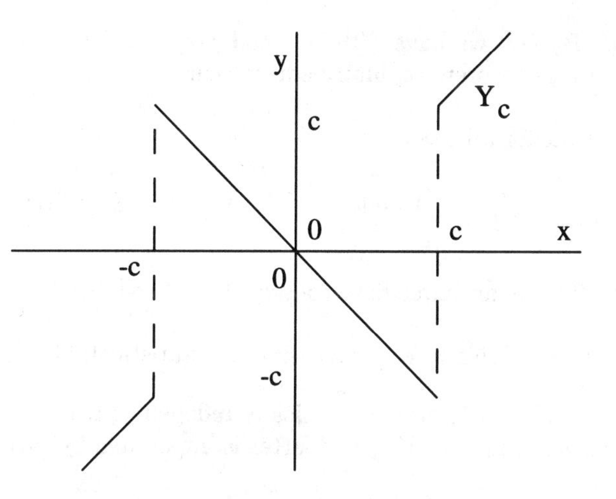

(ii) Prove that $Y_0 = X$ and $\lim_{c \to +\infty} Y_c = -X$.

Proof. It is trivial that $Y_0 = X$. Moreover, for every fixed x we have $Y_c(x) = -x = -X(x)$ for all $c > x$. Hence for all x

$$\lim_{c \to +\infty} Y_c(x) = -x = -X(x).$$

□

(iii) Prove that for all $c > 0$ the variable Y_c is $N(0,1)$-distributed.

Proof. Suppose $c > 0$. Looking at the figure we see that

$$\mathbb{P}[Y_c \leq z] = \mathbb{P}[(-\infty, z]] = \Phi(z) \quad \text{for } |z| \geq c.$$

For $|z| < c$ we have $\mathbb{P}[Y_c \leq z] = \mathbb{P}[(-\infty, -c] \cup [-z, c)]$. However, by a substitution $s := -t$ it follows that

$$\mathbb{P}\{[-z, c)\} = \int_{-z}^{c} \frac{1}{\sqrt{2\pi}} e^{-\frac{1}{2}t^2}\, dt = \int_{-c}^{z} \frac{1}{\sqrt{2\pi}} e^{-\frac{1}{2}s^2}\, ds = \mathbb{P}[(-c, z]].$$

Consequently $\mathbb{P}[(-\infty, -c] \cup [-z, c)] = \mathbb{P}[(-\infty, z]] = \Phi(z)$. □

(iv) Assume that the function $f : c \mapsto \text{cov}(X, Y_c)$ is continuous. Determine $f(0)$ and $\lim_{c \to +\infty} f(c)$.

Solution. We have $f(0) = \text{cov}(X, Y_0) = \text{var}(X) = 1$. Furthermore

$$\begin{aligned} \lim_{c\to\infty} f(c) &= \lim_{c\to\infty} \text{cov}(X, Y_c) = \text{cov}(X, \lim_{c\to\infty} Y_c) \\ &= \text{cov}(X, -X) = -\text{var}(X) = -1. \end{aligned}$$

□

(v) Prove that there is an element $a \in \mathbb{R}$ such that $f(a) = 0$.

Proof. By (iv) we have $f(0) > 0$ and $f(c) < 0$ for some c. Now apply the Weierstrass theorem on intermediate values. □

(vi) Prove that for all $c > 0$

$$\mathbb{P}(|X| \leq 1 \text{ and } |Y_c| \leq 1) \neq \mathbb{P}(|X| \leq 1)\, \mathbb{P}(|Y_c| \leq 1).$$

Proof. This is an immediate consequence of the fact that $|Y_c| = |X|$. □

(vii) Prove that X and Y_c can impossibly be statistically independent.

Proof. If X and Y_c were statistically independent, then so were (by Proposition I.4.2) $|X|$ and $|Y_c|$. The latter is impossible by (vi). □

(viii) At this point, the pair X, Y_a satisfies our three conditions. Isn't this in contradiction to Theorem I.6.7 ?

Solution. Here we have two $N(0,1)$-distributed stochastic variables X, Y_a satisfying $\text{cov}(X, Y_a) = 0$, which are statistically dependent. Theorem I.6.7 does not at all exclude this phenomenon, for Y_a is not in the linear span of X. □

Exercise 42

Given is a set of probability densities $f_1, \dots, f_n$. Prove that there exists a statistically independent system $X_1, \dots, X_n$ of stochastic variables such that X_i has probability density f_i (for all i).

Proof. Define $f : \mathbb{R}^n \to [0, +\infty)$ by $f := f_1 \otimes \cdots \otimes f_n$. It is easy to see that f presents a probability density. Now set $\Omega := \mathbb{R}^n$, $\mathbb{P} := f\lambda$ and $X_i := \pi_i$, the projection onto the ith co-ordinate (all i). Then

$$\begin{aligned} \mathbb{P}(X_i \in A) = \mathbb{P}(\pi_i \in A) &= \mathbb{P}[\pi_i^{-1}(A)] = \mathbb{P}[\mathbb{R} \times \cdots \times A \times \cdots \times \mathbb{R}] \\ &= \cdots (\text{Fubini}) \cdots = \int_A f_i(x)\, dx. \end{aligned}$$

It follows from this that $X_i = \pi_i$ has probability density f_i. By Theorem I.3.2 the system $X_1, \dots, X_n$ is statistically independent. □

Exercise 43

Suppose X and Y are stochastic variables that enjoy absolutely continuous distributions. Does this imply that (X, Y) also enjoys an absolutely continuous distribution?

Solution. No. Consider for example a stochastic variable X which is uniformly distributed on the interval $[0, 1]$. Then X enjoys an absolutely continuous distribution whereas (X, X) does not. Namely, the assumption that (X, X) has a probability density f leads to the contradiction that for $A := \{(x, x) : x \in [0, 1]\}$ we would have

$$1 = \mathbb{P}_{(X,X)}(A) = \int_A f(x, y)\, d(x, y) = \int_0^1 \int_x^x f(x, y)\, dy\, dx = \int_0^1 0\, dx = 0. \qquad \square$$

Exercise 44

Prove that the distribution function F_X of a stochastic variable X always satisfies the following conditions.

(i) $\lim_{x \to -\infty} F_X(x) = 0$ and $\lim_{x \to +\infty} F_X(x) = 1$,

(ii) F_X is increasing,

(iii) F_X is right-continuous.

Proof. By definition $F_X(x) = \mathbb{P}_X[(-\infty, x]]$ for all $x \in \mathbb{R}$. We know that $\mathbb{P}_X$ is a Borel measure on $\mathbb{R}$. We start by proving (ii).

(ii) If $x < y$, then $(-\infty, x] \subset (-\infty, y]$. Now apply Exercise 8 and conclude that $F_X(x) \leq F_X(y)$.

(i) For all $k = 1, 2, \ldots$ we write $A_k := (-\infty, -k]$. Then $A_1 \supset A_2 \supset \cdots$ and $\mathbb{P}_X(A_1) \leq 1 < +\infty$. Using Exercise 10 we see that

$$\lim_{k\to\infty} \mathbb{P}_X((-\infty, -k]) = \mathbb{P}_X\left(\bigcap_{k=1}^{\infty} A_k\right) = \mathbb{P}_X(\emptyset) = 0.$$

It follows from this and (ii) that $\lim_{x\to-\infty} F_X(x) = 0$. Similarly, define $B_k := (-\infty, k]$ for all $k = 1, 2, \ldots$. Then we have the disjoint decomposition

$$\mathbb{R} = B_1 \cup \left(\bigcup_{k=1}^{\infty} (B_{k+1} \setminus B_k)\right).$$

Consequently

$$\begin{aligned} 1 = \mathbb{P}_X(\mathbb{R}) &= \mathbb{P}_X(B_1) + \sum_{k=1}^{\infty} (\mathbb{P}_X(B_{k+1}) - \mathbb{P}_X(B_k)) \\ &\overset{(*)}{=} \lim_{k\to\infty} \mathbb{P}_X(B_k) = \lim_{k\to\infty} F_X(k). \end{aligned}$$

(Check the validity of $(*)$.) It follows that $\lim_{x\to+\infty} F_X(x) = 1$.

(iii) In order to prove that F_X is right-continuous in x, let $x_1, x_2, \ldots$ be any decreasing sequence such that $x_k \to x$. Then

$$\bigcap_{k=1}^{\infty} (-\infty, x_k] = (-\infty, x].$$

Applying Exercise 10 we learn that

$$\lim_{k\to\infty} \mathbb{P}_X((-\infty, x_k]) = \mathbb{P}_X((-\infty, x]).$$

This is the same as saying that $F_X(x_k) \to F_X(x)$. Consequently F_X is right-continuous in x. □

Exercise 45

By contour integration in the complex plane it is easily verified that for all $a \in \mathbb{R}$

$$\int_{-\infty}^{+\infty} e^{-(x+ia)^2}\, dx = \int_{-\infty}^{+\infty} e^{-x^2}\, dx \qquad (*)$$

Deduce from this that the characteristic function of a $N(\mu, \sigma^2)$-distributed variable X is given by

$$\chi_X(t) = e^{i\mu t - \frac{1}{2}\sigma^2 t^2} \quad (\text{all } t \in \mathbb{R}).$$

Solution. Set $Y := (X-\mu)/\sigma$. Then Y is (Theorem I.6.1) $N(0,1)$-distributed and by Proposition I.8.6 we have $\chi_X(t) = e^{i\mu t}\,\chi_Y(t\sigma)$. For $\chi_Y(t)$ we can write

$$\begin{aligned}
\chi_Y(t) = \mathbb{E}(e^{itY}) &= \int_{-\infty}^{+\infty} e^{ity}\,\frac{1}{\sqrt{2\pi}}\,e^{-\frac{1}{2}y^2}\,dy \\
(y := \sqrt{2}\,x) &= \int_{-\infty}^{+\infty} e^{it\sqrt{2}\,x}\,\frac{1}{\sqrt{2\pi}}\,e^{-\frac{1}{2}(\sqrt{2}\,x)^2}\,\sqrt{2}\,dx \\
&= \frac{1}{\sqrt{\pi}}\int_{-\infty}^{+\infty} e^{-(x-\frac{1}{2}\sqrt{2}\,it)^2}e^{-\frac{1}{2}t^2}\,dx \\
(\text{using } (*)) &= (e^{-\frac{1}{2}t^2}/\sqrt{\pi})\int_{-\infty}^{+\infty} e^{-x^2}\,dx \\
(\text{see Exercise 35(i)}) &= e^{-\frac{1}{2}t^2}.
\end{aligned}$$

Hence, $\chi_X(s) = e^{i\mu s}\,\chi_Y(s\sigma) = e^{i\mu s-\frac{1}{2}\sigma^2 s^2}$. This completes the proof. □

Exercise 46

Suppose that X and Y are statistically independent stochastic variables. Let $S := X+Y$. Prove that

1. $M_S(t) = M_X(t)\,M_Y(t)$ for all t for which these expressions make sense,
2. $\chi_S(t) = \chi_X(t)\,\chi_Y(t)$ for all t.

Proof.

1. For all t the stochastic variables e^{tX} and e^{tY} are statistically independent (Proposition I.4.2). Therefore
$$M_S(t) = \mathbb{E}(e^{tS}) = \mathbb{E}(e^{t(X+Y)}) = \mathbb{E}(e^{tX})\,\mathbb{E}(e^{tY}) = M_X(t)\,M_Y(t).$$

2. For a complex-valued stochastic variable X we can always write $X = X_1 + iX_2$, where X_1 and X_2 are real-valued. Now $\mathbb{E}(X)$ is defined by
$$\mathbb{E}(X) := \mathbb{E}(X_1) + i\,\mathbb{E}(X_2).$$
If $Y = Y_1 + iY_2$ is another complex-valued variable then X and Y are said to be statistically independent if the two 2-vectors (X_1, X_2) and (Y_1, Y_2) are so. In that case we have
$$\begin{aligned}
\mathbb{E}(XY) &= \mathbb{E}\left[(X_1+iX_2)(Y_1+iY_2)\right] \\
&= \mathbb{E}\left[X_1Y_1 - X_2Y_2 + i\,(X_1Y_2 + X_2Y_1)\right] \\
&= \mathbb{E}(X_1Y_1) - \mathbb{E}(X_2Y_2) + i\,\mathbb{E}(X_1Y_2) + i\,\mathbb{E}(X_2Y_1) \\
&= \mathbb{E}(X_1)\,\mathbb{E}(Y_1) - \mathbb{E}(X_2)\,\mathbb{E}(Y_2) + i\,\mathbb{E}(X_1)\,\mathbb{E}(Y_2) + i\,\mathbb{E}(X_2)\,\mathbb{E}(Y_1) \\
&= (\mathbb{E}(X_1) + i\,\mathbb{E}(X_2))\,(\mathbb{E}(Y_1) + i\,\mathbb{E}(Y_2)) = \mathbb{E}(X)\,\mathbb{E}(Y).
\end{aligned}$$

After these preparations, using Proposition I.4.2 again, we conclude that $e^{itX} = \cos(tX) + i\sin(tX)$ and $e^{itY} = \cos(tY) + i\sin(tY)$ are statistically independent. Therefore we can write

$$\chi_S(t) = \mathbb{E}(e^{itS}) = \mathbb{E}(e^{it(X+Y)}) = \mathbb{E}(e^{itX})\,\mathbb{E}(e^{itY}) = \chi_X(t)\,\chi_Y(t). \qquad \square$$

Exercise 47

If two statistically independent variables X and Y are Poisson distributed, then so is $X+Y$. Prove this.

Proof.

Solution 1. Suppose X and Y are statistically independent and Poisson distributed with parameters λ and μ respectively $(\lambda, \mu > 0)$. Then for all $k \in \mathbb{N}$ we have

$$\mathbb{P}(X = k) = \frac{\lambda^k}{k!}\,e^{-\lambda} \quad \text{and} \quad \mathbb{P}(Y = k) = \frac{\mu^k}{k!}\,e^{-\mu}.$$

Therefore

$$\begin{aligned}
\mathbb{P}(X+Y=k) &= \sum_{l=0}^{k} \mathbb{P}(X = l \text{ and } Y = k-l) \\
&= \sum_{l=0}^{k} \mathbb{P}(X = l)\,\mathbb{P}(Y = k-l) \\
&= \sum_{l=0}^{k} \frac{\lambda^l \mu^{k-l}}{l!(k-l)!}\,e^{-(\lambda+\mu)} \\
&= e^{-(\lambda+\mu)}\,\frac{1}{k!}\sum_{l=0}^{k}\binom{k}{l}\lambda^l\mu^{k-l} \\
&= \frac{(\lambda+\mu)^k}{k!}\,e^{-(\lambda+\mu)}.
\end{aligned}$$

This shows that $X+Y$ is Poisson distributed with parameter $\lambda+\mu$.

Solution 2. First we notice that the characteristic function of a variable X that is Poisson distributed with parameter λ is given by

$$\chi_X(t) = \sum_{k=0}^{\infty} \frac{\lambda^k}{k!} e^{itk-\lambda}.$$

In our case we get

$$\chi_{X+Y}(t) = \chi_X(t)\,\chi_Y(t) = \cdots = \sum_{k=0}^{\infty} \frac{(\lambda+\mu)^k}{k!}\,e^{itk-(\lambda+\mu)}.$$

We see that the characteristic function of $X+Y$ belongs to a Poisson distribution with parameter $\lambda+\mu$. $\square$

Exercise 48

Suppose X and Y are statistically independent and binomially distributed. Find conditions under which the sum is also binomially distributed.

Solution. Let X be a binomially distributed variable with parameters n, θ and let Y binomially distributed with parameters m, θ. If X and Y are statistically independent, then $X+Y$ is binomially distributed with parameters $n+m$ and θ.

To prove this we consider the characteristic function of a binomially distributed variable:

$$\begin{aligned} \chi_X(t) = \mathbb{E}(e^{itX}) &= \sum_{k=0}^{n} e^{itk} \binom{n}{k} \theta^k (1-\theta)^{n-k} \\ &= \sum_{k=0}^{n} \binom{n}{k} (\theta e^{it})^k (1-\theta)^{n-k} \\ &= (\theta e^{it} + (1-\theta))^n. \end{aligned}$$

Using Proposition I.8.8 we see that

$$\begin{aligned} \chi_{X+Y}(t) &= \chi_X(t)\,\chi_Y(t) \\ &= (\theta e^{it} + (1-\theta))^n\,(\theta e^{it} + (1-\theta))^m = (\theta e^{it} + (1-\theta))^{n+m}, \end{aligned}$$

so $X+Y$ has the characteristic function belonging to a binomial distribution with parameters $n+m$ and θ. □

Exercise 49

Suppose that X and Y are statistically independent Cauchy distributed stochastic variables. Prove that $X+Y$ also enjoys a Cauchy distribution.

Proof. Applying the result in Appendix D we may write

$$\chi_X(t) = e^{i\alpha_1 t} e^{-\beta_1 |t|} \quad \text{and} \quad \chi_Y(t) = e^{i\alpha_2 t} e^{-\beta_2 |t|}$$

for some $\alpha_1, \alpha_2, \beta_1, \beta_2 \in \mathbb{R}$. Because X and Y are independent we have (Proposition I.8.8):

$$\chi_{X+Y}(t) = \chi_X(t)\,\chi_Y(t) = e^{i(\alpha_1+\alpha_2)t}\, e^{-(\beta_1+\beta_2)|t|}.$$

This is, however, the characteristic function belonging to a Cauchy distribution with parameters $\alpha_1 + \alpha_2$ and $\beta_1 + \beta_2$. □

Exercise 50

Suppose X and Y are statistically independent stochastic variables having probability densities f_X and f_Y that are bounded and piecewise continuous. Assume there exists a constant $a > 0$ with $f_X(x) = f_Y(x) = 0$ for $|x| \geq a$, and define the *convolution product* $f_X * f_Y : \mathbb{R} \to [0, +\infty)$ by

$$(f_X * f_Y)(s) := \int_{-\infty}^{+\infty} f_X(x) f_Y(s - x)\, dx.$$

(i) Prove that this definition makes sense and that $f_X * f_Y$ presents a continuous function.

Proof.
There exists a number $K > 0$ such that $|f_X(x)|, |f_Y(x)| \leq K$ for all $x \in \mathbb{R}$. Furthermore $f_X(x) f_Y(s-x) = 0$ for all $|x| \geq a$. Keeping s fixed, the function $x \mapsto f_X(x) f_Y(s - x)$ is piecewise continuous and zero outside the interval $[-a, a]$. Such functions are integrable. This shows that the definition of $(f_X * f_Y)(s)$ makes sense.
Again keep s fixed. In order to prove that $f_X * f_Y$ is continuous, choose a sequence $s_1, s_2, \ldots \to s$. If f_Y is continuous in the point $s - x$, then we have

$$\lim_{n\to\infty} f_X(x) f_Y(s_n - x) = f_X(x) f_Y(s - x).$$

Define h_n by $h_n : x \mapsto f_X(x) f_Y(s_n - x)$. Now the function

$$h(x) := \lim_{n\to\infty} h_n(x)$$

is defined everywhere with the possible exception of a finite number of points. Moreover we have $|h_n| \leq g$, where $g : x \mapsto K^2\, 1_{[-a;a]}(x)$, satisfying $\int g(x)\, dx = K^2\, 2a < +\infty$. Applying Lebesgue's dominated convergence theorem (see Appendix A) we see that

$$\lim_{n\to\infty} \int_{-\infty}^{+\infty} f_X(x) f_Y(s_n - x)\, dx = \int_{-\infty}^{+\infty} h(x)\, dx = \int_{-\infty}^{+\infty} f_X(x) f_Y(s - x)\, dx.$$

This proves that $f_X * f_Y$ is continuous in s. □

(ii) For the variable $S := X + Y$ we have $f_S = f_X * f_Y$. Prove this.

Proof. Because X and Y are independent we have (Theorem I.3.2) that

$$\begin{aligned}
F_S(s) = \mathbb{P}_S((-\infty, s]) &= \mathbb{P}(X + Y \leq s) \\
&= \int_{-\infty}^{+\infty} \int_{-\infty}^{s-x} f_X(x) f_Y(y)\, dy\, dx \\
(y := t - x) \quad &= \int_{-\infty}^{+\infty} \int_{-\infty}^{s} f_X(x) f_Y(t - x)\, dt\, dx \\
(\text{Fubini}) \quad &= \int_{-\infty}^{s} \int_{-\infty}^{+\infty} f_X(x) f_Y(t - x)\, dt\, dx \\
&= \int_{-\infty}^{s} (f_X * f_Y)(t)\, dt.
\end{aligned}$$

This shows that $f_X * f_Y$ is the probability density of S. □

(iii) Prove that $(f_X * f_Y)(s) = 0$ if $|s| \geq 2a$.

Proof. If $|s| \geq 2a$ and $x \in [-a, a]$, then $|s - x| \geq ||s| - |x|| = |s| - |x| \geq 2a - a = a$ so $f_Y(s - x) = 0$. Consequently for $|s| \geq 2a$ we have

$$\begin{aligned}(f_X * f_Y)(s) &= \int_{-\infty}^{+\infty} f_X(x) f_Y(s - x)\, dx \\ &= \int_{-a}^{a} f_X(x) f_Y(s - x)\, dx \\ &= \int_{-a}^{a} f_X(x)\, 0\, dx = 0.\end{aligned}$$

□

(iv) Formulate and prove a discrete analogue of the above.

Solution. Suppose X and Y are discretely distributed stochastic variables with (discrete) probability densities f_X and f_Y. Let $B := \{b_1, b_2, \dots\}$ be the union of the ranges of X and Y. Define $f_X * f_Y : \mathbb{R} \to [0, +\infty)$ by

$$(f_X * f_Y)(s) := \sum_{k=1}^{\infty} f_X(b_k) f_Y(s - b_k).$$

The series on the right side is always convergent, for we have

$$0 \leq f_X(b_k) f_Y(s - b_k) \leq f_X(b_k) \quad \text{and} \quad \sum_k f_X(b_k) = 1.$$

Now let $S := X + Y$. The (discrete) probability density f_S of S is given by

$$\begin{aligned}f_S(s) = \mathbb{P}(X + Y = s) &= \sum_{k=1}^{\infty} \mathbb{P}(X = b_k \text{ and } Y = s - b_k) \\ &= \sum_{k=1}^{\infty} \mathbb{P}(X = b_k)\, \mathbb{P}(Y = s - b_k) \\ &= \sum_{k=1}^{\infty} f_X(b_k) f_Y(s - b_k) = (f_X * f_Y)(s).\end{aligned}$$

Hence $f_S = f_X * f_Y$. □

Exercise 51

Our goal is to visualize the central limit theorem.

Let X_1, X_2, X_3 be a statistically independent system of stochastic variables, all of them uniformly distributed on the interval $[-\frac{1}{2}, +\frac{1}{2}]$. Write $S_1 := X_1$, $S_2 := X_1 + X_2$ and $S_3 := X_1 + X_2 + X_3$.

(i) For S_1 we have $\mathbb{E}(S_1) = 0$ and $\operatorname{var}(S_1) = \frac{1}{12}$. Sketch the probability density of S_1 and that of the $N(0, 1/12)$-distribution in one figure.

Solution.

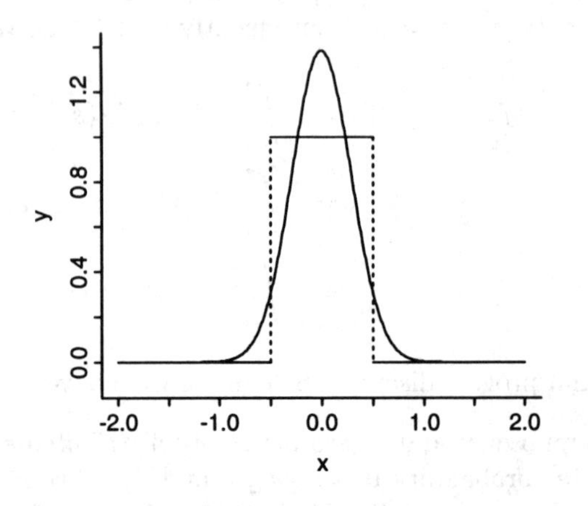

(ii) Determine the probability density of S_2.

Solution. Applying Exercise 50 (iii), we see that $f_{S_2}(s) = 0$ if $|s| \geq 1$. For $|s| < 1$ we get

$$\begin{aligned}
f_{S_2}(s) = (f_{X_1} * f_{X_2})(s) &= \int_{-\infty}^{+\infty} f_{X_1}(x) f_{X_2}(s-x)\, dx \\
&= \int_{-\frac{1}{2}}^{+\frac{1}{2}} f_{X_2}(s-x)\, dx \\
&= \int_{-\frac{1}{2}}^{+\frac{1}{2}} 1_{[-\frac{1}{2}, +\frac{1}{2}]}(s-x)\, dx \\
&= \int_{-\frac{1}{2}}^{+\frac{1}{2}} 1_{[s-\frac{1}{2}, s+\frac{1}{2}]}(x)\, dx.
\end{aligned}$$

In this way we obtain for $s \in [0, 1)$

$$f_{S_2}(s) = \int_{-\frac{1}{2}}^{+\frac{1}{2}} 1_{[s-\frac{1}{2}, s+\frac{1}{2}]}(x)\, dx = \int_{s-\frac{1}{2}}^{+\frac{1}{2}} 1\, dx = \tfrac{1}{2} - (s - \tfrac{1}{2}) = 1 - s$$

and for $s \in (-1, 0)$

$$\begin{aligned}
f_{S_2}(s) &= \int_{-\frac{1}{2}}^{+\frac{1}{2}} 1_{[s-\frac{1}{2}, s+\frac{1}{2}]}(x)\, dx \\
&= \int_{-\frac{1}{2}}^{s+\frac{1}{2}} 1\, dx = (s + \tfrac{1}{2}) - (-\tfrac{1}{2}) = 1 + s.
\end{aligned}$$

Summarizing, $f_{S_2}(s) = (1 - |s|)\,1_{(-1,1)}(s)$ for all $s \in \mathbb{R}$. □

(iii) Using (ii) (or Proposition I.5.1 and I.5.9) we see that $\mathbb{E}(S_2) = 0$ and $\mathrm{var}(S_2) = \frac{1}{6}$. Sketch the probability density of S_2 and the one belonging to the $N(0, 1/6)$-distribution in one figure.

Solution.

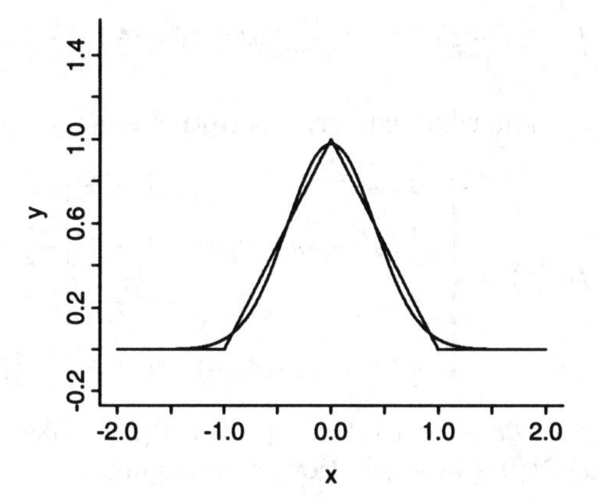

(iv) Determine the probability density of S_3.

Solution. The variables $S_2 = X_1 + X_2$ and X_3 are statistically independent and we have $S_3 = S_2 + X_3$. Applying Exercise 50 we get

$$\begin{aligned} f_{S_3}(s) &= (f_{S_2} * f_{X_3})(s) \\ &= \int_{-\infty}^{+\infty} f_{S_2}(x) f_{X_3}(s-x)\,dx \\ &= \int_{-1}^{1} (1-|x|)\,1_{[-\frac{1}{2},+\frac{1}{2}]}(s-x)\,dx \\ &= \int_{-1}^{1} (1-|x|)\,1_{[s-\frac{1}{2},s+\frac{1}{2}]}(x)\,dx. \end{aligned}$$

If $|s| > 3/2$, then the intervals $[s-\frac{1}{2}, s+\frac{1}{2}]$ and $[-1,+1]$ are disjoint, consequently $f_{S_3}(s) = 0$. For $s \in [-\frac{3}{2}, -\frac{1}{2}]$ we obtain

$$\begin{aligned} \int_{-1}^{s+\frac{1}{2}} (1-|x|)\,dx &= \int_{-1}^{s+\frac{1}{2}} (1+x)\,dx = \left[x + \tfrac{1}{2}\,x^2\right]_{-1}^{s+\frac{1}{2}} \\ &= (s+\tfrac{1}{2}) + \tfrac{1}{2}(s^2 + s + \tfrac{1}{4}) - (-\tfrac{1}{2}) \\ &= \tfrac{1}{2}\,(s^2 + 3s + \tfrac{9}{4}). \end{aligned}$$

For $s \in [-\frac{1}{2}, +\frac{1}{2}]$ we get

$$\begin{aligned}\int_{s-\frac{1}{2}}^{s+\frac{1}{2}} (1-|x|)\,dx &= \int_{s-\frac{1}{2}}^{0} (1+x)\,dx + \int_{0}^{s+\frac{1}{2}} (1-x)\,dx \\ &= [x+\tfrac{1}{2}x^2]_{s-\frac{1}{2}}^{0} + [x-\tfrac{1}{2}x^2]_0^{s+\frac{1}{2}} = \tfrac{3}{4} - s^2.\end{aligned}$$

For $s \in [\frac{1}{2}, \frac{3}{2}]$ we have

$$\int_{s-\frac{1}{2}}^{1} (1-x)\,dx = [x-\tfrac{1}{2}x^2]_{s-\frac{1}{2}}^{1} = \tfrac{1}{2}(s^2 - 3s + \tfrac{9}{4}).$$

Thus for f_{S_3} the following explicit description emerges:

$$f_{S_3}(s) = \begin{cases} 0 & \text{if } s \notin [-\frac{3}{2}, \frac{3}{2}], \\ \frac{1}{2}(s^2+3s+9/4) & \text{if } s \in [-\frac{3}{2}, -\frac{1}{2}], \\ \frac{3}{4} - s^2 & \text{if } s \in [-\frac{1}{2}, \frac{1}{2}], \\ \frac{1}{2}(s^2-3s+9/4) & \text{if } s \in [\frac{1}{2}, \frac{3}{2}]. \end{cases}$$

□

(v) We have $\mathbb{E}(S_3) = 0$ and $\mathrm{var}(S_3) = \frac{1}{4}$ (see also (iii)). Sketch the density of S_3 and that of the $N(0, \frac{1}{4})$-distribution in one figure.

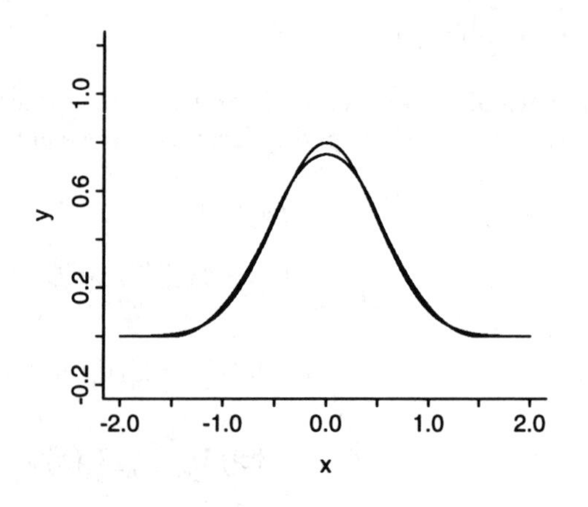

Exercise 52

Given are two statistically independent variables X and Y, having a common probability density of (exponential) type

$$f(x,\theta) = \frac{1}{\theta}\, e^{-x/\theta} \quad \text{if } x \geq 0,$$

and $f(x,\theta) = 0$ elsewhere. Here θ is a fixed positive number.

(i) Prove that $X+Y$ and Y/X are statistically independent.

Proof. Denote $V := X+Y$ and $W := Y/X$. We will prove that $F_{V,W}(a,b) = F_V(a)\,F_W(b)$ for all $a,b \in \mathbb{R}$. Evidently we may restrict ourselves to the case that $a,b > 0$.

Using the methods of I.10, we see that

$$\begin{aligned}
F_V(a) &= \mathbb{P}(X+Y \le a) = \mathbb{P}((X,Y) \in G) \\
&= \iint_G \frac{1}{\theta^2} e^{-(x+y)/\theta}\, dx\, dy \\
&= \int_0^a \int_0^{a-y} \frac{1}{\theta^2} e^{-(x+y)/\theta}\, dx\, dy \\
&= \int_0^a \frac{1}{\theta} e^{-y/\theta} \left[-e^{-x/\theta}\right]_0^{a-y} dy \\
&= \int_0^a \frac{1}{\theta} e^{-y/\theta} \left(1 - e^{(y-a)/\theta}\right) dy \\
&= \left[-e^{-y/\theta}\right]_0^a - \frac{a}{\theta} e^{-a/\theta} = 1 - \left(1 + \frac{a}{\theta}\right) e^{-a/\theta},
\end{aligned}$$

with G as depicted.

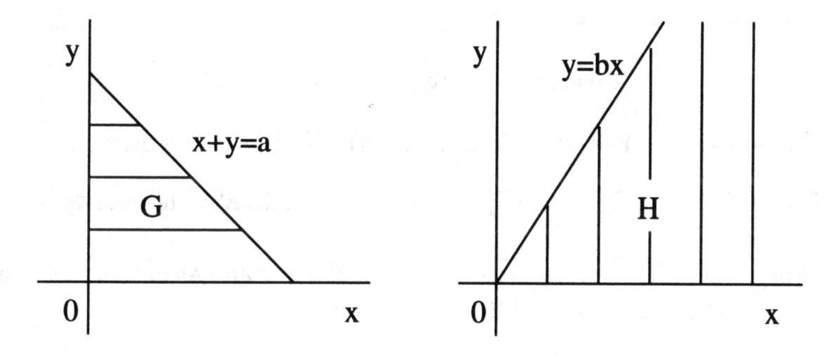

On the other hand (H is as in the picture), in a similar way

$$\begin{aligned}
F_W(b) &= \mathbb{P}(Y/X \le b) = \mathbb{P}((X,Y) \in H) \\
&= \int_0^{+\infty} \int_0^{bx} \frac{1}{\theta^2} e^{-(x+y)/\theta}\, dy\, dx \\
&= \int_0^{+\infty} \frac{1}{\theta} e^{-x/\theta} \left[-e^{-y/\theta}\right]_0^{bx} dx \\
&= \left[-e^{-x/\theta}\right]_0^{+\infty} - \int_0^{+\infty} \frac{1}{\theta} e^{-(1+b)x/\theta}\, dx \\
&= 1 + \frac{1}{1+b} \left[e^{-(1+b)x/\theta}\right]_0^{+\infty} = 1 - \frac{1}{1+b} = \frac{b}{1+b}.
\end{aligned}$$

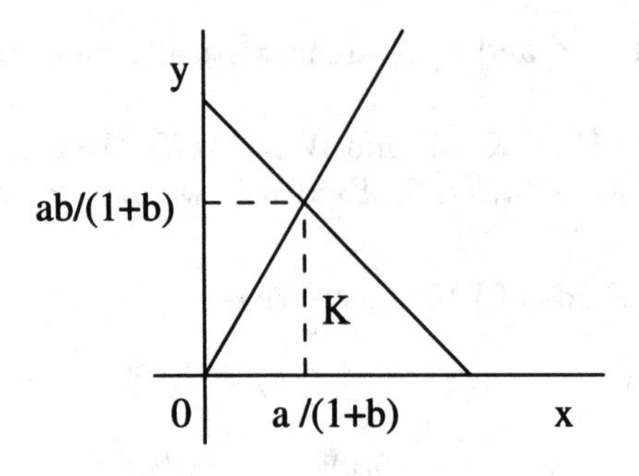

Finally, taking K as the intersection of G and H (see the second figure),

$$\begin{aligned} F_{V,W}(a,b) &= \mathbb{P}(V \le a, W \le b) = \mathbb{P}((V,W) \in K) \\ &= \iint_K \frac{1}{\theta^2} e^{-(x+y)/\theta}\, dx\, dy \\ &= \int_0^{ab/(1+b)} \int_{y/b}^{a-y} \frac{1}{\theta^2} e^{-(x+y)/\theta}\, dx\, dy \\ &= \cdots = \frac{b}{1+b} \left\{1 - e^{-a/\theta} \left(1 + \frac{a}{\theta}\right)\right\}. \end{aligned}$$

It follows that indeed

$$F_{V,W}(a,b) = F_V(a)\, F_W(b) \quad \text{for all } a, b \in \mathbb{R}.$$

Therefore $X + Y$ and Y/X are statistically independent. □

(ii) Prove that $X + Y$ and $X/(X + Y)$ are statistically independent.

Proof. As $X/(X+Y) = (1+Y/X)^{-1}$, the statement is a direct consequence of (i) and Proposition I.4.2. □

Chapter 2
Statistics and their distributions, estimation theory

2.1 Summary of Chapter II

2.1.1 Introduction (II.1)

A *sample* is understood to be a statistically independent sequence $X_1, \ldots, X_n$ of stochastic variables, all of them enjoying one and the same probability distribution. This common distribution is called the *(probability) distribution of the population.*

If $g : \mathbb{R}^n \to \mathbb{R}$ is a Borel function and $X_1, \ldots, X_n$ a sample, then the stochastic variable $g(X_1, \ldots, X_n)$ is said to be a *statistic.* An elementary, but important example of a statistic is the *sample mean* $\overline{X}$, which is obtained by setting $g(x_1, \ldots, x_n) := \frac{1}{n}(x_1 + \cdots + x_n)$:

$$\overline{X} := g(X_1, \ldots, X_n) = \frac{X_1 + \cdots + X_n}{n}.$$

If the population has mean μ and variance σ^2, we have $\mathbb{E}(\overline{X}) = \mu$ and $\operatorname{var}(\overline{X}) = \sigma^2/n$. It follows from this that in the case of a $N(\mu, \sigma^2)$-distributed population the statistic $\overline{X}$ is $N(\mu, \sigma^2/n)$-distributed.

The *sample variance* S^2 of a sample $X_1, \ldots, X_n$ is defined by

$$S^2 := \frac{1}{n-1} \sum_{i=1}^{n} (X_i - \overline{X})^2.$$

We have: $\mathbb{E}(S^2) = \sigma^2$. For this reason outcomes of S^2 are frequently interpreted as estimations of the (unknown) variance σ^2 of the population. Outcomes of $\overline{X}$ are often considered as being estimations of the (unknown) mean μ of the population.

2.1.2 The gamma distribution and the χ^2-distribution (II.2)

The *gamma function* $\Gamma : (0, +\infty) \to (0, +\infty)$ is defined by

$$\Gamma(x) := \int_0^{+\infty} t^{x-1} e^{-t}\, dt.$$

For all $x > 0$ we have $\Gamma(x+1) = x\,\Gamma(x)$. Because $\Gamma(1) = 1$, it follows from this that $\Gamma(k+1) = k!$ for all $k = 1, 2, \ldots$. Furthermore we have $\Gamma(\frac{1}{2}) = \sqrt{\pi}$.

A stochastic variable X is (by definition) *gamma-distributed* with parameters α and β, where $\alpha, \beta > 0$, if its probability density function is given by

$$f_X(x) = \frac{1}{\beta^\alpha \Gamma(\alpha)} x^{\alpha-1} e^{-x/\beta} \quad \text{if } x \geq 0$$

and $f_X(x) = 0$ elsewhere. If so, then the moment generating function of X is defined on the interval $(-\infty, 1/\beta)$ and is given by $M_X(t) = (1-\beta t)^{-\alpha}$. In such a case we have $\mathbb{E}(X) = \alpha\beta$ and $\operatorname{var}(X) = \alpha\beta^2$.

A stochastic variable X is said to be *χ^2-distributed with n degrees of freedom* (shortly: $\chi^2(n)$-distributed) if X enjoys a gamma-distribution with $\alpha = \frac{1}{2}n$ and $\beta = 2$. For such X we have $\mathbb{E}(X) = n$ and $\operatorname{var}(X) = 2n$. If $X_1, \ldots, X_n$ is a sample from a $N(0,1)$-distributed population, then the variable $X_1^2 + \cdots + X_n^2$ is χ^2-distributed with n degrees of freedom.

The χ^2-distribution enjoys the following two 'permanence properties'. If X and Y are statistically independent variables which are χ^2-distributed with m and n degrees of freedom, then $X+Y$ is χ^2-distributed with $m+n$ degrees of freedom. On the other hand, if $V = X + Y$ where X and Y are independent, V is $\chi^2(n)$-distributed, X is $\chi^2(m)$-distributed and $m < n$, then Y is $\chi^2(n-m)$-distributed.

If $X_1, \ldots, X_n$ is a sample from a $N(\mu, \sigma^2)$-distributed population, then the statistics S^2 and $\overline{X}$ are statistically independent. Moreover, the statistic $(n-1)S^2/\sigma^2$ is $\chi^2(n-1)$-distributed (Theorem II.2.12). For large n the $\chi^2(n)$-distribution is approximately the same as a $N(n, 2n)$-distribution.

2.1.3 The *t*-distribution (II.3)

A stochastic variable X is said to be *t-distributed with n degrees of freedom* if its probability density function is given by

$$f(x) = \frac{\Gamma\left(\frac{1}{2}(n+1)\right)}{\sqrt{\pi n}\,\Gamma\left(\frac{1}{2}n\right)} \left(1 + \frac{x^2}{n}\right)^{-\frac{1}{2}(n+1)} \quad (\text{all } x \in \mathbb{R}).$$

In a natural way this probability distribution arises as that of a variable of type $Z/\sqrt{Y/n}$ where Z and Y are statistically independent, Z enjoying a $N(0,1)$-distribution and Y a $\chi^2(n)$-distribution. If X is t-distributed with n degrees of freedom, then $\mathbb{E}(X) = 0$ if $n \geq 2$ and $\operatorname{var}(X) = n/(n-2)$ if $n \geq 3$.

The main result in this section is Theorem II.3.2: if $X_1, \ldots, X_n$ is a sample from a $N(\mu, \sigma^2)$-distributed population, then the statistic

$$\frac{\overline{X} - \mu}{S/\sqrt{n}}$$

enjoys a t-distribution with $n-1$ degrees of freedom.

2.1.4 Statistics to measure differences in mean (II.4)

Let $X_1, \ldots, X_m$ and $Y_1, \ldots, Y_n$ be two statistically independent samples from a $N(\mu_X, \sigma_X^2)$ and a $N(\mu_Y, \sigma_Y^2)$-distributed population. Then the statistic $\overline{X} - \overline{Y}$ enjoys a $N(\mu_X - \mu_Y, \sigma_X^2/m + \sigma_Y^2/n)$-distribution. Usually σ_X^2 and σ_Y^2 are unknown; we then replace them by S_X^2 and S_Y^2 respectively. If both m and n are large (≥ 30), the statistic

$$T = \frac{(\overline{X} - \overline{Y}) - (\mu_X - \mu_Y)}{\sqrt{S_X^2/m + S_Y^2/n}}$$

is approximately $N(0,1)$-distributed. If m or n is small then, concerning the distribution of T, a problem arises: the 'Behrens–Fisher-problem'.

Given two independent samples $X_1, \ldots, X_m$ and $Y_1, \ldots, Y_n$ the *pooled variance* S_p^2 is defined by

$$S_p^2 := \frac{(m-1)S_X^2 + (n-1)S_Y^2}{m+n-2}.$$

If $X_1, \ldots, X_m$ and $Y_1, \ldots, Y_n$ are independent samples from respectively a $N(\mu_X, \sigma^2)$- and a $N(\mu_Y, \sigma^2)$-distributed population (that is, if $\sigma_X^2 = \sigma_Y^2$), then the Behrens–Fisher problem can be solved in an elementary way. Namely, under these conditions the statistic

$$\frac{(\overline{X} - \overline{Y}) - (\mu_X - \mu_Y)}{S_p\sqrt{1/m + 1/n}}$$

is t-distributed with $m + n - 2$ degrees of freedom (Theorem II.4.5).

A *paired sample* $(X_1, Y_1), \ldots, (X_n, Y_n)$ is understood to be a statistically independent sequence of stochastic 2-vectors such that the differences $D_i := Y_i - X_i$ all have a common probability distribution. We denote the mean of the sample $D_1, \ldots, D_n$ by $\overline{D}$ and its sample variance by S_D^2. If for all i the statistic D_i is $N(\Delta, \sigma^2)$-distributed, then the statistic

$$\frac{\overline{D} - \Delta}{S_D/\sqrt{n}}$$

is t-distributed with $n - 1$ degrees of freedom.

2.1.5 The *F*-distribution (II.5)

A stochastic variable X is by definition *F-distributed* with m and n degrees of freedom in the numerator and the denominator respectively (more compactly: F_n^m-distributed), if its probability density is given by

$$f(x) = \frac{\Gamma(\frac{1}{2}(m+n))}{\Gamma(\frac{1}{2}m)\Gamma(\frac{1}{2}n)} \left(\frac{m}{n}\right)^{\frac{1}{2}m} x^{\frac{1}{2}m-1} \left(1 + \frac{m}{n}x\right)^{-\frac{1}{2}(m+n)} \quad \text{if } x \geq 0$$

and $f(x) = 0$ elsewhere.

F-distributions arise in the following natural way. If U and V are independent and $\chi^2(m)$ and $\chi^2(n)$-distributed respectively, then the statistic

$$\frac{U/m}{V/n}$$

is F_n^m-distributed (Theorem II.5.2). If X is F_n^m-distributed then $1/X$ enjoys a F_m^n-distribution.

If $X_1, \ldots, X_m$ and $Y_1, \ldots, Y_n$ are independent samples from a $N(\mu_X, \sigma_X^2)$ and a $N(\mu_Y, \sigma_Y^2)$-distributed population respectively, then the statistic

$$T = \frac{S_X^2/\sigma_X^2}{S_Y^2/\sigma_Y^2}$$

enjoys a F_{n-1}^{m-1}-distribution. This statistic is frequently used when comparing σ_X^2 and σ_Y^2.

2.1.6 The beta-distribution (II.6)

The *beta-function* $B : (0, +\infty) \times (0, +\infty) \to (0, +\infty)$ is defined by

$$B(p, q) := \int_0^1 t^{p-1}(1-t)^{q-1}\, dt \qquad (p, q > 0).$$

This integral can be represented in many different ways (see Propositions II.6.2-II.6.5). The beta-function is symmetric: $B(p, q) = B(q, p)$ for all $p, q > 0$. For all such p, q we have moreover

$$B(p, q) = \frac{\Gamma(p)\Gamma(q)}{\Gamma(p+q)}.$$

If the probability density function f of a variable X is given by

$$f(x) = \frac{1}{B(p, q)}\, x^{p-1}(1-x)^{q-1} \quad \text{on } [0, 1]$$

and $f(x) = 0$ elsewhere, then X is said to be *beta-distributed* with parameters p and q. For such stochastic variables we have by Proposition II.6.7:

$$\mathbb{E}(X) = \frac{p}{p+q} \quad \text{and} \quad \operatorname{var}(X) = \frac{pq}{(p+q)^2(p+q+1)}.$$

Now let X_β be beta-distributed with as its parameters the positive integers p and q, and let X_b be binomially distributed with parameters $p+q-1$ and $\theta \in (0, 1)$: then we have $\mathbb{P}(X_\beta \le \theta) = \mathbb{P}(X_b \ge p)$. So we can use statistical tables for the binomial distribution when dealing with a beta-distribution.

2.1.7 Populations that are not normally distributed (II.7)

If a stochastic variable can assume the values 0 and 1 only, then X is said to be *Bernoulli distributed.* The fraction $\theta = \mathbb{P}(X = 1)$ is called the parameter of the Bernoulli distribution. It is easily seen that $\mathbb{E}(X) = \theta$ and $\text{var}(X) = \theta(1-\theta)$. Now let $X_1, \ldots, X_n$ be a sample from a Bernoulli distributed population with parameter θ. Then, by the central limit theorem, for large n the statistic

$$T = \frac{\overline{X} - \theta}{\sqrt{\theta(1-\theta)/n}}$$

is approximately $N(0,1)$-distributed.

2.1.8 Bayesian estimation (II.8)

In this section the *Bayesian point of view* is discussed. We assume that, carrying out an experiment, it is known *beforehand* that the parameter to be estimated enjoys a certain probability distribution. If the experiment shows outcomes that are compatible with this probability distribution, then, using our foreknowledge together with the information supplied by these outcomes, sharper estimations can be made.

Now let X and Y be stochastic variables that are discretely distributed or enjoy an absolutely continuous distribution. The *conditional probability density of X, given $Y = y$,* is defined by

$$f_X(x \,|\, Y = y) := \frac{f_{X,Y}(x,y)}{f_Y(y)} \quad \text{if } f_Y(y) > 0$$

and $f_X(x \,|\, Y = y) := 0$ if $f_Y(y) = 0$.

The next equalities then apply:

$$f_{X,Y}(x,y) = f_X(x \,|\, Y = y) f_Y(y) = f_Y(y \,|\, X = x) f_X(x),$$

and

$$\begin{cases} f_X(x) = \int_{-\infty}^{+\infty} f_X(x \,|\, Y = y) f_Y(y)\, dy, \\ f_Y(y) = \int_{-\infty}^{+\infty} f_Y(y \,|\, X = x) f_X(x)\, dy. \end{cases}$$

Suppose that X is binomially distributed with parameters n and θ. Look upon θ as being the outcome of a stochastic variable Θ which is beta-distributed with parameters p and q. Then the variable Θ is, given $X = k$, beta-distributed with parameters $p' = p + k$ and $q' = q + n - k$.

Another result of this kind is the following. Let X be a $N(\mu, \sigma^2)$-distributed variable, where σ^2 is known and μ the outcome of a variable M which enjoys a $N(\mu_0, {\sigma_0}^2)$-distribution. Then the variable M, given $X = x$, is $N(\mu_1, {\sigma_1}^2)$-distributed where

$$\mu_1 = \frac{\sigma_0^2 x + \sigma^2 \mu_0}{\sigma_0^2 + \sigma^2} \quad \text{and} \quad \frac{1}{\sigma_1^2} = \frac{1}{\sigma_0^2} + \frac{1}{\sigma^2}.$$

2.1.9 Minimum variance estimation (II.9)

Let Θ be any set and let $f : \mathbb{R} \times \Theta \to [0, \infty)$ be a function such that for all θ the map $x \mapsto f(x, \theta)$ presents a probability density (denoted by $f(\bullet, \theta)$). Then we talk about a *family of probability densities* with parameter $\theta \in \Theta$. In connection to this the set Θ is called the *parameter space*. Systematically this section is dealing with populations enjoying a probability density which is a member of such a family $(f(\bullet, \theta))_{\theta \in \Theta}$.

By a *characteristic* of the population we mean a function $\kappa : \Theta \to \mathbb{R}$ (or $\mathbb{C}$). If κ is constant then we talk about a trivial characteristic.

In general we do not know the (exact) value of $\kappa(\theta)$. In order to get an impression of the numerical value of $\kappa(\theta)$ we draw a sample from the population. If we look upon the outcome of a statistic $T = g(X_1, \dots, X_n)$ as being an estimation of $\kappa(\theta)$, then we call T an *estimator of κ based on the sample* $X_1, \dots, X_n$. Generally the expectation value of T depends on θ, therefore we shall frequently write $\mathbb{E}_\theta(T)$ instead of $\mathbb{E}(T)$. T is said to be an *unbiased estimator* of κ if $\mathbb{E}_\theta(T) = \kappa(\theta)$ for *all* $\theta \in \Theta$. Unbiased estimators may not exist. If they do, they are not always useful. By Proposition II.9.1 we always have the following 'trichotomy'. Concerning a characteristic there exists either no unbiased estimator, exactly one unbiased estimator or there exist infinitely many unbiased estimators. If there are infinitely many, there may be a desire to select among them estimators of minimal variance. In a certain sense, estimators of minimal variance can be seen as the best ones.

An estimator T based on a sample $X_1, \dots, X_n$ is called *linear* if it is a linear combination of the variables $X_1, \dots, X_n$. Suppose that we are dealing with a population with a probability density from the family $\{f(\bullet, \theta) : \theta \in \Theta\}$. If for all $\theta \in \Theta$ the mean μ and the variance σ^2 of the population exist, and if μ is a non-trivial characteristic, then $\overline{X}$ is among the unbiased linear estimators of μ the one of minimal variance. $\overline{X}$ is unique in this respect.

Two variables T_1 and T_2 are called *essentially equal* with respect to the family of densities $\{f(\bullet, \theta) : \theta \in \Theta\}$ if for all $\theta \in \Theta$ we have $\mathbb{P}_\theta(T_1 = T_2) = 1$. If so, then T_1 and T_2 are identically distributed.

Next, let $\mathfrak{C}$ be a collection statistics in $X_1, \dots, X_n$, all of them of finite variance. For any characteristic κ we set

$$\mathfrak{C}_\kappa := \{T \in \mathfrak{C} : \mathbb{E}_\theta(T) = \kappa(\theta) \text{ for all } \theta \in \Theta\}.$$

$\mathfrak{C}$ is called *convex* if

$$T_1, T_2 \in \mathfrak{C}\,,\ \lambda \in [0,1] \quad \Longrightarrow \quad \lambda T_1 + (1 - \lambda) T_2 \in \mathfrak{C}.$$

If $\mathfrak{C}$ is convex, then so is $\mathfrak{C}_\kappa$. For all $\theta \in \Theta$ we define $m(\theta)$ by

$$m(\theta) := \inf \{\mathrm{var}_\theta(T) \ : \ T \in \mathfrak{C}_\kappa\}.$$

If for $T_1, T_2 \in \mathfrak{C}_\kappa$ we have

$$\mathrm{var}_\theta(T_1) = m(\theta) = \mathrm{var}_\theta(T_2) \quad \text{for all } \theta \in \Theta,$$

then (Theorem II.9.4) T_1 and T_2 are essentially equal.

Let $\mathfrak{P}_n$ be the set of all permutations of $\{1, 2, \ldots, n\}$. For $\pi \in \mathfrak{P}_n$ we define the orthogonal linear operator $\mathbf{Q}(\pi) : \mathbb{R}^n \to \mathbb{R}^n$ by $(x_1, \ldots, x_n) \mapsto (x_{\pi(1)}, \ldots, x_{\pi(n)})$; such a map is called an *exchange of coordinates.* For all functions $g : \mathbb{R}^n \to \mathbb{R}$ the map $g \circ \mathbf{Q}(\pi)$ will be denoted by g^π. We say that g is *symmetric* if $g = g^\pi$ for all $\pi \in \mathfrak{P}_n$. For any statistic $T = g(X_1, \ldots, X_n)$ we denote $T^\pi := g^\pi(X_1, \ldots, X_n)$. Now the variables T and T^π are always identically distributed. The statistic T is said to be *essentially symmetric* if for all $\pi \in \mathfrak{P}_n$ the statistics T and T^π are essentially equal. A collection $\mathfrak{C}$ of statistics is called *symmetric* if

$$T \in \mathfrak{C} \implies T^\pi \in \mathfrak{C} \quad \text{for all } \pi \in \mathfrak{P}_n.$$

We now state Theorem II.9.6. Suppose $X_1, \ldots, X_n$ is a sample from a population with probability density $f(\bullet, \theta)$ where $\theta \in \Theta$. Let $\mathfrak{C}$ be a symmetric collection of statistics (based on $X_1, \ldots, X_n$), all of them of finite variance. Furthermore, let κ be a characteristic of the population. If $T_{\min}$ is an element in $\mathfrak{C}_\kappa$ such that

$$\mathrm{var}_\theta(T_{\min}) = \inf \{\mathrm{var}_\theta(T) \ : \ T \in \mathfrak{C}_\kappa\} = m(\theta) \quad \text{for all } \theta \in \Theta,$$

then $T_{\min}$ is essentially symmetric.

The content of Theorem II.9.7 is a version of the *information inequality* (also *Cramér–Rao inequality*). It is assumed that Θ is an open interval in $\mathbb{R}$ and that $\kappa : \Theta \to \mathbb{R}$ is a differentiable function. If T is an unbiased estimator of the characteristic κ, then we have (under certain regularity conditions) that

$$(\kappa'(\theta))^2 \ \le \ n \, \mathrm{var}_\theta(T) \, \mathbb{E}_\theta \left[\left(\frac{\partial}{\partial \theta} \log f(X, \theta) \right)^2 \right]$$

for all $\theta \in \Theta$ and $X \in \{X_1, \ldots, X_n\}$. Here the expression

$$\mathbb{E}_\theta \left[\left(\frac{\partial}{\partial \theta} \log f(X, \theta) \right)^2 \right]$$

is called the *Fisher information* supplied by a single measurement. If the Fisher information is non-zero, then the information inequality gives us a lower bound for $\mathrm{var}_\theta(T)$.

2.1.10 Maximum likelihood estimation, sufficiency (II.10)

Instead of searching for unbiased estimators of minimal variance it is also useful to consider *maximum likelihood estimators.* Let $X_1, \ldots, X_n$ be a sample from a population with probability density f. The *likelihood function* of the sample is understood to be the probability density of the stochastic vector $(X_1, \ldots, X_n)$, that is to say, the likelihood function is the map $L : \mathbb{R}^n \to [0, +\infty)$ defined by

$$L(x_1, \ldots, x_n) := f(x_1) \cdots f(x_n).$$

Outcomes $(x_1, \ldots, x_n)$ for which L assumes a relatively large value are considered 'likely'. When dealing with a family $\{f(\bullet, \theta) \; : \; \theta \in \Theta\}$, we denote L_θ instead of L. Given an outcome $(x_1, \ldots, x_n)$ we maximize the function $\theta \mapsto L_\theta(x_1, \ldots, x_n)$. We assume that there is exactly one point $\hat{\theta} \in \Theta$ for which the maximum is realized; such a point $\hat{\theta}$ is called the *maximum likelihood estimation* of θ. Because $\hat{\theta}$ depends generally on $(x_1, \ldots, x_n)$ we shall also write $\hat{\theta}(x_1, \ldots, x_n)$ instead of $\hat{\theta}$, thus looking upon $\hat{\theta}$ as a map $\hat{\theta} : \mathbb{R}^n \to \Theta$. Given a characteristic $\kappa : \Theta \to \mathbb{R}$, the statistic $T = \kappa\,(\hat{\theta}(X_1, \ldots, X_n))$ is said to be the *maximum likelihood estimator* of κ.

An unbiased estimator of minimal variance is not automatically a maximum likelihood estimator (Example 12). Maximum likelihood of an estimator is preserved when carrying out transformations on the estimator; this in contrast to unbiasedness, which is in general lost when applying transformations. Maximum likelihood estimators are always (essentially) symmetric.

An important concept in mathematical statistics is that of sufficiency. Suppose $X_1, \ldots, X_n$ is a sample from a population having a probability density $f(\bullet, \theta)$, where $\theta \in \Theta$. Let $g : \mathbb{R}^n \to \mathbb{R}^p$ be any Borel function and let $\mathbf{T} := g(X_1, \ldots, X_n)$ be the associated (vectorial) statistic. This statistic is said to be *sufficient* for the parameter θ if for all $\mathbf{t}$ the conditional probability distribution of the stochastic n-vector $(X_1, \ldots, X_n)$, given $\mathbf{T} = \mathbf{t}$, does not depend on θ.

A very useful criterion for sufficiency is the so-called *factorization theorem* (Theorem II.10.3). This theorem states that $\mathbf{T}$ is sufficient if and only if the likelihood function L can be factorized in the following way:

$$L(x_1, \ldots, x_n) = h(x_1, \ldots, x_n)\, \varphi(\theta, g(x_1, \ldots, x_n)),$$

where $\varphi : \Theta \times \mathbb{R}^p \to [0, +\infty)$ and where $h : \mathbb{R}^n \to [0, +\infty)$ does not depend on θ. We know that, when dealing with $N(\mu, \sigma^2)$-distributed populations, the vectorial statistic $(\overline{X}, S^2)$ is sufficient for the (vectorial) parameter (μ, σ).

2.2 Exercises to Chapter II

Exercise 1

Prove that $\Gamma\,(x + 1) = x\,\Gamma\,(x)$ for all $x > 0$.

Proof.

$$\begin{aligned}\Gamma(x+1) &= \int_0^{+\infty} t^x e^{-t}\,dt = \int_0^{+\infty} -t^x\,de^{-t} \\ &= \left[-t^x e^{-t}\right]_0^{+\infty} + \int_0^{+\infty} x t^{x-1} e^{-t}\,dt \\ &= x\int_0^{+\infty} t^{x-1}e^{-t}\,dt = x\,\Gamma(x).\end{aligned}$$

□

Exercise 2

Prove that $\Gamma\left(\frac{1}{2}\right) = \sqrt{\pi}$.

Proof.

$$\begin{aligned}\Gamma\left(\tfrac{1}{2}\right) &= \int_0^{+\infty} t^{-\frac{1}{2}} e^{-t}\,dt = \int_0^{+\infty} \left(\tfrac{1}{2}s^2\right)^{-\frac{1}{2}} e^{-\frac{1}{2}s^2}\,d\left(\tfrac{1}{2}s^2\right) \\ &= \sqrt{2}\int_0^{+\infty} e^{-\frac{1}{2}s^2}\,ds = \tfrac{1}{2}\sqrt{2}\int_{-\infty}^{+\infty} e^{-\frac{1}{2}s^2}\,ds \\ &= \tfrac{1}{2}\sqrt{2}\,\sqrt{2\pi} = \sqrt{\pi}.\end{aligned}$$

□

Exercise 3

If X is gamma distributed with parameters α and β, then $\mathbb{E}(X) = \alpha\beta$ and $\mathrm{var}(X) = \alpha\beta^2$. Prove this.

Proof. By Proposition II.2.4 we have $M_X(t) = (1-\beta t)^{-\alpha}$. Consequently

$$M_X'(t) = \alpha\beta(1-\beta t)^{-\alpha-1} \quad \text{and} \quad M_X''(t) = \alpha\beta^2(\alpha+1)(1-\beta t)^{-\alpha-2}.$$

Applying Proposition I.8.2 we get $\mathbb{E}(X) = M_X'(0) = \alpha\beta$ and $\mathbb{E}(X^2) = M_X''(0) = \alpha^2\beta^2 + \alpha\beta^2$. So $\mathrm{var}(X) = \alpha^2\beta^2 + \alpha\beta^2 - \alpha^2\beta^2 = \alpha\beta^2$. □

Exercise 4

Let $X_1, \ldots, X_n$ be a sample from a $N(0,1)$-distributed population. Set $S_n := \sum_{i=1}^n X_i^2$.

(i) Prove that for all $a \in \mathbb{R}$ we have

$$\lim_{n\to\infty} \mathbb{P}\left(\frac{S_n - n}{\sqrt{2n}} \leq a\right) = \Phi(a),$$

where Φ is the distribution function belonging to the $N(0,1)$-distribution.

Proof. Applying Proposition II.2.7 we see that the variables ${X_1}^2, \dots, {X_n}^2$ are all $\chi^2(1)$-distributed. Consequently, by Proposition II.2.8, for all i we have

$$\mathbb{E}(X_i^2) = 1 \quad \text{and} \quad \operatorname{var}(X_i^2) = 2.$$

The variables ${X_1}^2, \dots, {X_n}^2$ constitute (Proposition I.4.2) a statistically independent system. Applying the central limit theorem we conclude that for all $a \in \mathbb{R}$

$$\lim_{n\to\infty} \mathbb{P}\left(\frac{\frac{1}{n}S_n - 1}{\sqrt{2}/\sqrt{n}} \le a\right) = \Phi(a).$$

□

(ii) If for all $n \in \mathbb{N}^*$ the variable T_n is $\chi^2(n)$-distributed, then

$$\lim_{n\to\infty} \mathbb{P}\left(\frac{T_n - n}{\sqrt{2n}} \le a\right) = \Phi(a) \quad \text{for all } a \in \mathbb{R}.$$

Prove this.

Proof. Let $X_1, \dots, X_n$ be a sample from a standard normal distributed population (§I.11, Exercise 42 guarantees the existence of such a system of stochastic variables). Set $S_n := \sum_i X_i^2$. Now S_n and T_n are identically distributed and we therefore have for $n = 1, 2, \dots$

$$\mathbb{P}\left(\frac{S_n - n}{\sqrt{2n}} \le a\right) = \mathbb{P}\left(\frac{T_n - n}{\sqrt{2n}} \le a\right)$$

for all $a \in \mathbb{R}$. Next apply (i). □

Exercise 5

If X is $\chi^2(100)$-distributed, then by Table IV we have $\mathbb{P}(X \le 118.50) = 0.90$. Apply Exercise 4 to approximate $\mathbb{P}(X \le 118.50)$.

Solution. By Exercise 4 we have

$$\mathbb{P}\left(\frac{X - 100}{\sqrt{2 \cdot 100}} \le a\right) \approx \Phi(a),$$

or equivalently, $\mathbb{P}(X \le 10a\sqrt{2} + 100) \approx \Phi(a)$. Setting $10a\sqrt{2} + 100 = 118.50$ provides $a = 1.850/\sqrt{2} = 1.308$. Therefore we get

$$\mathbb{P}(X \le 118.50) \approx \Phi(1.308) = 0.904.$$

We see that this approximation is accurate in two decimals. □

Exercise 6

Let $X_1, \ldots, X_n$ be a sample from a $N(\mu, \sigma^2)$-distributed population. Prove that $\text{var}(S^2) = 2\sigma^4/(n-1)$.

Proof. We know (Theorem II.2.12) that $(n-1)S^2/\sigma^2$ is $\chi^2(n-1)$-distributed. Therefore (Proposition II.2.8) we have

$$\left(\frac{n-1}{\sigma^2}\right)^2 \text{var}(S^2) = \text{var}\left(\frac{n-1}{\sigma^2}S^2\right) = 2(n-1). \qquad \square$$

Exercise 7

Given is a sample X_1, X_2 from a population that is uniformly distributed on the interval $(0, 1)$.

(i) Show that $\mathbb{E}(X_1) = \frac{1}{2}$ and $\text{var}(X_1) = \frac{1}{12}$.

Solution. See §I.11, Exercise 19. $\square$

(ii) Show that $S^2 = \frac{1}{2}(X_1 - X_2)^2$.

Solution.

$$\begin{aligned} 2S^2 &= 2 \cdot \tfrac{1}{2-1} \textstyle\sum_{i=1}^{2} (X_i - \overline{X})^2 = 2\left[(X_1 - \overline{X})^2 + (X_2 - \overline{X})^2\right] \\ &= 2\left[(\tfrac{1}{4}X_1{}^2 - \tfrac{1}{2}X_1X_2 + \tfrac{1}{4}X_2{}^2) + (\tfrac{1}{4}X_1{}^2 - \tfrac{1}{2}X_1X_2 + \tfrac{1}{4}X_2{}^2)\right] \\ &= X_1{}^2 - 2X_1X_2 + X_2{}^2 = (X_1 - X_2)^2. \end{aligned} \qquad \square$$

(iii) Prove that $\mathbb{E}(S^4) = \frac{1}{60}$.

Proof. From (i) we know that $\mathbb{E}(X_1) = \frac{1}{2}$ and $\mathbb{E}(X_1{}^2) = \frac{1}{3}$. Moreover,

$$\mathbb{E}(X_1{}^3) = \int_0^1 x^3 \cdot 1\,dx = \tfrac{1}{4} \quad \text{and} \quad \mathbb{E}(X_1{}^4) = \int_0^1 x^4 \cdot 1\,dx = \tfrac{1}{5}.$$

For X_2 of course the same holds. Furthermore $S^4 = \frac{1}{4}(X_1 - X_2)^4$, hence

$$\begin{aligned} \mathbb{E}(S^4) &= \tfrac{1}{4}\,\mathbb{E}[(X_1 - X_2)^4] \\ &= \tfrac{1}{4}\,\mathbb{E}(X_1{}^4 - 4X_1{}^3X_2 + 6X_1{}^2X_2{}^2 - 4X_1X_2{}^3 + X_2{}^4) \\ &= \tfrac{1}{4}\left\{\mathbb{E}(X_1{}^4) - 4\,\mathbb{E}(X_1{}^3)\mathbb{E}(X_2) + \cdots + \mathbb{E}(X_2{}^4)\right\} \\ &= \tfrac{1}{4}\left\{\ldots\right\} = \tfrac{1}{60}. \end{aligned} \qquad \square$$

(iv) *Question.* Does the result of Exercise 6, in which $\text{var}(S^2) = 2\sigma^4/(n-1)$, apply in the present situation?

Solution. Here $2\sigma^4/(n-1) = 2\left(\frac{1}{12}\right)^2 = \frac{1}{72}$ whereas $\text{var}(S^2) = \mathbb{E}(S^4) - \mathbb{E}(S^2)^2 = \frac{1}{60} - \left(\frac{1}{12}\right)^2 = \frac{7}{720}$. This shows that the result in Exercise 6 does not apply when dealing with uniformly distributed populations. $\square$

(v) Determine the numerical value of $\mathbb{P}(\overline{X} \leq \frac{1}{4})$ and $\mathbb{P}\left(S^2 \geq \frac{1}{8}\right)$.

Solution.

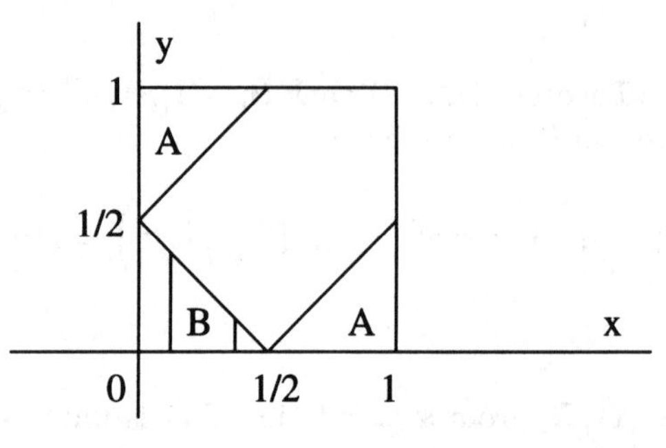

For the density f_{X_1,X_2} of the 2-vector (X_1, X_2) we have

$$f_{X_1,X_2}(x_1,x_2) = \begin{cases} 1 & \text{if } x_1, x_2 \in [0,1] \times [0,1], \\ 0 & \text{elsewhere.} \end{cases}$$

Consequently

$$\mathbb{P}(\overline{X} \leq \tfrac{1}{4}) = \mathbb{P}\left(X_1 + X_2 \leq \tfrac{1}{2}\right) = [\text{area of } B] = \tfrac{1}{8},$$

with B as in the figure. In the same way we find (A as in the figure):

$$\begin{aligned} \mathbb{P}\left(S^2 \geq \tfrac{1}{8}\right) &= \mathbb{P}\left[(X_1 - X_2)^2 \geq \tfrac{1}{4}\right] \\ &= \mathbb{P}\left(|X_1 - X_2| \geq \tfrac{1}{2}\right) = [\text{area of } A] = 2 \cdot \tfrac{1}{8} = \tfrac{1}{4}. \end{aligned}$$

□

(vi) Determine $\mathbb{P}\left(S^2 \geq \frac{1}{8} \text{ and } \overline{X} \leq \frac{1}{4}\right)$.

Solution. We have (see the figure above):

$$\mathbb{P}\left(S^2 \geq \tfrac{1}{8} \text{ and } \overline{X} \leq \tfrac{1}{4}\right) = \mathbb{P}\left[(X_1, X_2) = (0, \tfrac{1}{2}) \text{ or } (X_1, X_2) = (\tfrac{1}{2}, 0)\right] = 0.$$

□

(vii) Prove that S^2 and $\overline{X}$ are statistically dependent.
Isn't this in contradiction to Theorem II.2.11, which states that S^2 and $\overline{X}$ are statistically independent?

Solution. By (v) we learn that $\mathbb{P}\left(S^2 \geq \frac{1}{8}\right) \mathbb{P}\left(\overline{X} \leq \frac{1}{4}\right) \neq 0$. Aside from this (by (vi)) we have $\mathbb{P}\left(S^2 \geq \frac{1}{8} \text{ and } \overline{X} \leq \frac{1}{4}\right) = 0$, which shows that S^2 and $\overline{X}$ are not statistically independent.

This phenomenon is not at all in contradiction to Theorem II.2.11, for this theorem is about *normally* distributed populations. □

Exercise 8

Suppose X is t-distributed with n degrees of freedom, where $n \geq 3$. Prove that $\mathbb{E}(X) = 0$ and $\mathrm{var}(X) = n/(n-2)$.

Proof. The probability density of X is given by

$$f_X(x) = f(x) = \frac{\Gamma\left(\frac{1}{2}(n+1)\right)}{\sqrt{\pi n}\,\Gamma\left(\frac{1}{2}n\right)}\left(1+\frac{x^2}{n}\right)^{-\frac{1}{2}(n+1)} \qquad (x \in \mathbb{R}).$$

It follows that

$$\begin{aligned}
\mathbb{E}(X^2) &= \frac{\Gamma\left(\frac{1}{2}(n+1)\right)}{\sqrt{\pi n}\,\Gamma\left(\frac{1}{2}n\right)} \int_{-\infty}^{+\infty} x^2\left(1+\frac{x^2}{n}\right)^{-\frac{1}{2}(n+1)} dx \\
(x := \sqrt{n}\,s) \quad &= \frac{\Gamma\left(\frac{1}{2}(n+1)\right)}{\sqrt{\pi n}\,\Gamma\left(\frac{1}{2}n\right)} \int_{-\infty}^{+\infty} (\sqrt{n}\,s)^2\left(1+\frac{(\sqrt{n}\,s)^2}{n}\right)^{-\frac{1}{2}(n+1)} d(\sqrt{n}\,s) \\
&= \frac{\Gamma\left(\frac{1}{2}(n+1)\right)}{\sqrt{\pi}\,\Gamma\left(\frac{1}{2}n\right)}\, 2n \int_0^{+\infty} s^2(1+s^2)^{-\frac{1}{2}(n+1)}\, ds \\
(\text{Prop. II.6.4}) \quad &= \frac{\Gamma\left(\frac{1}{2}(n+1)\right)}{\Gamma\left(\frac{1}{2}\right)\Gamma\left(\frac{1}{2}n\right)}\, n\, B\left(\frac{3}{2}, \frac{n-2}{2}\right) \\
(\text{Prop. II.6.6}) \quad &= n\, \frac{\Gamma\left(\frac{1}{2}(n+1)\right)}{\Gamma\left(\frac{1}{2}\right)\Gamma\left(\frac{1}{2}n\right)}\, \frac{\Gamma\left(\frac{3}{2}\right)\Gamma\left(\frac{1}{2}n-1\right)}{\Gamma\left(\frac{1}{2}(n+1)\right)} \\
(\text{Prop. II.2.1}) \quad &= n\, \frac{\frac{1}{2}\Gamma\left(\frac{1}{2}\right)\Gamma\left(\frac{1}{2}n-1\right)}{\Gamma\left(\frac{1}{2}\right)\left(\frac{1}{2}n-1\right)\Gamma\left(\frac{1}{2}n-1\right)} = \frac{n}{n-2} \qquad (*)
\end{aligned}$$

Part of the conclusion is that $\mathbb{E}(X^2)$ exists and therefore also $\mathbb{E}(X)$ (see §I.11 Exercise 18). The function $g : x \mapsto x f_X(x)$ enjoys the property that $g(-x) = -g(x)$. Therefore

$$\mathbb{E}(X) = \int_{-\infty}^{+\infty} x f_X(x)\, dx = \int_{-\infty}^{+\infty} g(x)\, dx = 0.$$

It follows from this, together with $(*)$, that $\mathrm{var}(X) = n/(n-2)$. □

Exercise 9

In this exercise our goal is to prove that for large n the t-distribution with n degrees of freedom is approximately the same as the $N(0,1)$-distribution. We take for granted that $\log\Gamma(x)$ presents a convex function.

(i) Prove that for all $x > 0$

$$\frac{\Gamma\left(x+\frac{1}{2}\right)}{\sqrt{x}\,\Gamma(x)} \leq 1.$$

Proof. Starting from the fact that $\log \Gamma(x)$ is a convex function, we may write

$$(\log \Gamma)(x + \tfrac{1}{2}) \leq \tfrac{1}{2}\{(\log \Gamma)(x) + (\log \Gamma)(x+1)\} \quad \text{for all } x > 0.$$

Hence

$$\begin{aligned}
\log(x^{\frac{1}{2}}\,\Gamma(x)) &= \log \Gamma(x) + \log x^{\frac{1}{2}} \\
&= \tfrac{1}{2}(2 \log \Gamma(x) + \log x) \\
&= \tfrac{1}{2}(\log \Gamma(x) + \log(x\Gamma(x))) \\
&= \tfrac{1}{2}(\log \Gamma(x) + \log \Gamma(x+1)) \\
&\geq (\log \Gamma)(x + \tfrac{1}{2}).
\end{aligned}$$

Consequently, for all $x > 0$

$$\log \frac{\Gamma\left(x + \frac{1}{2}\right)}{x^{\frac{1}{2}}\,\Gamma(x)} = \log\, \Gamma\left(x + \tfrac{1}{2}\right) - \log\, x^{\frac{1}{2}}\Gamma(x) \;\leq\; 0 = \log 1.$$

Because $x \mapsto \log x$ is an increasing function this completes the proof. □

(ii) Prove that

$$\sqrt{\frac{x - \frac{1}{2}}{x}} \;\leq\; \frac{\Gamma\left(x + \frac{1}{2}\right)}{\sqrt{x}\,\Gamma(x)} \;\leq\; 1 \quad \text{for all } x > \tfrac{1}{2}.$$

Proof. The second inequality was proved in (i). The first one is equivalent to

$$\sqrt{x - \tfrac{1}{2}}\;\Gamma(x) \;\leq\; \Gamma\left(x + \tfrac{1}{2}\right),$$

which, in turn (by Proposition II.2.1), is equivalent to

$$\sqrt{x - \tfrac{1}{2}}\;\Gamma(x) \;\leq\; \left(x - \tfrac{1}{2}\right)\Gamma\left(x - \tfrac{1}{2}\right).$$

Replacing $x - \frac{1}{2}$ by x, the latter is exactly (i). □

(iii) Show that

$$\lim_{n \to \infty} \frac{\Gamma\left(\frac{1}{2}(n+1)\right)}{\sqrt{\pi n}\,\Gamma\left(\frac{1}{2}n\right)} = \frac{1}{\sqrt{2\pi}}.$$

Solution. By (ii) we have for $n = 2, 3, \ldots$ that

$$1 \;\geq\; \frac{\Gamma\left(\frac{1}{2}(n+1)\right)}{\sqrt{\frac{1}{2}n}\;\Gamma\left(\frac{1}{2}n\right)} \;\geq\; \sqrt{\frac{n-1}{n}}.$$

We see that the expression in the middle is 'sandwiched' between two expressions having limit 1 for $n \to \infty$. The result follows from this. □

(iv) Let f_n be the probability density of the t-distribution with n degrees of freedom. Prove that

$$\lim_{n\to\infty} f_n(x) = \frac{1}{\sqrt{2\pi}} e^{-\frac{1}{2}x^2} \quad \text{for all } x \in \mathbb{R}.$$

Proof. By definition of the t-distribution, the function f_n is given by

$$f_n(x) = \frac{\Gamma(\frac{1}{2}(n+1))}{\sqrt{\pi n}\,\Gamma(\frac{1}{2}n)} \left(1 + x^2/n\right)^{-\frac{1}{2}(n+1)}.$$

For all $x \in \mathbb{R}$ we have

$$\begin{aligned}
\lim_{n\to\infty} (1 + x^2/n)^{-\frac{1}{2}(n+1)} &= \left[\lim_{n\to\infty} (1 + x^2/n)^{n+1}\right]^{-\frac{1}{2}} \\
&= \left[\lim_{n\to\infty} (1 + x^2/n)^{n}\right]^{-\frac{1}{2}} \left[\lim_{n\to\infty} (1 + x^2/n)\right]^{-\frac{1}{2}} \\
&= e^{-\frac{1}{2}x^2}.
\end{aligned}$$

This, together with (iii), proves the statement in the exercise. □

(v) Let F_n be the distribution function belonging to the t-distribution with n degrees of freedom and let Φ be that of the $N(0,1)$-distribution. Prove that for all $x \in \mathbb{R}$

$$\lim_{n\to\infty} F_n(x) = \Phi(x).$$

Proof. We can write

$$\begin{aligned}
\lim_{n\to\infty} F_n(x) &= \lim_{n\to\infty} \int_{-\infty}^{x} f_n(y)\,dy = \int_{-\infty}^{x} \lim_{n\to\infty} f_n(y)\,dy \\
&\overset{\text{(iv)}}{=} \int_{-\infty}^{x} \frac{1}{\sqrt{2\pi}} e^{-\frac{1}{2}y^2}\,dy = \Phi(x).
\end{aligned}$$

Here the second equality has to be justified. This can be done by applying Lebesgue's dominated convergence theorem (see Appendix A). Define $g : x \mapsto 2/(2+x^2)$. Then $|f_n| \le g$ and $\int g(x)\,dx < +\infty$. This guarantees that

$$\lim_{n\to\infty} \int_{-\infty}^{x} f_n(y)\,dy = \int_{-\infty}^{x} \lim_{n\to\infty} f_n(y)\,dy.$$ □

Exercise 10

Let F_n and Φ be as in Exercise 9(v). Determine $|F_{20}(1.325) - \Phi(1.325)|$ by means of some statistical table.

Solution. Reading in the Tables II and III we learn that

$$|F_{20}(1.325) - \Phi(1.325)| = |0.90 - 0.907| = 0.01.$$

We see that already for $n = 20$ the numerical value of $F_n(1.325)$ can be approximated by $\Phi(1.325)$ in a fairly accurate way. □

Exercise 11

Suppose that X is F_n^m-distributed. Determine $\mathbb{E}(X)$ (for $n \geq 3$) and $\operatorname{var}(X)$ (for $n \geq 5$).

Solution. For $n \geq 3$ we see, carrying out a substitution $y = \frac{m}{n}x$, that

$$\begin{aligned}
&\int_{-\infty}^{+\infty} |x| f_X(x)\,dx \\
&= \int_0^{+\infty} \frac{\Gamma\left(\frac{1}{2}(m+n)\right)}{\Gamma\left(\frac{1}{2}m\right)\Gamma\left(\frac{1}{2}n\right)} \left(\frac{m}{n}\right)^{\frac{1}{2}m} x^{\frac{1}{2}m} \left(1+\frac{m}{n}x\right)^{-\frac{1}{2}(m+n)} dx \\
&= \frac{n}{m} \frac{\Gamma\left(\frac{1}{2}(m+n)\right)}{\Gamma\left(\frac{1}{2}m\right)\Gamma\left(\frac{1}{2}n\right)} \int_0^{+\infty} y^{\frac{1}{2}m}(1+y)^{-\frac{1}{2}(m+n)}\,dy \\
&\overset{(*)}{=} \frac{n}{m} \frac{\Gamma\left(\frac{1}{2}(m+n)\right)}{\Gamma\left(\frac{1}{2}m\right)\Gamma\left(\frac{1}{2}n\right)} B\left(\tfrac{1}{2}m+1, \tfrac{1}{2}n-1\right) \\
&\overset{(**)}{=} \frac{n}{m} \frac{\Gamma\left(\frac{1}{2}(m+n)\right)}{\Gamma\left(\frac{1}{2}m\right)\left(\frac{1}{2}n-1\right)\Gamma\left(\frac{1}{2}n-1\right)} \frac{\frac{1}{2}m\Gamma\left(\frac{1}{2}m\right)\Gamma\left(\frac{1}{2}n-1\right)}{\Gamma\left(\frac{1}{2}(m+n)\right)} \\
&= \frac{n}{n-2} < +\infty.
\end{aligned}$$

This proves that $\mathbb{E}(X)$ exists and that $\mathbb{E}(X) = n/(n-2)$. At $(*)$ we used Proposition II.6.2; at $(**)$ Theorem II.6.6.

For $n \geq 5$ we have

$$\mathbb{E}(X^2) = \frac{\Gamma\left(\frac{1}{2}(m+n)\right)}{\Gamma\left(\frac{1}{2}m\right)\Gamma\left(\frac{1}{2}n\right)} \left(\frac{m}{n}\right)^{\frac{1}{2}m} \int_0^{+\infty} x^{\frac{1}{2}m+1}\left(1+\frac{m}{n}x\right)^{-\frac{1}{2}(m+n)} dx.$$

Via a substitution $y := \frac{m}{n}x$ we see that

$$\begin{aligned}
\mathbb{E}(X^2) &= \frac{\Gamma\left(\frac{1}{2}(m+n)\right)}{\Gamma\left(\frac{1}{2}m\right)\Gamma\left(\frac{1}{2}n\right)} \left(\frac{n}{m}\right)^2 \int_0^{+\infty} y^{\frac{1}{2}m+1}(1+y)^{-\frac{1}{2}(m+n)}\,dy \\
&= \frac{\Gamma\left(\frac{1}{2}(m+n)\right)}{\Gamma\left(\frac{1}{2}m\right)\Gamma\left(\frac{1}{2}n\right)} \left(\frac{n}{m}\right)^2 B\left(\tfrac{1}{2}m+2, \tfrac{1}{2}n-2\right) \\
&= \left(\frac{n}{m}\right)^2 \frac{\left(\frac{1}{2}m+1\right)\left(\frac{1}{2}m\right)}{\left(\frac{1}{2}n-1\right)\left(\frac{1}{2}n-2\right)} = \frac{n^2(m+2)}{m(n-2)(n-4)} < +\infty.
\end{aligned}$$

It thus appears that

$$\begin{aligned}\operatorname{var}(X) &= \frac{n^2(m+2)}{m(n-2)(n-4)} - \frac{n^2}{(n-2)^2} \\ &= \frac{n^2(m+2)(n-2) - n^2m(n-4)}{m(n-2)^2(n-4)} \\ &= \frac{n^3m + 2n^3 - 2n^2m - 4n^2 - n^3m + 4n^2m}{m(n-2)^2(n-4)} \\ &= \frac{2n^2(m+n-2)}{m(n-2)^2(n-4)}.\end{aligned}$$

□

Exercise 12

Suppose X is F_8^4-distributed.

(i) Show that

$$f_X(x) = \begin{cases} 5x(1+\frac{1}{2}x)^{-6} & \text{if } x \geq 0, \\ 0 & \text{elsewhere.} \end{cases}$$

Solution. By definition of the F-distribution, $f_X(x) = 0$ for $x < 0$. Moreover, for $x \geq 0$ we have

$$\begin{aligned} f_X(x) &= \frac{\Gamma(6)}{\Gamma(4)\Gamma(2)} \left(\tfrac{1}{2}\right)^2 x\left(1+\tfrac{1}{2}x\right)^{-6} \\ &= \tfrac{5!}{3!1!}\,\tfrac{1}{4}x\left(1+\tfrac{1}{2}x\right)^{-6} = 5x\left(1+\tfrac{1}{2}x\right)^{-6}. \end{aligned}$$

□

(ii) Compute $\int_{2.81}^{+\infty} f_X(x)\,dx$.

Solution.

$$\begin{aligned} \int_{2.81}^{+\infty} f_X(x)\,dx &= \int_{2.81}^{+\infty} 5x\left(1+\tfrac{1}{2}x\right)^{-6} dx = \int_{2.81}^{+\infty} -2x\,d\left(1+\tfrac{1}{2}x\right)^{-5} \\ &= \left[-2x\left(1+\tfrac{1}{2}x\right)^{-5}\right]_{2.81}^{+\infty} + 2\int_{2.81}^{+\infty} \left(1+\tfrac{1}{2}x\right)^{-5} dx \\ &= \left[-2x\left(1+\tfrac{1}{2}x\right)^{-5}\right]_{2.81}^{+\infty} - \left[\left(1+\tfrac{1}{2}x\right)^{-4}\right]_{2.81}^{+\infty} \\ &= 0.0698 + 0.0299 = 0.100. \end{aligned}$$

This shows that with a probability of 90% the variable X will assume a value in the interval $[0, 2.81]$. □

(iii) Compare the result of (ii) to the result you get by using the table describing the F-distribution.

Solution. In Table V, describing the F-distribution we find that $F(x) = 0.90$ for $x = 2.81$. □

Exercise 13

If X is t-distributed with n degrees of freedom then the variable $Y = X^2$ is F_n^1-distributed. Prove this.

Proof. Let Z_1, Z_2 be two statistically independent variables such that Z_1 is $N(0,1)$-distributed and Z_2 is $\chi^2(n)$-distributed (for the existence of such a pair see §I.11 Exercise 42). Now X and $Z_1/\sqrt{Z_2/n}$ are identically distributed. By Proposition II.2.7 the variable ${Z_1}^2$ is $\chi^2(1)$-distributed. Therefore (by Theorem II.5.1) the variable

$$\left(\frac{Z_1}{\sqrt{Z_2/n}}\right)^2 = \frac{{Z_1}^2/1}{{Z_2}^2/n}$$

is F_n^1-distributed. Applying §I.11 Exercise 27, we conclude that $Y = X^2$ is also F_n^1-distributed. □

Exercise 14

Suppose X is beta-distributed with parameters $p = \frac{1}{2}m$ and $q = \frac{1}{2}n$. Prove that the variable $Y := nX \,/\, m(1-X)$ is F_n^m-distributed.

Proof. Define $\varphi : (0,1) \to (0,+\infty)$ by $\varphi(x) := nx/m(1-x)$; then

$$\varphi'(x) = \frac{n}{m(1-x)^2} > 0 \quad \text{and} \quad \varphi^{-1}(y) = my/(n+my).$$

Now we have

$$F_Y(y) = \mathbb{P}(Y \le y) = \mathbb{P}(X \le \varphi^{-1}(y)) = F_X(\varphi^{-1}(y)).$$

By differentiation we obtain (see also §I.11 Exercise 33)

$$f_Y(y) = f_X(\varphi^{-1}(y))\,(\varphi^{-1})'(y) = f_X(\varphi^{-1}(y))\,/\varphi'(\varphi^{-1}(y))\,.$$

Consequently

$$f_Y(y) = f_X\left(\frac{my}{n+my}\right)\frac{m}{n}\left(1-\frac{my}{n+my}\right)^2 = f_X\left(\frac{my}{n+my}\right)\frac{nm}{(n+my)^2}$$

if $y > 0$ (for $y \le 0$ of course $F_Y(y) = f_Y(y) = 0$).

By hypothesis X is beta-distributed with parameters $p = \frac{1}{2}m$ and $q = \frac{1}{2}n$, so

$$f_X(x) = \frac{x^{\frac{1}{2}m-1}(1-x)^{\frac{1}{2}n-1}}{B\left(\frac{1}{2}m, \frac{1}{2}n\right)} \qquad \text{if } x \in (0,1)$$

and $f_X(x) = 0$ elsewhere.

Summarizing we see that $f_Y(y) = 0$ if $y \le 0$ and that for all $y > 0$

$$\begin{aligned} f_Y(y) &= \frac{nm}{(n+my)^2} \frac{\left(\frac{my}{n+my}\right)^{\frac{1}{2}m-1}\left(1-\frac{my}{n+my}\right)^{\frac{1}{2}n-1}}{B\left(\frac{1}{2}m,\frac{1}{2}n\right)} \\ &= \frac{n^{\frac{1}{2}n}m^{\frac{1}{2}m}y^{\frac{1}{2}m-1}}{(n+my)^{\frac{1}{2}(m+n)}B\left(\frac{1}{2}m,\frac{1}{2}n\right)} \\ &= \frac{\Gamma\left(\frac{1}{2}(m+n)\right)}{\Gamma\left(\frac{1}{2}m\right)\Gamma\left(\frac{1}{2}n\right)}\left(\frac{m}{n}\right)^{\frac{1}{2}m} y^{\frac{1}{2}m-1}\left(1+\frac{m}{n}y\right)^{-\frac{1}{2}(m+n)}. \end{aligned}$$ □

Exercise 15

Given is a stochastic variable X which is beta-distributed with parameters 4 and 7.

(i) Determine the numerical value of $\mathbb{P}(X \le 0.2)$.

Solution. Let X_b be a binomially distributed variable with parameters 10 and 0.2. Then, applying Proposition II.6.9 we arrive at $\mathbb{P}(X \le 0.2) = \mathbb{P}(X_b \ge 4) = 1 - \mathbb{P}(X_b \le 3) = 0.12$. □

(ii) Sketch the graph of F_X.

Solution. Compute, in the way we did in (i), the numerical value of $F_X(x) = \mathbb{P}(X \le x)$ for $x = 0.1, 0.2, \ldots, 0.9$. This leads to the following figure:

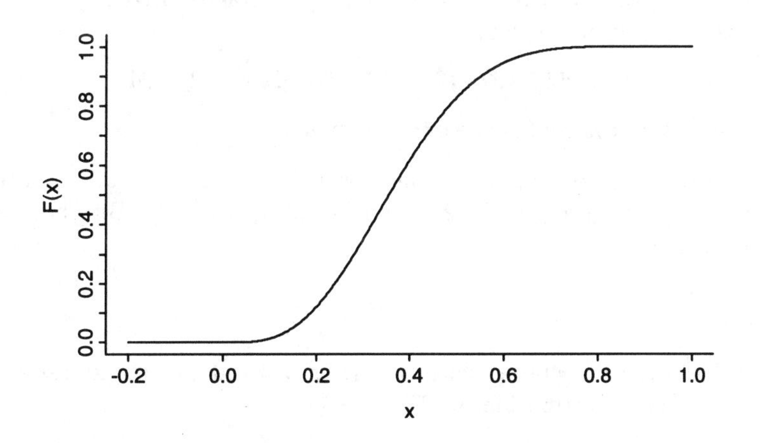

(iii) The *median* is the number m for which $\mathbb{P}(X \le m) = 0.5$. Determine the median of X.

Solution. If X_b is binomially distributed with parameters $n = 10$ and $\theta \approx 0.35$ then $\mathbb{P}(X_b \le 3) \approx 0.50$. It follows from this that $m \approx 0.35$ (see also the figure above). □

(iv) *Question.* Do we have $\mathbb{E}(X) = m$?

Solution. By Proposition II.6.7

$$\mathbb{E}(X) = \tfrac{4}{4+7} = 0.364.$$

Hence the median and the expectation value of X are not quite the same, though they do not differ very much here. □

(v) Are there stochastic variables such that m is not uniquely determined by the probability equation $\mathbb{P}(X \leq m) = 0.5$?

Solution. Yes. If X has a distribution function as plotted in the figure below, then $\mathbb{P}(X \leq m) = 0.5$ for all $m \in [a, b]$.

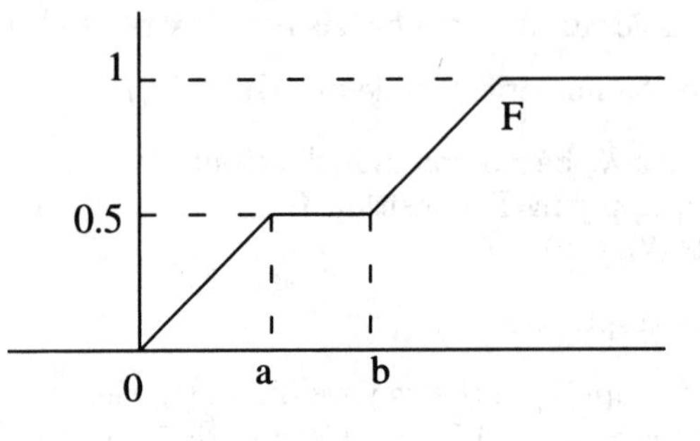

Exercise 16

Suppose $T_1, \ldots, T_9$ is a sample from a $N(\mu_1, {\sigma_1}^2)$-distributed population. The outcome is given by the sequence

220, 205, 192, 198, 201, 207, 195, 201, 204.

(i) Construct a 95% confidence interval for μ_1.

Solution. Elementary calculations show that the outcomes of $\overline{T}$ and the sample variance S_1^2 are given by $\overline{T} = 202.56$ and $S_1 = 8.1103$. By Theorem II.3.2 the variable

$$Y := \frac{\overline{T} - \mu_1}{S_1/\sqrt{9}}$$

is (a priori) t-distributed with 8 degrees of freedom. Using Table III, describing the t-distribution, we learn that

$$\mathbb{P}(|Y| \leq 2.306) = 0.95.$$

Substituting the outcomes of $\overline{T}$ and S_1 in the inequality

$$-2.306 \leq \frac{\overline{T} - \mu_1}{S_1/\sqrt{9}} \leq 2.306$$

we obtain (196.3, 208.8) as a 95% confidence interval for μ_1. □

(ii) Construct a 95% confidence interval for σ_1.

Solution. By Theorem II.2.12 the variable $8S_1^2/\sigma_1^2$ is $\chi^2(8)$-distributed. Cutting off a 2.5% left and a 2.5% right tail we find that

$$\mathbb{P}\left(2.18 \leq 8S_1^2/\sigma_1^2 \leq 17.53\right) = 0.95.$$

Substituting the outcome of S_1 in the inequality

$$2.18 \;\leq\; 8S_1^2/\sigma_1^2 \;\leq\; 17.53$$

we obtain (5.48, 15.54) as a 95% confidence interval for σ_1. □

(iii) A second sample $V_1, \ldots, V_7$ from a $N(\mu_2, {\sigma_2}^2)$-distributed population provides $\overline{V} = 207.714$ and $S_2^2 = 36.238$. Construct a 90% confidence interval for σ_1/σ_2.

Solution. By Theorem II.5.3 the variable $(\sigma_2^2/\sigma_1^2)\,(S_1^2/S_2^2)$ enjoys (a priori) a F_6^8-distribution. It follows from this that

$$\mathbb{P}\left(\tfrac{1}{3.58} \leq \;(\sigma_2^2/\sigma_1^2)\,(S_1^2/S_2^2) \;\leq\; 4.15\right) = 0.90.$$

Substituting the outcomes of S_1^2 and S_2^2 in this inequality we obtain (0.66, 2.55) as a 90% confidence interval for σ_1/σ_2. □

(iv) Suppose $\sigma_1 = \sigma_2 = \sigma$. Find a 95% confidence interval for $\mu_2 - \mu_1$.

Solution. By Proposition II.4.5 the variable

$$Y := \frac{\overline{T} - \overline{V} - (\mu_1 - \mu_2)}{S_p\sqrt{\frac{1}{9} + \frac{1}{7}}}$$

is (a priori) t-distributed with 14 degrees of freedom. We have outcomes $\overline{T} - \overline{V} = -5.154$ and $S_p^2 = (8S_1^2 + 6S_2^2)/14 = 53.117$ so that $S_p = 7.288$. Using Table III it can be read off that

$$\mathbb{P}(|Y| \leq 2.145) = 0.95.$$

Substituting the outcomes of $\overline{T} - \overline{V}$ and S_p in the inequality

$$-2.145 \;<\; \frac{\overline{T} - \overline{V} - (\mu_1 - \mu_2)}{S_p\sqrt{\frac{1}{9} + \frac{1}{7}}} \;<\; 2.145$$

we obtain (−2.7, 13.0) as a 95% confidence interval for $\mu_2 - \mu_1$. □

Exercise 17

Let X be beta-distributed with parameters $p = \frac{1}{2}m$ and $q = \frac{1}{2}n$. Prove that

$$\mathrm{var}(X) = \frac{pq}{(p+q)^2(p+q+1)}.$$

Proof. We have (Definition II.6.2, Theorem II.6.6 and Proposition II.2.1):

$$\begin{aligned} \mathbb{E}(X^2) &= \int_0^1 x^2 \left(\frac{x^{p-1}(1-x)^{q-1}}{B(p,q)} \right) dx \\ &= \frac{B(p+2,q)}{B(p,q)} = \frac{(p+1)p}{(p+q)(p+q+1)}. \end{aligned}$$

Using Proposition II.6.7 (i), we conclude that

$$\begin{aligned} \mathrm{var}(X) = \mathbb{E}(X^2) - \mathbb{E}(X)^2 &= \frac{(p+1)p}{(p+q)(p+q+1)} - \left(\frac{p}{p+q} \right)^2 \\ &= \frac{pq}{(p+q)^2(p+q+1)}. \end{aligned}$$

□

Exercise 18

A sample $X_1, \ldots, X_{500}$ is drawn from a Bernoulli distributed population with unknown parameter θ. The outcome shows: $\sum_i X_i = 108$. Construct a 90% confidence interval for the parameter θ.

Solution. For all i we have $\mathbb{E}(X_i) = p$ and $\mathrm{var}(X_i) = p(1-p)$. By the central limit theorem the variable

$$Z := \sqrt{500}\, \frac{|\overline{X} - \theta|}{\sqrt{\theta(1-\theta)}}$$

is approximately $N(0,1)$-distributed, so that $\mathbb{P}(|Z| \leq 1.645) \approx 0.90$. Substituting $\overline{X} = 108/500 = 0.216$ in the inequality

$$\sqrt{500}\, \frac{|\overline{X} - \theta|}{\sqrt{\theta(1-\theta)}} \leq 1.645$$

we get the following inequality in θ:

$$\left| \frac{\theta - 0.216}{\sqrt{\theta(1-\theta)}} \right| \leq 0.07357 \quad \text{or} \quad \frac{(\theta - 0.216)^2}{\theta(1-\theta)} \leq 0.00541.$$

This is equivalent to the inequality $(1+0.00541)\theta^2 - (0.432+0.00541)\theta + 0.216^2 \leq 0$. Solving this elementary quadratic inequality provides (0.195, 0.240) as a 90% confidence interval for θ.

□

Exercise 19

Given is a population that is Cauchy distributed with parameters $\alpha = 0$ and $\beta = 1$, so the probability density is given by

$$f(x) = \frac{1}{\pi}\frac{1}{x^2+1}.$$

From this population we draw a sample $X_1, \dots, X_n$.

(i) Prove that $\chi_{X_k} = \chi_{\overline{X}}$ for all k.

Proof. The characteristic function of X_k is the same for all k. Therefore it suffices to prove that $\chi_{X_1} = \chi_{\overline{X}}$. We have (see §I.8 and Appendix D):

$$\chi_{X_1}(t) = \int_{-\infty}^{+\infty} e^{itx}\frac{1}{\pi}\frac{1}{x^2+1}\,dx = \frac{1}{\pi}\pi e^{-|t|} = e^{-|t|}.$$

Furthermore, setting $S_n := \sum_{k=1}^{n} X_k$ we can write (by a generalized version of Proposition I.8.8):

$$\chi_{S_n}(t) = \prod_{k=1}^{n}\chi_{X_k}(t) = e^{-n|t|}.$$

Using Proposition I.8.6 we deduce from this that

$$\chi_{\overline{X}}(t) = \chi_{S_n}(t/n) = e^{-n\,|t/n|} = e^{-|t|}.$$

It follows that $\chi_{X_k} = \chi_{\overline{X}}$ for all k. □

(ii) *Question.* What can we say about the distribution of $\overline{X}$?

Solution. The probability distribution of a stochastic variable is completely characterized by its characteristic function. Therefore, by (i), we conclude that $\overline{X}$ and X_1 are identically distributed. □

(iii) For both the variables $X_1 + X_2$ and $X_1 - X_2$ the characteristic function is given by $\chi(t) = e^{-2|t|}$. Prove this.

Proof. The variables X_1, X_2 being independent, we have by Proposition I.8.8 $\chi_{X_1+X_2}(t) = \chi_{X_1}(t)\,\chi_{X_2}(t) = e^{-2|t|}$. The variable $-X_2$ enjoys the same distribution as X_2, and the variables $X_1, -X_2$ are (Proposition I.4.2) statistically independent. For this reason we may write

$$\chi_{X_1-X_2}(t) = \chi_{X_1+(-X_2)}(t) = \chi_{X_1}(t)\,\chi_{-X_2}(t) = e^{-2|t|}.$$ □

(iv) Show that $X_1 + X_2$ and $X_1 - X_2$ are dependent.

Solution. If $X_1 + X_2$ and $X_1 - X_2$ were statistically independent, then on the one hand we would have for all $t \in \mathbb{R}$ that

$$\chi_{2X_1}(t) = \chi_{(X_1+X_2)+(X_1-X_2)}(t) = \chi_{X_1+X_2}(t)\,\chi_{X_1-X_2}(t) = e^{-4|t|} \qquad (*)$$

On the other hand, however, we have (Proposition I.8.6) that

$$\chi_{2X_1}(t) = \chi_{X_1}(2t) = e^{-2|t|}.$$

This is in contradiction to (*). Conclusion: $X_1 + X_2$ and $X_1 - X_2$ are not statistically independent. □

Exercise 20

Denote the velocity of a randomly chosen molecule in an ideal gas by $\mathbf{V}$. Starting from the fact that $\|\mathbf{V}\|^2/\sigma^2$ is $\chi^2(3)$-distributed, deduce an explicit form for the velocity density of molecules in an ideal gas. In other words, find the density of $\|\mathbf{V}\|$.

Solution. We have

$$\mathbb{P}\left(\|\mathbf{V}\|^2/\sigma^2 \leq z\right) = \int_0^z \frac{1}{2^{\frac{3}{2}}\Gamma\left(\frac{3}{2}\right)}\, x^{\frac{1}{2}}\, e^{-\frac{1}{2}x}\, dx = \int_0^z \frac{1}{\sqrt{2\pi}}\, x^{\frac{1}{2}}\, e^{-\frac{1}{2}x}\, dx$$

for $z \geq 0$ and $\mathbb{P}\left(\|\mathbf{V}\|^2/\sigma^2 \leq z\right) = 0$ for $z < 0$. It follows that for all $v > 0$

$$\begin{aligned}
\mathbb{P}(\|\mathbf{V}\| \leq v) &= \int_0^{v^2/\sigma^2} \frac{1}{\sqrt{2\pi}}\, x^{\frac{1}{2}} e^{-\frac{1}{2}x}\, dx \\
(x := t^2/\sigma^2) &= \int_0^v \frac{1}{\sqrt{2\pi}}\, (t^2/\sigma^2)^{\frac{1}{2}} e^{-\frac{1}{2}t^2/\sigma^2}\, d(t^2/\sigma^2) \\
&= \int_0^v \frac{2}{\sqrt{2\pi}\,\sigma^3}\, t^2 e^{-\frac{1}{2}t^2/\sigma^2}\, dt.
\end{aligned}$$

Summarizing, the probability density ('velocity density') f of $\|\mathbf{V}\|$ is given by

$$f(v) = \frac{2}{\sqrt{2\pi}\,\sigma^3} v^2 e^{-\frac{1}{2}v^2/\sigma^2} \quad \text{for } v \geq 0$$

and $f = 0$ elsewhere. □

Exercise 21

We are measuring the velocity $\mathbf{V} = (V_x, V_y)$ of some High Speed Ants, which move on a flat, homogeneous disc, on which a coordinate system has been put. We assume the components V_x and V_y to be statistically independent.

(i) Describe the probability distribution of $\|\mathbf{V}\|^2$.

Solution. By arguments concerning symmetry we may assume that the probability distribution of $\mathbf{V}$ is rotation invariant. Applying §I.11 Exercise 39 it then follows that V_x and V_y both share some $N(0, \sigma^2)$-distribution. Therefore the variable

$$\frac{\| \mathbf{V} \|^2}{\sigma^2} = \frac{V_x^2}{\sigma^2} + \frac{V_y^2}{\sigma^2} = \left(\frac{V_x}{\sigma}\right)^2 + \left(\frac{V_y}{\sigma}\right)^2$$

enjoys a $\chi^2(2)$-distribution.

Now set $F := F_{\|\mathbf{V}\|^2}$. Of course $F(z) = 0$ for $z < 0$. For $z \geq 0$ we have

$$\begin{aligned} F(z) &= \mathbb{P}\left(\| \mathbf{V} \|^2 \leq z\right) = \mathbb{P}\left(\frac{\| \mathbf{V} \|^2}{\sigma^2} \leq \frac{z}{\sigma^2}\right) \\ &= \int_0^{z/\sigma^2} \tfrac{1}{2} e^{-\frac{1}{2}x}\, dx = [-e^{-\frac{1}{2}x}]_0^{z/\sigma^2} = 1 - e^{-z/(2\sigma^2)}, \end{aligned}$$

thus obtaining an explicit form for F. □

(ii) Give an explicit expression for $f_{\|\mathbf{V}\|}$.

Solution. Let G be the distribution function of $\| \mathbf{V} \|$. Of course $G(v) = 0$ for all $v < 0$. For $v \geq 0$ we have

$$G(v) = \mathbb{P}(\| \mathbf{V} \| \leq v) = \mathbb{P}\left(\| \mathbf{V} \|^2 \leq v^2\right) = 1 - e^{-v^2/(2\sigma^2)}.$$

Differentiation of G shows that

$$f_{\|\mathbf{V}\|}(v) = \begin{cases} \frac{1}{\sigma^2}\, v\, e^{-v^2/2\sigma^2} & \text{if } v \geq 0, \\ 0 & \text{elsewhere.} \end{cases}$$

□

(iii) *Question.* What do you think about the assumption that V_x and V_y are statistically independent?

Solution. Assuming the velocity of the ants will not be near the velocity of light, it is quite reasonable to presume statistical independence of V_x and V_y. Measurement of V_x does not provide any information about V_y and vice versa. There is no link between V_x and V_y. □

Exercise 22

Given is a population enjoying an absolutely continuous distribution with probability density f, for which

$$\begin{cases} f(x) > 0 & \text{if } x > 0, \\ f(x) = 0 & \text{elsewhere.} \end{cases}$$

We draw a sample X_1, X_2 from this population.

(i) Prove that $S^2 = \frac{1}{2}(X_1 - X_2)^2$.

Solution. See Exercise 7 (ii). □

(ii) Prove that $\mathbb{P}(S^2 \geq 2a^2) \neq 0$ and $\mathbb{P}(\overline{X} \leq a) \neq 0$ for all $a > 0$.

Proof. With probability 1 we have $X_1, X_2 > 0$. Consequently, for $a > 0$, we can write (see the figure for $(*)$)

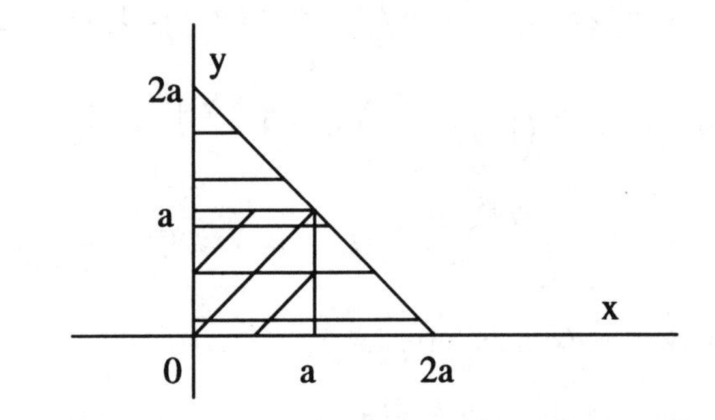

$$\begin{aligned}\mathbb{P}(\overline{X} \leq a) &= \mathbb{P}(X_1 + X_2 \leq 2a) \overset{(*)}{\geq} \mathbb{P}(X_1 \leq a \text{ and } X_2 \leq a) \\ &= (\mathbb{P}[X_1 \leq a])^2 = \left(\int_0^a f(x)\,dx\right)^2 > 0.\end{aligned}$$

The last inequality can be justified as follows: f is strictly positive and continuous on $[\frac{1}{2}a, a]$. Hence f attains a minimum $k > 0$ on this interval and we therefore have $\int_0^a f(x)\,dx \geq \frac{1}{2}a\,k > 0$. Furthermore (the figure below shows the region B):

$$\mathbb{P}(S^2 \geq 2a^2) \overset{(i)}{=} \mathbb{P}(|X_1 - X_2| \geq 2a) = \mathbb{P}[(X_1, X_2) \in B].$$

Similar arguments show that this probability too is positive. □

(iii) Determine the numerical value of $\mathbb{P}(S^2 \geq 2a^2 \text{ and } \overline{X} \leq a)$.

Solution.

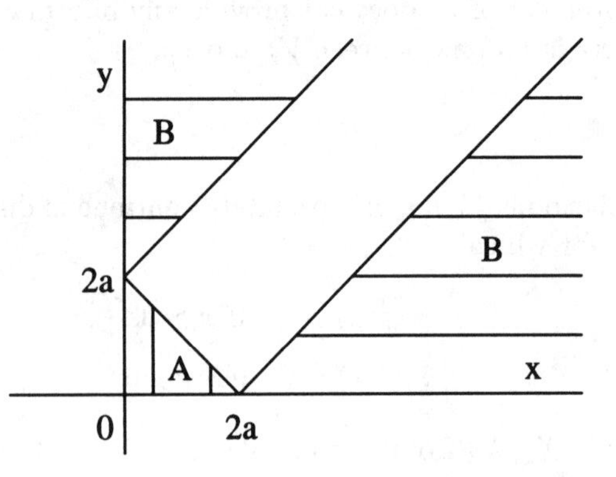

Let A and B be as in the figure. In the notations of (ii) we may write

$$\mathbb{P}\,(S^2 \geq 2a^2 \ \text{ and } \ \overline{X} \leq a) = \mathbb{P}\,[(X_1, X_2) \in B \ \text{ and } \ (X_1, X_2) \in A] = 0,$$

as $A \cap B = \emptyset$. □

(iv) *Question.* Are $\overline{X}$ and S^2 statistically independent?

Solution. Because of (ii) and (iii) we have

$$\mathbb{P}\,(S^2 \geq 2a^2 \ \text{ and } \ \overline{X} \leq a) \ \neq \ \mathbb{P}(S^2 \geq 2a^2)\ \mathbb{P}\left(\overline{X} \leq a\right).$$

This shows that $\overline{X}$ and S^2 are not statistically independent. □

Exercise 23

Given is a population with a probability density f, satisfying $f(x) = f(-x)$ for all $x \in \mathbb{R}$. The mean μ and variance σ^2 of this population exist.

(i) Prove that

$$\chi(t) = \int_{-\infty}^{+\infty} f(x) e^{itx}\, dx = \int_{-\infty}^{+\infty} f(x) \cos(tx)\, dx.$$

Proof. We can write

$$\chi(t) = \int_{-\infty}^{+\infty} f(x) e^{itx}\, dx = \int_{-\infty}^{+\infty} f(x) \cos(tx)\, dx + i \int_{-\infty}^{+\infty} f(x) \sin(tx)\, dx.$$

Because $\int_{-\infty}^{+\infty} |f(x) \sin(tx)|\, dx \leq \int_{-\infty}^{+\infty} f(x) \cdot 1\, dx = 1$, the integral

$$\int_{-\infty}^{+\infty} f(x) \sin(tx)\, dx$$

exists. Moreover,

$$\begin{aligned} \int_0^{+\infty} f(x) \sin(tx)\, dx &= \int_0^{-\infty} f(-y) \sin(-ty)\, d(-y) \\ &= -\int_{-\infty}^{0} f(y) \sin(ty)\, dy, \end{aligned}$$

which shows that $\int_{-\infty}^{+\infty} f(x) \sin(tx)\, dx = 0$. □

(ii) Show that $\chi(0) = 1,\ \chi'(0) = 0$ and $\chi''(0) = -\sigma^2$.

Solution. Trivially we have $\chi(0) = \int_{-\infty}^{+\infty} f(x)\cdot 1\,dx = 1$. Applying Lebesgue's dominated convergence theorem (see Appendix A) we can justify differentiation under the integral sign. Thus we obtain

$$\begin{aligned}\chi'(t) &= \frac{\partial}{\partial t}\int_{-\infty}^{+\infty} f(x)\cos(tx)\,dx \\ &= \int_{-\infty}^{+\infty} \frac{\partial}{\partial t} f(x)\cos(tx)\,dx = \int_{-\infty}^{+\infty} f(x)\cdot -x\,\sin(tx)\,dx.\end{aligned}$$

This shows that $\chi'(0) = 0$. In the same way we may write

$$\chi''(t) = \int_{-\infty}^{+\infty} \frac{\partial}{\partial t}(-x\,f(x)\sin(tx))\,dx = \int_{-\infty}^{+\infty} -x^2 f(x)\cos(tx)\,dx,$$

which shows that $\chi''(0) = -\int_{-\infty}^{+\infty} x^2\,f(x)\,dx$. Because f is symmetric with respect to the origin we have $\mu = 0$; consequently $\chi''(0) = -\sigma^2$. □

(iii) Next we draw a sample X_1, X_2 and we set $\overline{X} := \frac{1}{2}(X_1 + X_2)$ and $\tilde{X} := \frac{1}{2}(X_1 - X_2)$. Prove that $\chi_{\overline{X}}(t) = \chi_{\tilde{X}}(t) = \{\chi(t/2)\}^2$.

Proof. The variables X_1 and X_2 are statistically independent. Applying Proposition I.4.2 it follows that X_1 and $-X_2$ are also independent. Using Proposition I.8.8 we reveal that $\chi_{2\overline{X}}(t) = \chi_{X_1}(t)\,\chi_{X_2}(t) = \{\chi(t)\}^2$ and

$$\begin{aligned}\chi_{2\tilde{X}}(t) &= \chi_{X_1}(t)\,\chi_{-X_2}(t) = \chi_{X_1}(t)\,\chi_{X_2}(-t) \\ &\overset{\text{(i)}}{=} \chi_{X_1}(t)\,\chi_{X_2}(t) = \{\chi(t)\}^2.\end{aligned}$$

Applying Proposition I.8.6 we see that

$$\chi_{\overline{X}}(t) = \chi_{\frac{1}{2}\cdot 2\overline{X}}(t) = \chi_{2\overline{X}}(t/2) = \{\chi(t/2)\}^2$$

and

$$\chi_{\frac{1}{2}\cdot 2\tilde{X}}(t) = \chi_{2\tilde{X}}(t/2) = \{\chi(t/2)\}^2.$$

This had to be proved. □

(iv) Verify that $\operatorname{cov}(\overline{X}, \tilde{X}) = 0$.

Solution. Obviously

$$\begin{aligned}\operatorname{cov}(\overline{X}, \tilde{X}) &= \tfrac{1}{4}\operatorname{cov}(X_1 + X_2, X_1 - X_2) \\ &= \tfrac{1}{4}\{\operatorname{cov}(X_1, X_1) - 0 + 0 - \operatorname{cov}(X_2, X_2)\} \\ &= \tfrac{1}{4}\{\operatorname{var}(X_1) - 0 + 0 - \operatorname{var}(X_2)\} = 0.\end{aligned}$$

□

(v) From now on we assume that $\overline{X}$ and $\tilde{X}$ are statistically independent. Prove that $\chi(t) = \{\chi(t/2)\}^4$ for all $t \in \mathbb{R}$.

Solution. Using Proposition I.8.8 together with (iii) we learn that on the one hand

$$\chi_{\overline{X}+\tilde{X}}(t) = \chi_{\overline{X}}(t)\,\chi_{\tilde{X}}(t) = \{\chi(t/2)\}^4,$$

whereas on the other hand

$$\chi_{\overline{X}+\tilde{X}}(t) = \chi_{X_1}(t) = \chi(t).$$

So $\chi(t) = \{\chi(t/2)\}^4$ for all $t \in \mathbb{R}$. □

(vi) Define $\varphi : \mathbb{R} \to \mathbb{R}$ by $\varphi(t) := \log \chi(t)$. Show that $\varphi(0) = 0$, $\varphi'(0) = 0$ and $\varphi''(0) = -\sigma^2$.

Solution. First we check whether the definition of φ makes sense. By (i) we have $\chi(t) \in \mathbb{R}$. By (iv) we have $\chi(t) = \{\chi(t/2)\}^4$, which shows that $\chi(t) \geq 0$. Finally, it cannot occur that $\chi(t) = 0$. To see this, suppose it does. Then by (iv) we also have $\chi(t/2) = 0$. Applying (iv) repeatedly we see that $\chi(t/2^n) = 0$ for all $n = 1, 2, \ldots$. Therefore we would have $1 = \chi(0) = \lim_{n\to\infty} \chi(t/2^n) = 0$. This contradiction implies that $\chi(t) > 0$ for all t.

By (ii) we have $\varphi(0) = \log \chi(0) = \log 1 = 0$. Furthermore $\varphi'(t) = \chi'(t)/\chi(t)$, which shows (again we use (ii)) that $\varphi'(0) = 0/1 = 0$. Finally $\varphi''(t) = (\chi(t)\chi''(t) - \chi'(t)^2)/\chi(t)^2$; this shows that $\varphi''(0) = \chi''(0) = -\sigma^2$. □

(vii) Set $\psi(t) := \varphi(t)/t^2$. Prove that

$$\lim_{t\to 0} \psi(t) = -\tfrac{1}{2}\sigma^2.$$

Proof. Applying l'Hôpital's rule twice, we get

$$\lim_{t\to 0} \frac{\varphi(t)}{t^2} = \lim_{t\to 0} \frac{\varphi'(t)}{2t} = \lim_{t\to 0} \frac{\varphi''(t)}{2} = -\tfrac{1}{2}\sigma^2.$$

□

(viii) Prove that $\psi(t) = \psi(t/2)$ for all $t \neq 0$ and deduce from this that $\psi(t) = -\frac{1}{2}\sigma^2$ for all $t \neq 0$.

Proof. For $t \neq 0$ we can write down the following chain of equivalent statements:

$$\begin{aligned}
\psi(t) = \psi(t/2) \quad &\Leftrightarrow \quad \varphi(t)/t^2 = \varphi(t/2)/(t/2)^2 \\
&\Leftrightarrow \quad \varphi(t) = 4\varphi(t/2) \\
&\Leftrightarrow \quad \log \chi(t) = 4 \log \chi(t/2) \\
&\Leftrightarrow \quad \chi(t) = \{\chi(t/2)\}^4.
\end{aligned}$$

So by (iv) we have $\psi(t) = \psi(t/2)$ for all $t \neq 0$. This implies of course that $\psi(t) = \psi(t/2) = \psi(t/4) = \cdots$ for all $t \neq 0$. Hence for any $t \neq 0$

$$\psi(t) = \lim_{k\to\infty} \psi(t/2^k) = \lim_{s\to 0} \psi(s) \overset{\text{(vi)}}{=} -\tfrac{1}{2}\sigma^2. \qquad \square$$

(ix) Show that the population in question is necessarily $N(0, \sigma^2)$-distributed.

Solution. By (vii) we have for all $t \neq 0$ that $\varphi(t) = -\sigma^2 t^2/2$. Because $\varphi(0) = 0$ this holds for *all* $t \in \mathbb{R}$. Hence $\chi(t) = e^{-\frac{1}{2}\sigma^2 t^2}$ for all $t \in \mathbb{R}$. This is, however, the characteristic function belonging to a $N(0, \sigma^2)$-distribution (Proposition I.8.7), so that the population is $N(0, \sigma^2)$-distributed. $\square$

Exercise 24

Given is a population with mean μ and variance σ^2. The population has a probability density which is symmetric around μ. Furthermore the population enjoys the following property: for all samples X_1, X_2 the variables $\tilde{X}$ and $\overline{X}$ (as defined in Exercise 23 (iii)) are statistically independent. Prove that the population is necessarily $N(\mu, \sigma^2)$-distributed.

Proof. Apply Exercise 23 to the variable $X - \mu$ and conclude that $X - \mu$ is $N(0, \sigma^2)$-distributed. Then apply Theorem I.6.1 and conclude that X is $N(\mu, \sigma^2)$-distributed. $\square$

Exercise 25

Suppose X_1, X_2 is a sample from a standard normal distributed population. The probability density of the 2-vector (X_1, X_2) is denoted by f. For f we have

$$f(x_1, x_2) = \frac{1}{(\sqrt{2\pi})^2}\, e^{-\frac{1}{2}x_1^2}\, e^{-\frac{1}{2}x_2^2} = \frac{1}{2\pi}\, e^{-\frac{1}{2}(x_1^2 + x_2^2)}.$$

Now look at the picture. A probability density $\tilde{f}$ is defined to be equal to $2f$ on the interior of the squares *I* and *III*, equal to zero on the interior of the squares *II* and *IV* and equal to f elsewhere.

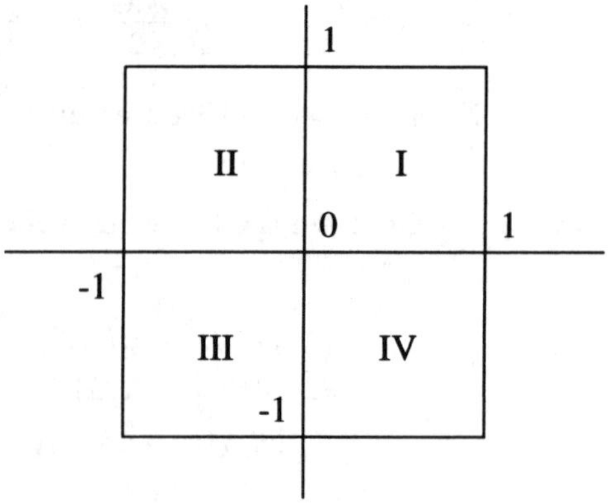

(i) Show that $\tilde{f}$ is indeed a probability density.

Solution.

1. The function $\tilde{f}$ is Borel. To see this, look at the following decomposition:

$$\tilde{f} = 2f\ 1_A + f\ 1_B,$$

where A and B are suitably chosen Borel sets. Now apply §I.11 Exercise 7.

2. It is trivial that $\tilde{f} \geq 0$.

3. Next we prove that $\int \tilde{f}\, d\lambda = 1$. Exploiting rotational invariance (Proposition I.6.2) we may write

$$\iint_{I\&III} f\ d\lambda = \iint_{\mathbf{Q}(I\&III)} f\ d\lambda = \iint_{II\&IV} f\ d\lambda,$$

where the matrix $[\mathbf{Q}]$ of $\mathbf{Q}$ is given by

$$[\mathbf{Q}] = \begin{pmatrix} 0 & -1 \\ 1 & 0 \end{pmatrix} = \begin{pmatrix} \cos\pi/2 & -\sin\pi/2 \\ \sin\pi/2 & \cos\pi/2 \end{pmatrix}.$$

The orthogonal map $\mathbf{Q}$ presents rotation over an angle $\frac{1}{2}\pi$, anticlockwise. We see that

$$\iint_{I\&III} 2f\ d\lambda = \iint_{I\&III} f\ d\lambda + \iint_{II\&IV} f\ d\lambda.$$

Hence

$$\iint_{I\&...\&IV} \tilde{f}\ d\lambda = \iint_{I\&III} 2f\ d\lambda = \iint_{I\&...\&IV} f\ d\lambda,$$

so that $\iint \tilde{f}\ d\lambda = \iint f\ d\lambda = 1$. □

(ii) Show that for all $x_1 \in \mathbb{R} \setminus \{-1, 0, 1\}$ we have

$$\int_{-\infty}^{+\infty} \tilde{f}(x_1, x_2)\, dx_2 = \frac{1}{\sqrt{2\pi}}\ e^{-\frac{1}{2}x_1^2}.$$

Solution. For $x_1 < -1$ or $x_1 > 1$ we have $\tilde{f}(x_1, x_2) = f(x_1, x_2)$, so

$$\begin{aligned} \int_{-\infty}^{+\infty} \tilde{f}(x_1, x_2)\, dx_2 &= \int_{-\infty}^{+\infty} f(x_1, x_2)\, dx_2 \\ &= \frac{1}{\sqrt{2\pi}}\ e^{-\frac{1}{2}x_1^2} \int_{-\infty}^{+\infty} \frac{1}{\sqrt{2\pi}}\ e^{-\frac{1}{2}x_2^2}\, dx_2 = \frac{1}{\sqrt{2\pi}}\ e^{-\frac{1}{2}x_1^2} \end{aligned}$$

for all $x_1 \notin [-1, +1]$. For $x_1 \in (-1, 0)$ we can write

$$\begin{aligned}
\int_{-\infty}^{+\infty} f(x_1, x_2)\, dx_2 - \int_{-\infty}^{+\infty} \tilde{f}(x_1, x_2)\, dx_2 \\
&= \int_{-\infty}^{+\infty} (f - \tilde{f})(x_1, x_2)\, dx_2 = \int_{-1}^{1} (f - \tilde{f})(x_1, x_2)\, dx_2 \\
&= \int_{-1}^{0} (f - 2f)(x_1, x_2)\, dx_2 + \int_{0}^{1} (f - 0)(x_1, x_2)\, dx_2 \\
&= \int_{-1}^{0} (-f)(x_1, x_2)\, dx_2 + \int_{-1}^{0} f(x_1, -y)\, dy \\
&= -\int_{-1}^{0} f(x_1, x_2)\, dx_2 + \int_{-1}^{0} f(x_1, y)\, dy = 0.
\end{aligned}$$

Consequently

$$\int_{-\infty}^{+\infty} \tilde{f}(x_1, x_2)\, dx_2 = \int_{-\infty}^{+\infty} f(x_1, x_2)\, dx_2 = \frac{1}{\sqrt{2\pi}}\, e^{-\frac{1}{2}x_1^2}.$$

For $x_1 \in (0, 1)$ similar arguments apply. □

(iii) Suppose the stochastic 2-vector $(\tilde{X}_1, \tilde{X}_2)$ has probability density $\tilde{f}$. Prove that $\tilde{X}_1$ and $\tilde{X}_2$ are both $N(0, 1)$-distributed.

Proof. Via (ii) we see that

$$\mathbb{P}\left(\tilde{X}_1 \leq x\right) = \int_{-\infty}^{x} \int_{-\infty}^{+\infty} \tilde{f}(x_1, x_2)\, dx_2\, dx_1 = \int_{-\infty}^{x} \frac{1}{\sqrt{2\pi}}\, e^{-\frac{1}{2}x_1^2}\, dx_1.$$

This shows that $\tilde{X}_1$ is $N(0, 1)$-distributed. Of course the same applies to $\tilde{X}_2$. □

(iv) Prove that $\tilde{X}_1$ and $\tilde{X}_2$ are not statistically independent.

Proof. First note that {square II} $= [-1, 0] \times [0, 1]$. By definition of $\tilde{f}$ we have on the one hand

$$\mathbb{P}\left[(\tilde{X}_1, \tilde{X}_2) \in \text{square } II\right] = 0.$$

On the other hand, however, we have

$$\mathbb{P}\left(\tilde{X}_1 \in [-1, 0]\right)\, \mathbb{P}\left(\tilde{X}_2 \in [0, 1]\right) > 0.$$

Summarizing, $\tilde{X}_1$ and $\tilde{X}_2$ cannot be independent. □

(v) Prove that $\tilde{X}_1^{\,2} + \tilde{X}_2^{\,2}$ enjoys a $\chi^2(2)$-distribution.

Proof. Denote square *I* by A_1, square *II* by A_2 and so on. Let $V := \mathbb{R}^2 \setminus \{A_1 \cup A_2 \cup A_3 \cup A_4\}$ and let $B_R = \{\mathbf{x} \in \mathbb{R}^2 : x_1^2 + x_2^2 < R^2\}$, where $R \geq 0$.

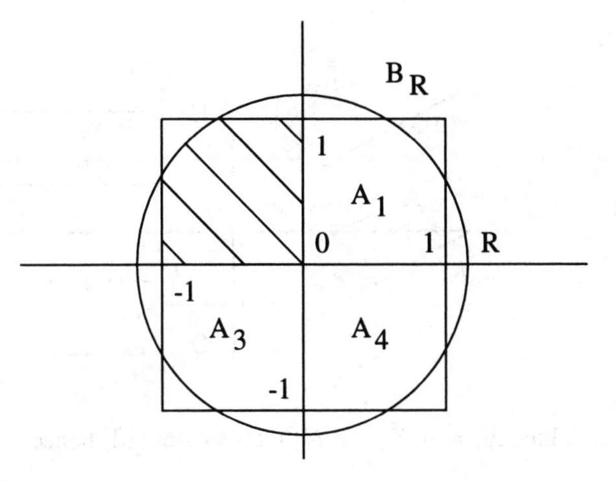

In these notations we can write (by rotational invariance):

$$\iint_{B_R \cap A_1} f\, d\lambda = \iint_{\mathbf{Q}(B_R \cap A_1)} f\, d\lambda = \iint_{B_R \cap A_2} f\, d\lambda,$$

with $\mathbf{Q}$ as in point 3 of (i). Similarly

$$\iint_{B_R \cap A_3} f\, d\lambda = \iint_{B_R \cap A_4} f\, d\lambda.$$

Consequently we may write:

$$\begin{aligned}
\mathbb{P}[\tilde{X}_1{}^2 + \tilde{X}_2{}^2 \le R^2] &= \iint_{B_R} \tilde{f}\, d\lambda \\
&= \iint_{B_R \cap V} \tilde{f}\, d\lambda + \sum_{i=1}^{4} \iint_{B_R \cap A_i} \tilde{f}\, d\lambda \\
&= \iint_{B_R \cap V} f\, d\lambda + \iint_{B_R \cap A_1} 2f\, d\lambda + \iint_{B_R \cap A_3} 2f\, d\lambda \\
&= \iint_{B_R \cap V} f\, d\lambda + \sum_{i=1}^{4} \iint_{B_R \cap A_i} f\, d\lambda \\
&= \iint_{B_R} f\, d\lambda = \mathbb{P}[X_1{}^2 + X_2{}^2 \le R^2].
\end{aligned}$$

This shows that $\tilde{X}_1{}^2 + \tilde{X}_2{}^2$ and $X_1{}^2 + X_2{}^2$ are identically distributed. The latter being $\chi^2(2)$-distributed, this completes the proof. □

(vi) Show that

$$\rho(\tilde{X}_1, \tilde{X}_2) = \frac{2}{\pi e}\, (\sqrt{e} - 1)^2.$$

Solution.

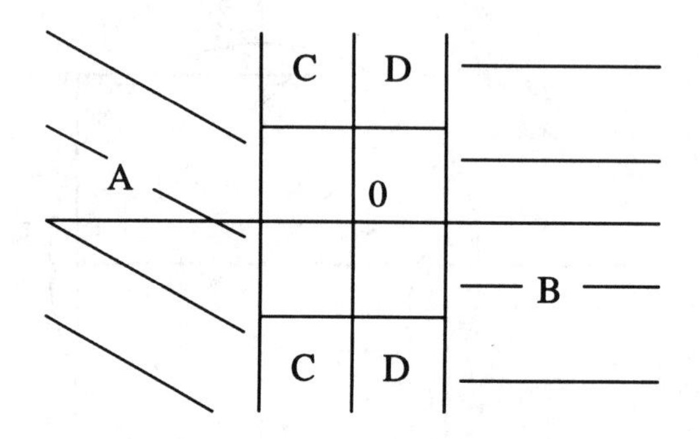

The variables $\tilde{X}_1$ and $\tilde{X}_2$ are $N(0,1)$-distributed, hence

$$\begin{aligned}\rho(\tilde{X}_1, \tilde{X}_2) &= \frac{\operatorname{cov}(\tilde{X}_1, \tilde{X}_2)}{\sigma_{\tilde{X}_1}\sigma_{\tilde{X}_2}} = \operatorname{cov}(\tilde{X}_1, \tilde{X}_2) = \mathbb{E}(\tilde{X}_1\tilde{X}_2) - 0 \cdot 0 \\ &= \int_{-\infty}^{+\infty}\int_{-\infty}^{+\infty} x_1 x_2 \tilde{f}(x_1, x_2)\, dx_2\, dx_1.\end{aligned}$$

Now set $A := (-\infty, -1) \times (-\infty, +\infty)$ and $B := (1, +\infty) \times (-\infty, +\infty)$. Define $\varphi : \mathbb{R} \to [0, +\infty)$ by $\varphi(x) = \frac{1}{\sqrt{2\pi}} e^{-x^2/2}$. Then $\tilde{f}(x_1, x_2) = f(x_1, x_2) = \varphi(x_1)\varphi(x_2)$ for all $(x_1, x_2) \in A \cup B$. In this notation we see that

$$\begin{aligned}\iint_A x_1 x_2 \tilde{f}(x_1, x_2)\, dx_2\, dx_1 &= \iint_A x_1 x_2\, \varphi(x_1)\varphi(x_2)\, dx_2\, dx_1 \\ &= \int_{-\infty}^{-1} x_1 \varphi(x_1) \left\{ \int_{-\infty}^{+\infty} x_2 \varphi(x_2)\, dx_2 \right\} dx_1 = 0,\end{aligned}$$

since the inner integral equals $\mathbb{E}(X_2) = 0$. In the same way of course

$$\iint_B x_1 x_2 \tilde{f}(x_1, x_2)\, dx_2\, dx_1 = 0.$$

Next, define the sets C and D in $\mathbb{R}^2$ (as in the figure) by

$$\begin{cases} C := [-1, 0] \times ((-\infty, -1) \cup (1, +\infty)), \\ D := [0, 1] \times ((-\infty, -1) \cup (1, +\infty)). \end{cases}$$

Now

$$\begin{aligned}\iint_C x_1 x_2 \tilde{f}(x_1, x_2)\, dx_2\, dx_1 &= \int_{-1}^{0}\int_{-\infty}^{-1} x_1 x_2\, \varphi(x_1)\varphi(x_2)\, dx_2\, dx_1 \\ &\quad + \int_{-1}^{0}\int_{1}^{+\infty} x_1 x_2\, \varphi(x_1)\varphi(x_2)\, dx_2\, dx_1.\end{aligned}$$

Substituting $y := -x_2$ in the first integral on the right side, we see that

$$\iint_C x_1 x_2 \tilde{f}(x_1, x_2)\, dx_2\, dx_1 = 0.$$

Analogously

$$\iint_D x_1 x_2 \tilde{f}(x_1, x_2)\, dx_2\, dx_1 = 0.$$

Consequently

$$\begin{aligned}
\rho(\tilde{X_1}, \tilde{X_2}) &= \int_{-1}^{1}\int_{-1}^{1} x_1 x_2\, \tilde{f}(x_1, x_2)\, dx_2\, dx_1 \\
&= \iint_{A_1} 2x_1 x_2\, f(x_1, x_2)\, dx_2\, dx_1 \\
&\quad + \iint_{A_3} 2x_1 x_2\, f(x_1, x_2)\, dx_2\, dx_1 \\
&= 4 \iint_{A_1} x_1 x_2\, \varphi(x_1)\varphi(x_2)\, dx_2\, dx_1.
\end{aligned}$$

Here again we exploited rotational invariance together with the fact that $\mathbf{Q}^2(A_3) = A_1$. It thus appears that

$$\begin{aligned}
\rho(\tilde{X_1}, \tilde{X_2}) &= 4 \int_0^1 \int_0^1 x_1 x_2\, f(x_1) f(x_2)\, dx_2\, dx_1 \\
&= 4 \left(\int_0^1 x_1 f(x_1)\, dx_1 \right)^2 = 4 \left(\int_0^1 \frac{1}{\sqrt{2\pi}}\, x e^{-\frac{1}{2}x^2}\, dx \right)^2 \\
&= 4 \left(\left[-\frac{1}{\sqrt{2\pi}} e^{-\frac{1}{2}x^2} \right]_0^1 \right)^2 = \cdots = \frac{2}{\pi e}\, (\sqrt{e} - 1)^2.
\end{aligned}$$

□

Exercise 26

Let $X_1, \ldots, X_n$ be a sample of size n from a Bernoulli distributed population with parameter θ. Prove that the variable $S := X_1 + \cdots + X_n$ is binomially distributed with parameters n and θ.

Proof. We take an arbitrary $k \in \{0, \ldots, n\}$. Now

$$\mathbb{P}(S = k) = \mathbb{P}(X_1 + \cdots + X_n = k) = \sum_{\boldsymbol{\alpha} \in A_k} \mathbb{P}(\mathbf{X} = \boldsymbol{\alpha})$$

where $\mathbf{X} := (X_1, \ldots, X_n)$ and $A_k := \{\boldsymbol{\alpha} \in \{0,1\}^n \;:\; |\boldsymbol{\alpha}| = k\}$ for $|\boldsymbol{\alpha}| := \alpha_1 + \cdots + \alpha_n$. For $|\boldsymbol{\alpha}| = k$ we have

$$\begin{aligned}
\mathbb{P}(\mathbf{X} = \boldsymbol{\alpha}) &= \mathbb{P}(X_1 = \alpha_1, \ \ldots \ , X_n = \alpha_n) \\
&= \mathbb{P}(X_1 = \alpha_1) \ \cdots \ \mathbb{P}(X_n = \alpha_n) \\
&= \theta^{\alpha_1} (1 - \theta)^{1-\alpha_1} \ \cdots \ \theta^{\alpha_n} (1 - \theta)^{1-\alpha_n} \\
&= \theta^{|\boldsymbol{\alpha}|} (1 - \theta)^{n-|\boldsymbol{\alpha}|} = \theta^k (1 - \theta)^{n-k}.
\end{aligned}$$

It follows that

$$\begin{aligned}\mathbb{P}(S=k) &= \sum_{\alpha\in A_k} \theta^k(1-\theta)^{n-k}\\ &= \#(A_k)\times\theta^k(1-\theta)^{n-k} = \binom{n}{k}\theta^k(1-\theta)^{n-k}.\end{aligned}$$

So S is binomially distributed with the required parameters. □

Exercise 27

Let X be a $N(\mu,\sigma^2)$-distributed stochastic variable, where σ^2 is known. Consider μ as the outcome of a stochastic variable M which is $N(\mu_0,{\sigma_0}^2)$-distributed.

(i) Express $f_X(x\,|\,M=\mu)$ in terms of x,μ and σ.

Solution. Evidently we have

$$f_X(x\,|\,M=\mu)=\frac{1}{\sigma\sqrt{2\pi}}\exp\left[-\frac{1}{2}\left(\frac{x-\mu}{\sigma}\right)^2\right].$$

□

(ii) Express $f_M(\mu)$ in terms of μ, μ_0 and σ_0.

Solution. We have:

$$f_M(\mu)=\frac{1}{\sigma_0\sqrt{2\pi}}\exp\left[-\frac{1}{2}\left(\frac{\mu-\mu_0}{\sigma_0}\right)^2\right].$$

□

(iii) Verify that $f_{X,M}(x,\mu)$ equals

$$\frac{1}{2\sigma_0\sigma\pi}\exp\left[-\frac{\sigma_0^2+\sigma^2}{2\sigma_0^2\sigma^2}\left\{\left(\mu-\frac{\sigma^2\mu_0+x\sigma_0^2}{\sigma_0^2+\sigma^2}\right)^2+\frac{\sigma_0^2\sigma^2(x-\mu_0)^2}{(\sigma_0^2+\sigma^2)^2}\right\}\right].\quad (*)$$

Solution. By Proposition II.8.1 we can write

$$\begin{aligned}f_{X,M}(x,\mu) &= f_X(x\,|\,M=\mu)f_M(\mu)\\ &= \frac{1}{2\sigma_0\sigma\pi}\exp\left[-\frac{1}{2\sigma_0^2\sigma^2}\left\{(x-\mu)^2\sigma_0^2+(\mu-\mu_0)^2\sigma^2\right\}\right].\end{aligned}$$

We have

$$\begin{aligned}&(x-\mu)^2\sigma_0^2+(\mu-\mu_0)^2\sigma^2\\ &\qquad=(\sigma_0^2+\sigma^2)\,\mu^2+(-2x\sigma_0^2-2\mu_0\sigma^2)\,\mu+(x^2\sigma_0^2+\mu_0^2\sigma^2).\end{aligned}$$

This expression is identical to

$$(\sigma_0^2+\sigma^2)\left\{\left(\mu-\frac{\sigma^2\mu_0+x\sigma_0^2}{\sigma_0^2+\sigma^2}\right)^2+\alpha\right\},\quad (**)$$

where

$$\alpha := -\left(\frac{\sigma^2\mu_0 + x\sigma_0^2}{\sigma_0^2 + \sigma^2}\right)^2 + \frac{\sigma^2\mu_0^2 + x^2\sigma_0^2}{\sigma_0^2 + \sigma^2}.$$

So we have

$$\begin{aligned}
&(\sigma_0^2 + \sigma^2)^2\alpha \\
&\quad = -(\sigma^2\mu_0 + x\sigma_0^2)^2 + (\sigma_0^2 + \sigma^2)(\sigma^2\mu_0^2 + x^2\sigma_0^2) \\
&\quad = -x^2{\sigma_0}^4 - 2x\mu_0\sigma_0^2\sigma^2 - \mu_0^2\sigma^4 + x^2{\sigma_0}^4 + \mu_0^2\sigma_0^2\sigma^2 + x^2\sigma_0^2\sigma^2 + \mu_0^2\sigma^4 \\
&\quad = \sigma_0^2\sigma^2(x^2 + \mu_0^2 - 2x\mu_0) = \sigma_0^2\sigma^2(x - \mu_0)^2.
\end{aligned}$$

Substituting this in (∗∗), we get (∗). □

(iv) Prove that

$$f_X(x) = \frac{1}{\sqrt{2\pi}\sqrt{\sigma_0^2 + \sigma^2}} \exp\left(-\frac{1}{2}\frac{(x - \mu_0)^2}{\sigma_0^2 + \sigma^2}\right).$$

Proof. First we notice that for all $\beta \in \mathbb{R}$ and all $\delta > 0$, we have

$$\int_{-\infty}^{+\infty} \frac{1}{\sqrt{2\pi}}\frac{1}{\delta} \exp\left[-\frac{1}{2}\left(\frac{x - \beta}{\delta}\right)^2\right] dx = 1, \qquad (*)$$

because this presents the integral of the density of a $N(\beta, \delta^2)$-distribution. Next we work out the equality

$$f_X(x) = \int_{-\infty}^{+\infty} f_{X,M}(x, \mu)\, d\mu.$$

Let α be as in (iii), that is:

$$\alpha = \frac{\sigma_0^2\sigma^2(x - \mu_0)^2}{(\sigma_0^2 + \sigma^2)^2}.$$

Furthermore, let μ_1 be given by

$$\mu_1 := \frac{\sigma^2\mu_0 + x\sigma_0^2}{\sigma_0^2 + \sigma^2}$$

and σ_1 by

$$\sigma_1 := \sqrt{\frac{\sigma_0^2\sigma^2}{\sigma_0^2 + \sigma^2}} = \frac{\sigma_0\sigma}{\sqrt{\sigma_0^2 + \sigma^2}}.$$

Then we can write

$$
\begin{aligned}
f_X(x) &= \int_{-\infty}^{+\infty} \frac{1}{2\sigma_0\sigma\pi} \exp\left[-\frac{1}{2\sigma_1^2}\{(\mu-\mu_1)^2+\alpha\}\right] d\mu \\
&= \frac{\exp\left(-\frac{1}{2}\alpha/\sigma_1^2\right)}{2\sigma_0\sigma\pi} \int_{-\infty}^{+\infty} \exp\left[-\frac{1}{2}\left(\frac{\mu-\mu_1}{\sigma_1}\right)^2\right] d\mu \\
\text{(see } (*)) \quad &= \frac{\sqrt{2\pi}\,\sigma_1 \exp\left(-\frac{1}{2}\alpha/\sigma_1^2\right)}{2\sigma_0\sigma\pi} \\
&= \frac{1}{\sqrt{2\pi}\sqrt{\sigma_0^2+\sigma^2}} \exp\left(-\frac{1}{2}\frac{(x-\mu_0)^2}{\sigma_0^2+\sigma^2}\right).
\end{aligned}
$$

□

(v) Prove that

$$
f_M(\mu \,|\, X=x) = \frac{1}{\sigma_1\sqrt{2\pi}}\, e^{-\frac{1}{2}(\mu-\mu_1)^2/\sigma_1^2}.
$$

Proof. By Definition II.8.1 we have

$$
f_M(\mu \,|\, X=x) = \frac{f_{X,M}(x,\mu)}{f_X(x)}.
$$

It follows, using (iii) and (iv), that

$$
\begin{aligned}
&f_M(\mu \,|\, X=x) \\
&= \sqrt{\frac{\sigma_0^2+\sigma^2}{2\pi\sigma_0^2\sigma^2}} \exp\left[-\frac{\sigma_0^2+\sigma^2}{2\sigma_0^2\sigma^2}\left\{(\mu-\mu_1)^2+\alpha\right\} + \frac{(x-\mu_0)^2}{2(\sigma_0^2+\sigma^2)}\right] \\
&= \frac{1}{\sqrt{2\pi}\,\sigma_1} \exp\left[-\frac{1}{2\sigma_1^2}\left\{(\mu-\mu_1)^2+\alpha-\sigma_1^2\frac{(x-\mu_0)^2}{\sigma_0^2+\sigma^2}\right\}\right].
\end{aligned}
$$

Because

$$
\sigma_1^2\frac{(x-\mu_0)^2}{\sigma_0^2+\sigma^2} = \frac{\sigma_0^2\sigma^2(x-\mu_0)^2}{(\sigma_0^2+\sigma^2)^2} = \alpha,
$$

we arrive at

$$
f_M(\mu \,|\, X=x) = \frac{1}{\sigma_1\sqrt{2\pi}} \exp\left[-\frac{1}{2}\left(\frac{\mu-\mu_1}{\sigma_1}\right)^2\right].
$$

□

Exercise 28

We assume the length of Dutch university students to be $N(\mu, 100)$-distributed. The parameter μ is regarded as an outcome of a $N(175, 60)$-distributed stochastic variable M. The sample mean of a sample $L_1, \ldots, L_{12}$ shows an outcome $\overline{L} = 178$.

(i) Construct a 95% confidence interval for μ, using the probability distribution of M only (ignoring the outcome of the sample).

Solution. Looking at Table II we see that

$$\mathbb{P}\left(\left|\frac{M-175}{\sqrt{60}}\right| \leq 1.96\right) = 0.95.$$

Thus the interval (160, 190) emerges as a 95% confidence interval for μ. □

(ii) Construct a 95% confidence interval for μ, using the probability distribution of M together with the outcome of the sample.

Solution. First notice that $\overline{L}$ is $N(\mu, \frac{100}{12})$-distributed. By Theorem II.8.3 the variable M, given the outcome $\overline{L} = 178$, is $N(\mu_1, {\sigma_1}^2)$-distributed, where

$$\mu_1 = \frac{60 \cdot 178 + \frac{100}{12} \cdot 175}{100/12 + 60} = 177.63 \qquad (*)$$

and

$$\sigma_1^2 = \frac{\sigma_0^2 \sigma^2}{\sigma_0^2 + \sigma^2} = \frac{60 \cdot 100/12}{60 + 100/12} = 7.32 \qquad (**)$$

We have (see Table II):

$$\mathbb{P}\left(\left|\frac{M-\mu_1}{\sigma_1}\right| \leq 1.96\right) = 0.95.$$

Substituting $(*)$ and $(**)$ in the inequality

$$\left|\frac{M-\mu_1}{\sigma_1}\right| \leq 1.96,$$

the interval $(172.3, 182.9)$ emerges as a 95% confidence interval for μ. □

(iii) Construct a 95% confidence interval for μ using the outcome of $\overline{L}$ only (ignoring the extra information that M is $N(175, 60)$-distributed).

Solution. The variable $\overline{L}$ is $N(\mu, 100/12)$-distributed. Therefore the variable

$$\frac{\overline{L}-\mu}{\sqrt{100/12}}$$

enjoys (a priori) a $N(0,1)$-distribution. Using Table II we see that

$$\mathbb{P}\left(\left|\frac{\overline{L}-\mu}{\sqrt{100/12}}\right| \leq 1.96\right) = 0.95.$$

We substitute the outcome $\overline{L} = 178$ in the inequality; this results in a 95% confidence interval given by $(172.3, 183.7)$. □

Exercise 29

Given is a stochastic variable X which is Poisson distributed with parameter λ. We look upon λ as being the outcome of a stochastic variable Λ which is gamma distributed with parameters α and β.

(i) Express $f_X(k \mid \Lambda = \lambda)$ in terms of k and λ.

Solution. By definition of the Poisson distribution we have

$$f_X(k \mid \Lambda = \lambda) = \frac{\lambda^k e^{-\lambda}}{k!} \quad \text{for all } k \in \mathbb{N}.$$ □

(ii) Show that

$$f_{\Lambda,X}(\lambda, k) = f_{X,\Lambda}(k, \lambda) = \frac{\lambda^{\alpha+k-1} e^{-\lambda(1+1/\beta)}}{\beta^\alpha \Gamma(k+1)\Gamma(\alpha)}.$$

Solution. Using Proposition II.2.1 we write

$$\begin{aligned} f_{X,\Lambda}(k,\lambda) = f_X(k \mid \Lambda = \lambda)\, f_\Lambda(\lambda) &= \frac{\lambda^k e^{-\lambda}}{k!} \frac{\lambda^{\alpha-1} e^{-\lambda/\beta}}{\beta^\alpha \Gamma(\alpha)} \\ &= \frac{\lambda^{\alpha+k-1} e^{-\lambda(1+1/\beta)}}{\beta^\alpha \Gamma(k+1)\Gamma(\alpha)} \end{aligned}$$

for all $k \in \mathbb{N}$ and $\lambda > 0$. □

(iii) Prove that, given $X = k$, the variable Λ is gamma distributed with parameters $\alpha + k$ and $\beta/(\beta+1)$.

Proof. First we notice (using (ii) and passing to the variable $\mu := (1+1/\beta)\,\lambda$) that

$$\begin{aligned} f_X(k) &= \int_{-\infty}^{+\infty} f_{X,\Lambda}(k,\lambda)\, d\lambda \\ &= \frac{1}{k!\,\beta^\alpha\,\Gamma(\alpha)} \int_0^{+\infty} \lambda^{\alpha+k-1} e^{-\lambda(1+1/\beta)}\, d\lambda \\ &= \frac{\int_0^{+\infty} \mu^{\alpha+k-1}\, e^{-\mu}\, d\mu}{k!\,\beta^\alpha\,\Gamma(\alpha)\,(1+1/\beta)^{\alpha+k}} \\ &= \frac{\Gamma(\alpha+k)\,(\beta/(\beta+1))^{\alpha+k}}{k!\,\beta^\alpha\,\Gamma(\alpha)}, \end{aligned}$$

where we used the fact that $\Gamma(x+1) = x\,\Gamma(x)$. Now we see that for all $k \in \mathbb{N}$ and $\lambda > 0$

$$\begin{aligned} f_\Lambda(\lambda \mid X = k) &= f_{\Lambda,X}(\lambda,k)/f_X(k) \\ &= \frac{\lambda^{\alpha+k-1} e^{-\lambda(1+1/\beta)}}{\beta^\alpha \Gamma(k+1)\Gamma(\alpha)} \Bigg/ \frac{\Gamma(\alpha+k)\,(\beta/(\beta+1))^{\alpha+k}}{k!\,\beta^\alpha\,\Gamma(\alpha)} \\ &= \frac{\lambda^{\alpha+k-1}\, e^{-\lambda(1+1/\beta)}}{\Gamma(\alpha+k)\,(\beta/(\beta+1))^{\alpha+k}}. \end{aligned}$$

This is, however, exactly the density belonging to a gamma distribution with parameters $\alpha + k$ and $\beta/(\beta+1)$. □

(iv) Express $\mathbb{E}(\Lambda \mid X = k)$ in terms of α, β and k.

Solution. Using (iii) and Proposition II.2.5 we see that

$$\mathbb{E}(\Lambda \mid X = k) = (\alpha + k)\,\frac{\beta}{\beta+1}.$$ □

Exercise 30

A sample $X_1, \ldots, X_n$ is drawn from a $N(0, \sigma^2)$-distributed population.

(i) Determine the Fisher information supplied by a single measurement.

Solution. By definition the Fisher information supplied by a single measurement equals

$$\begin{aligned}
&\mathbb{E}_\sigma\left[\left(\frac{\partial}{\partial\sigma}\log\left(\frac{1}{\sqrt{2\pi}\,\sigma}\,e^{-\frac{1}{2}X_i^2/\sigma^2}\right)\right)^2\right] \\
&= \mathbb{E}_\sigma\left[\left(\frac{\partial}{\partial\sigma}\{-\tfrac{1}{2}\log(2\pi) - \log\sigma - \tfrac{1}{2}X_i^2/\sigma^2\}\right)^2\right] \\
&= \mathbb{E}_\sigma\left[(-1/\sigma + X_i^2/\sigma^3)^2\right] = \mathbb{E}_\sigma\left[\sigma^{-6}(X_i^2 - \sigma^2)^2\right] \\
&= \sigma^{-6}\{\mathbb{E}_\sigma\left[X_i^4\right] - 2\sigma^2\mathbb{E}_\sigma\left[X_i^2\right] + \sigma^4\} \\
&= \sigma^{-6}\{\mu_4(\sigma) - \sigma^4\},
\end{aligned}$$

where $\mu_4(\sigma)$ is the fourth moment of the $N(0, \sigma^2)$-distribution.
By Proposition I.8.2 we have $\mu_4(\sigma) = M^{(4)}(0)$. Using Proposition I.8.7 we may write $M(t) = \exp(\frac{1}{2}\sigma^2 t^2)$, $M'(t) = M^{(1)}(t) = \sigma^2 t\exp(\frac{1}{2}\sigma^2 t^2)$,

$$\begin{aligned}
M^{(2)}(t) &= (\sigma^4 t^2 + \sigma^2)\exp(\tfrac{1}{2}\sigma^2 t^2) \quad \text{and} \\
M^{(3)}(t) &= (2\sigma^4 t + \sigma^2 t\,(\sigma^4 t^2 + \sigma^2))\exp(\tfrac{1}{2}\sigma^2 t^2) \\
&= (\sigma^6 t^3 + 3\sigma^4 t)\exp(\tfrac{1}{2}\sigma^2 t^2).
\end{aligned}$$

Finally we arrive at the following expression for $M^{(4)}(t)$:

$$\begin{aligned}
M^{(4)}(t) &= (3\sigma^6 t^2 + 3\sigma^4 + \sigma^2 t\,(\sigma^6 t^3 + 3\sigma^4 t))\exp(\tfrac{1}{2}\sigma^2 t^2) \\
&= (\sigma^8 t^4 + 6\sigma^6 t^2 + 3\sigma^4)\exp(\tfrac{1}{2}\sigma^2 t^2).
\end{aligned}$$

It follows from this that $\mu_4(\sigma) = M^{(4)}(0) = 3\sigma^4$. Summarizing, the Fisher information supplied by a single measurement equals $2\sigma^{-2}$. □

(ii) Prove that $(X_1^2 + \cdots + X_n^2)/n$ is a variable which brings about equality in the information inequality.

Proof. Here $\kappa(\sigma) = \sigma^2$, so $\kappa'(\sigma) = 2\sigma$. We have to prove that

$$\operatorname{var}_\sigma\left((X_1^2 + \cdots + X_n^2)/n\right) = \frac{(2\sigma)^2}{n\,2\sigma^{-2}} = \frac{2\sigma^4}{n}.$$

We can write

$$\operatorname{var}_\sigma\left(\frac{1}{n}\sum_{i=1}^{n} X_i^2\right) = \frac{1}{n^2}\sum_{i=1}^{n}\operatorname{var}_\sigma\left(X_i^2\right) = \frac{n}{n^2}\operatorname{var}_\sigma\left(X_1^2\right).$$

In order to determine $\operatorname{var}_\sigma\left(X_1^2/\sigma^2\right)$ we note that X_1/σ is $N(0,1)$-distributed. Consequently X_1^2/σ^2 is $\chi^2(1)$-distributed (Proposition II.2.7). Using Proposition II.2.8 we see that $\operatorname{var}_\sigma\left(X_1^2/\sigma^2\right) = 2$. (Note that this argument provides an alternative proof of the fact that $\mu_4(\sigma) = \operatorname{var}_\sigma\left(X_1^2\right) + \mu_2(\sigma)^2 = 3\sigma^4$.) It follows from this that $\operatorname{var}_\sigma\left(X_1^2\right) = 2\sigma^4$. □

Exercise 31

Given is an exponentially distributed population with parameter $\theta > 0$, that is to say, we are dealing with a family of probability densities given by

$$f(x,\theta) = \frac{1}{\theta}\,e^{-x/\theta}\,1_{[0,+\infty)}(x).$$

(i) Show that the nth moment of such a probability distribution is given by $n!\,\theta^n$.

Solution. Denoting the nth moment by $\mu_n(\theta)$ we can write

$$\begin{aligned}\mu_n(\theta) = \int_0^{+\infty} x^n\,\frac{1}{\theta}\,e^{-x/\theta}\,dx &= \theta^n\int_0^{+\infty}(x/\theta)^n\,e^{-x/\theta}\,d(x/\theta)\\ &= \theta^n\,\Gamma(n+1) = n!\,\theta^n\end{aligned}$$

(see for example Proposition II.2.1). □

(ii) Determine the Fisher information supplied by a single measurement.

Solution. First we note that

$$\frac{\partial}{\partial\theta}\log f(X,\theta) = \frac{\partial}{\partial\theta}(-\log\theta - X/\theta) = (X-\theta)/\theta^2.$$

Therefore (see (i))

$$\begin{aligned}&\mathbb{E}_\theta\left[\left(\frac{\partial}{\partial\theta}\log f(X,\theta)\right)^2\right]\\ &\quad= \mathbb{E}_\theta\left[(X-\theta)^2/\theta^4\right] = \theta^{-4}\left\{\mu_2(\theta) - 2\theta\,\mu_1(\theta) + \theta^2\right\}\\ &\quad= \theta^{-4}\left\{2\theta^2 - 2\theta\,\theta + \theta^2\right\} = \theta^{-2}.\end{aligned}$$

□

(iii) Determine var (S^2).

Solution. In a fairly straightforward way the variable S^2 can be decomposed as follows:

$$S^2 = \frac{1}{n}\sum_{i=1}^{n} X_i^2 - \frac{1}{n(n-1)}\sum_{i\neq j} X_i X_j.$$

Therefore

$$\begin{aligned}
\operatorname{var}(S^2) &= \operatorname{cov}(S^2, S^2) \\
&= \frac{1}{n^2}\operatorname{var}\left(\sum_{i=1}^{n} X_i^2\right) - \frac{2}{n^2(n-1)}\operatorname{cov}\left(\sum_{i=1}^{n} X_i^2\,,\,\sum_{i\neq j} X_i X_j\right) \\
&\quad + \frac{1}{n^2(n-1)^2}\operatorname{cov}\left(\sum_{i\neq j} X_i X_j\,,\,\sum_{i\neq j} X_i X_j\right).
\end{aligned}$$

In order to express this in terms of n and θ, making use of (i) and symmetry arguments, we compute:

$$\operatorname{var}\left(\sum_{i=1}^{n} X_i^2\right) = n\operatorname{var}(X_1^2) = n\,\left(\mu_4(\theta) - \mu_2(\theta)^2\right) = 20\,n\,\theta^4.$$

In the same way,

$$\begin{aligned}
\operatorname{cov}\left(\sum_{i=1}^{n} X_i^2\,,\,\sum_{i\neq j} X_i X_j\right) &= n\operatorname{cov}\left(X_1^2\,,\,\sum_{i\neq j} X_i X_j\right) \\
&= n\left(\operatorname{cov}(X_1^2\,,\,\textstyle\sum_{j=2}^{n} X_1 X_j) + \operatorname{cov}(X_1^2\,,\,\textstyle\sum_{i=2}^{n} X_i X_1\,)\right) \\
&= 2n(n-1)\operatorname{cov}(X_1^2, X_1X_2) = 8n(n-1)\,\theta^4
\end{aligned}$$

and

$$\begin{aligned}
\operatorname{cov}\left(\sum_{i\neq j} X_i X_j, \sum_{i\neq j} X_i X_j\right) &= n(n-1)\operatorname{cov}\left(X_1X_2, \sum_{i\neq j} X_i X_j\right) \\
&= n(n-1)\left\{\operatorname{cov}(X_1X_2, \textstyle\sum_{j=3}^{n} X_1X_j) + \operatorname{cov}(X_1X_2, \textstyle\sum_{j=3}^{n} X_2X_j)\right. \\
&\qquad + \operatorname{cov}(X_1X_2, \textstyle\sum_{i=3}^{n} X_iX_1) + \operatorname{cov}(X_1X_2, \textstyle\sum_{i=3}^{n} X_iX_2) \\
&\qquad \left. + \operatorname{cov}(X_1X_2, X_1X_2) + \operatorname{cov}(X_1X_2, X_2X_1\,)\right\} \\
&= n(n-1)\,\{4(n-2)\operatorname{cov}(X_1X_2, X_1X_3) + 2\operatorname{cov}(X_1X_2, X_1X_2)\} \\
&= n(n-1)\,\{4(n-2)\,(\mu_2(\theta)\mu_1(\theta)^2 - \mu_1(\theta)^4)\,+2\,(\mu_2(\theta)^2 - \mu_1(\theta)^4)\} \\
&= n(n-1)\,\{4(n-2)\,\theta^4 + 6\,\theta^4\} = 2n(n-1)(2n-1)\,\theta^4.
\end{aligned}$$

Summarizing, we arrive at

$$\begin{aligned}\mathrm{var}(S^2) &= \frac{20n\,\theta^4}{n^2} - \frac{2\cdot 8n(n-1)\,\theta^4}{n^2(n-1)} + \frac{2n(n-1)(2n-1)\,\theta^4}{n^2(n-1)^2} \\ &= \frac{2\theta^4}{n(n-1)}\,\{10(n-1) - 8(n-1) + (2n-1)\} \\ &= \frac{2(4n-3)}{n(n-1)}\,\theta^4.\end{aligned}$$

□

(iv) Does S^2 achieve equality in the information inequality?

Solution. The variance of the population (see (i)) equals $\mu_2(\theta) - \mu_1^2(\theta) = 2\theta^2 - \theta^2 = \theta^2$, so the characteristic associated with the population variance is given by $\kappa : \theta \mapsto \theta^2$. By (ii) it follows from this that the lower bound in the information inequality is $4\theta^2/(n\,\theta^{-2}) = 4\theta^4/n$. This is (see (iii)) strictly less than the variance of S^2. □

Exercise 32

Given is a population with a probability density $f(\bullet, \theta)$ defined by

$$f(x,\theta) \;:=\; 1_{(\theta-\frac{1}{2},\theta+\frac{1}{2})}(x).$$

for all $x \in \mathbb{R}$ and $\theta > 0$. A sample X_1, X_2, X_3 is drawn from this population. We set $G := \max\{X_1, X_2, X_3\}$, $K := \min\{X_1, X_2, X_3\}$ and $T := \frac{1}{2}(G + K)$.

(i) Show that $\mathbb{E}(T) = \theta$, the mean of the population.

Solution. For $i = 1, 2, 3$ we define the variables Z_i by $Z_i := X_i - \theta$. Furthermore we set

$$\begin{cases} \tilde{G} := \max\{Z_1, Z_2, Z_3\} = G - \theta, \\ \tilde{K} := \min\{Z_1, Z_2, Z_3\} = K - \theta, \\ \tilde{T} := \frac{1}{2}(\tilde{G} + \tilde{K}) = T - \theta. \end{cases}$$

The variables Z_1, Z_2, Z_3 are statistically independent and they are all uniformly distributed on the interval $(-\frac{1}{2}, \frac{1}{2})$. Next define

$$\begin{aligned} U_1 &:= \{(z_1, z_2, z_3) : -\tfrac{1}{2} \le z_1 \le z_2 \le z_3 \le \tfrac{1}{2}\}, \\ U_2 &:= \{(z_1, z_2, z_3) : -\tfrac{1}{2} \le z_2 \le z_1 \le z_3 \le \tfrac{1}{2}\} \end{aligned}$$

etcetera. There are 3!, that is 6, sets of this type: $U_1, \ldots, U_6$. Note that the mutual intersections all have measure zero, hence these intersections are of no importance in our examinations. We have

$$\bigcup_{i=1}^{6} U_i = [-\tfrac{1}{2}, \tfrac{1}{2}] \times [-\tfrac{1}{2}, \tfrac{1}{2}] \times [-\tfrac{1}{2}, \tfrac{1}{2}].$$

Next we compute $\mathbb{E}(\tilde{G})$ and $\mathbb{E}(\tilde{K})$. For $\mathbf{Z} := (Z_1, Z_2, Z_3)$ we can write

$$
\begin{aligned}
\mathbb{E}(\tilde{G}) &= \int \max\{z_1, z_2, z_3\} f_{\mathbf{Z}}(\mathbf{z})\, d\mathbf{z} \\
&= \sum_{i=1}^{6} \int_{U_i} \max\{z_1, z_2, z_3\}\, d\mathbf{z} = 6 \int_{U_1} \max\{z_1, z_2, z_3\}\, d\mathbf{z} \\
&= 6 \int_{-\frac{1}{2}}^{\frac{1}{2}} \int_{-\frac{1}{2}}^{z_3} \int_{-\frac{1}{2}}^{z_2} z_3 \; dz_1\, dz_2\, dz_3 = 6 \int_{-\frac{1}{2}}^{\frac{1}{2}} \int_{-\frac{1}{2}}^{z_3} z_3\, (z_2 + \tfrac{1}{2})\, dz_2\, dz_3 \\
&= 6 \int_{-\frac{1}{2}}^{\frac{1}{2}} z_3\, \left[\tfrac{1}{2} z_2^2 + \tfrac{1}{2} z_2\right]_{-\frac{1}{2}}^{z_3} dz_3 = 3 \int_{-\frac{1}{2}}^{\frac{1}{2}} z_3^3 + z_3^2 - (\tfrac{1}{4} - \tfrac{1}{2})\, z_3\, dz_3 \\
&= 3\, \left[\tfrac{1}{4} z_3^4 + \tfrac{1}{3} z_3^3 + \tfrac{1}{8} z_3^2\right]_{-\frac{1}{2}}^{\frac{1}{2}} = \left[z_3^3\right]_{-\frac{1}{2}}^{\frac{1}{2}} = 2\,(\tfrac{1}{2})^3 = \tfrac{1}{4}.
\end{aligned}
$$

In the same way it appears that $\mathbb{E}(\tilde{K}) = -\frac{1}{4}$. It follows that

$$
\mathbb{E}(\tilde{T}) = \tfrac{1}{2}(\mathbb{E}(\tilde{G}) + \mathbb{E}(\tilde{K})) = 0 \quad \text{and} \quad \mathbb{E}(T) = \mathbb{E}(\tilde{T} + \theta) = \theta.
$$

□

(ii) Determine $\mathrm{var}(G)$, $\mathrm{var}(K)$ and $\mathrm{cov}(G, K)$.

Solution. Both 'variance' and 'covariance' are invariant under translations. Therefore $\mathrm{var}(G) = \mathrm{var}(\tilde{G})$, $\mathrm{var}(K) = \mathrm{var}(\tilde{K})$ and $\mathrm{cov}(G, K) = \mathrm{cov}(\tilde{G}, \tilde{K})$. As in (i) we deduce that

$$
\begin{aligned}
\mathbb{E}(\tilde{G}^2) &= 6 \int_{-\frac{1}{2}}^{\frac{1}{2}} \int_{-\frac{1}{2}}^{z_3} \int_{-\frac{1}{2}}^{z_2} z_3^2\, dz_1\, dz_2\, dz_3 \\
&= 3 \int_{-\frac{1}{2}}^{\frac{1}{2}} z_3^4 + z_3^3 - (\tfrac{1}{4} - \tfrac{1}{2}) z_3^2\, dz_3 \\
&= 3\, [\tfrac{1}{5} z_3^5 + \tfrac{1}{4} z_3^4 + \tfrac{1}{12} z_3^3]_{-\frac{1}{2}}^{\frac{1}{2}} \\
&= 3\, (\tfrac{2}{5}\, (\tfrac{1}{2})^5 + \tfrac{2}{12}\, (\tfrac{1}{2})^3) = \tfrac{3}{80} + \tfrac{1}{16} \;=\; \tfrac{1}{10}.
\end{aligned}
$$

It follows that $\mathrm{var}(\tilde{G}) = \mathbb{E}(\tilde{G}^2) - \mathbb{E}(\tilde{G})^2 = \frac{1}{10} - \left(\frac{1}{4}\right)^2 = \frac{3}{80}$. In a similar way we obtain $\mathrm{var}(\tilde{K}) = \frac{3}{80}$. Furthermore, proceeding as before,

$$
\begin{aligned}
\mathbb{E}(\tilde{G}\tilde{K}) &= 6 \int_{U_1} \max\{z_1, z_2, z_3\} \min\{z_1, z_2, z_3\}\, d\mathbf{z} \\
&= 6 \int_{-\frac{1}{2}}^{\frac{1}{2}} \int_{-\frac{1}{2}}^{z_3} \int_{-\frac{1}{2}}^{z_2} z_3 z_1\, dz_1\, dz_2\, dz_3 =
\end{aligned}
$$

$$
\begin{aligned}
&= 6\int_{-\frac12}^{\frac12} z_3 \int_{-\frac12}^{z_3} \left[\tfrac12 z_1^2\right]_{-\frac12}^{z_2} dz_2\, dz_3 = 6\int_{-\frac12}^{\frac12} z_3 \left[\tfrac16 z_2^3 - \tfrac18 z_2\right]_{-\frac12}^{z_3} dz_3 \\
&= 6\int_{-\frac12}^{\frac12} \tfrac16 z_3^4 - \tfrac18 z_3^2 - (-\tfrac{1}{48} + \tfrac{1}{16}) z_3\, dz_3 \\
&= 6\left[\tfrac{1}{30} z_3^5 - \tfrac{1}{24} z_3^3 - \tfrac{1}{48} z_3^2\right]_{-\frac12}^{\frac12} = \left[\tfrac15 z_3^5 - \tfrac14 z_3^3 - \tfrac18 z_3^2\right]_{-\frac12}^{\frac12} \\
&= \tfrac25 \left(\tfrac12\right)^5 - \tfrac12 \left(\tfrac12\right)^3 = \tfrac{1}{80} - \tfrac{1}{16} = -\tfrac{4}{80} = -\tfrac{1}{20}.
\end{aligned}
$$

Hence $\operatorname{cov}(\tilde G, \tilde K) = \mathbb{E}(\tilde G \tilde K) - \mathbb{E}(\tilde G)\mathbb{E}(\tilde K) = -\frac{1}{20} - \left(\frac14 \cdot -\frac14\right) = \frac{-4+5}{80} = \frac{1}{80}$. Summarizing, we have

$$\operatorname{cov}(G,K) = \tfrac{1}{80}, \ \operatorname{var}(K) = \tfrac{3}{80} \quad \text{and} \quad \operatorname{var}(G) = \tfrac{3}{80}. \qquad \square$$

(iii) Show that $\operatorname{var}(T) < \operatorname{var}(\overline{X})$.

Solution. The variance of T is given by

$$
\begin{aligned}
\operatorname{var}(T) = \operatorname{var}\left(\tfrac12 (G+K)\right) &= \tfrac14 \operatorname{cov}(G+K, G+K) \\
&= \tfrac14 (\operatorname{var}(G) + 2\operatorname{cov}(G,K) + \operatorname{var}(K)) \\
&= \tfrac14 \left(\tfrac{3}{80} + \tfrac{2}{80} + \tfrac{3}{80}\right) = \tfrac{1}{40}.
\end{aligned}
$$

The variance of $\overline{X}$ is given by $\operatorname{var}(\overline{X}) = \frac13 \operatorname{var}(X_1) = \frac13 \operatorname{var}(Z_1)$. In §I.11, Exercise 19 we learnt that $\operatorname{var}(Z_1) = \frac{1}{12}$: consequently $\operatorname{var}(\overline{X}) = \frac{1}{36}$. This shows that $\operatorname{var}(T) < \operatorname{var}(\overline{X})$. $\square$

(iv) Let $D := \{(x,\theta) \in \mathbb{R}^2 : |x - \theta| \neq \frac12\}$. Prove that f is continuously differentiable on D.

Proof. Setting $A := \{(x,\theta) \in \mathbb{R}^2 : |x-\theta| < \frac12\}$, we have $f|_D = 1_A$. The function f is constant on the open set A. It follows from this that $f'(\mathbf{y}) = 0$ for all $\mathbf{y} \in A$. Furthermore, f equals zero on $D \setminus A$, which shows that $f'(\mathbf{y}) = 0$ for all $\mathbf{y} \in D \setminus A$. Consequently f is differentiable on D and $f'(\mathbf{y}) = 0$ for all $\mathbf{y} \in D$. $\square$

(v) Determine $\frac{\partial}{\partial\theta} f$ on D.

Solution. A direct consequence of (iv) is that $\frac{\partial}{\partial\theta} f = 0$ on D. $\square$

(vi) *Question.* For which continuous functions $g : \mathbb{R} \to \mathbb{R}$ do we have

$$\frac{\partial}{\partial\theta} \int g(x)\, f(x,\theta)\, dx = \int g(x)\, \frac{\partial}{\partial\theta} f(x,\theta)\, dx \ ?$$

Solution. This kind of differentiation under the integral sign is correct (see (v)) if and only if

$$\frac{\partial}{\partial\theta} \int g(x)\, 1_{(\theta - \frac12, \theta + \frac12)}(x)\, dx = 0.$$

Applying the fundamental theorem of calculus we see that the above is equivalent to

$$\frac{\partial}{\partial\theta}\int_{\theta-\frac{1}{2}}^{\theta+\frac{1}{2}} g(x)\,dx = g(\theta+\tfrac{1}{2}) - g(\theta-\tfrac{1}{2}) = 0.$$

In turn this is equivalent to $g(\theta+\frac{1}{2}) = g(\theta-\frac{1}{2})$ for all $\theta \in \mathbb{R}$, that is: g is a periodic function with period 1. □

(vii) *Question.* Do the regularity conditions that are mentioned in the information inequality apply in the current situation?

Solution. No. In the notation of (iv) the function f is constant on A and on $D \setminus A$. It follows that

$$\begin{aligned}\mathbb{E}_\theta\left[\left(\frac{\partial}{\partial\theta}\log f(X,\theta)\right)^2\right] &= \int_{-\infty}^{+\infty}\left(\frac{\partial}{\partial\theta}\log f(x,\theta)\right)^2 f(x,\theta)\,dx \\ &= \int_{A_\theta}\left(\frac{\partial}{\partial\theta}\log f(x,\theta)\right)^2 f(x,\theta)\,dx \\ &\quad + \int_{D\setminus A_\theta}\left(\frac{\partial}{\partial\theta}\log f(x,\theta)\right)^2 f(x,\theta)\,dx = 0,\end{aligned}$$

where by A_θ we mean $\{x \in \mathbb{R} : (x,\theta) \in A\}$. Therefore, if the information inequality would apply, for $\kappa : \theta \mapsto \theta$ we would have

$$1 = (\kappa'(\theta))^2 \leq n \operatorname{var}_\theta(T)\,\mathbb{E}_\theta\left[\left(\frac{\partial}{\partial\theta}\log f(X,\theta)\right)^2\right] = 0,$$

which is of course false. This is due to the fact that it is not allowed to bluntly differentiate under the integral sign. The regularity conditions in the information inequality are clearly not satisfied. □

Exercise 33

A sample $X_1, \ldots, X_n$ is drawn from a Poisson distributed population with parameter $\lambda \geq 0$. Here we are dealing with a discrete probability density $f(\bullet,\lambda)$, given by

$$f(k,\lambda) = e^{-\lambda}\frac{\lambda^k}{k!} \quad \text{for all } k \in \mathbb{N}.$$

Determine the maximum likelihood estimator of λ.

Solution. Here the likelihood function L_λ assumes the form

$$L_\lambda(k_1,\ldots,k_n) = e^{-n\lambda}\frac{\lambda^{k_1+\cdots+k_n}}{k_1!\cdots k_n!} \quad \text{if } k_1,\ldots,k_n \in \mathbb{N}.$$

Given an outcome $(k_1, \ldots, k_n) \in \mathbb{N}^n$, we have to maximize the function $f : \lambda \mapsto \lambda^{|\mathbf{k}|} e^{-n\lambda}/\mathbf{k}!$, where $\mathbf{k}! := k_1! \cdots k_n!$ and $|\mathbf{k}| := k_1 + \cdots + k_n$.

For $|\mathbf{k}| \neq 0$, we can write

$$\frac{d}{d\lambda} f(\lambda) = \frac{|\mathbf{k}|}{\mathbf{k}!} \lambda^{|\mathbf{k}|-1} e^{-n\lambda} - \frac{n}{\mathbf{k}!} \lambda^{|\mathbf{k}|} e^{-n\lambda} = \frac{\lambda^{|\mathbf{k}|-1} e^{-n\lambda}(|\mathbf{k}| - n\lambda)}{\mathbf{k}!}.$$

This derivative vanishes for $\lambda = \frac{1}{n}|\mathbf{k}| = \overline{k}$. It is easy to see that $\lambda \mapsto f(\lambda)$ indeed attains a maximum for this value of λ. It follows that the maximum likelihood estimator for λ is $\overline{X}$.

If $|\mathbf{k}| = 0$, then $k_1 = \cdots = k_n = 0$. In that case the strictly decreasing function $\lambda \mapsto e^{-n\lambda}$ has to be maximized. This leads to $\lambda = 0$, which can also be read as $\lambda = \overline{x}$. Summarizing, $\overline{X}$ is the maximum likelihood estimator of λ. □

Exercise 34

Define for all $p = 1, 2, \ldots$ the compact set S_p by

$$S_p := \{\boldsymbol{\theta} \in [0,1]^p : \theta_1 + \cdots + \theta_p = 1\}.$$

Suppose there is given a population with a discrete probability density $f(\bullet, \boldsymbol{\theta})$, where $\boldsymbol{\theta} \in \Theta$ and $f(i, \boldsymbol{\theta}) := \theta_i$ $(i = 1, \ldots, p)$. A sample $X_1, \ldots, X_n$ is drawn from this population and the outcome is $(i_1, \ldots, i_n)$. We denote the frequency of occurence of the number j in the sequence $(i_1, \ldots, i_n)$ by k_j. In this way the 'frequency vector' $(k_1, \ldots, k_p)$ emerges. The likelihood function $L_{\boldsymbol{\theta}}$ can now be expressed as

$$L_{\boldsymbol{\theta}}((i_1, \ldots, i_n)) = \theta_{i_1} \cdots \theta_{i_n} = \theta_1^{k_1} \cdots \theta_p^{k_p}.$$

(i) Prove that the function $f_p : \boldsymbol{\theta} \mapsto \theta_1^{k_1} \cdots \theta_p^{k_p}$ attains on S_p a global maximum in the point $\hat{\boldsymbol{\theta}} = (k_1/n, \ldots, k_p/n)$ (only).

Proof. The proof is split up into two steps, applying induction on p.

Step 1. The case $p = 2$. In this case we are dealing with the function

$$(\theta_1, \theta_2) \mapsto \theta_1^{k_1} \theta_2^{k_2}$$

which is to be maximized under the constraint $\theta_1 + \theta_2 = 1$.

Write $k := k_1$, $\theta := \theta_1$. Then $\theta_2 = 1 - \theta$ and $k_2 = n - k$; thus we arrive at the problem to find a maximum of the function

$$\theta \mapsto \theta^k (1 - \theta)^{n-k}.$$

This function is maximal (check this) in the point $\theta = k/n$, that is, in the point $(\theta_1, \theta_2) = (k_1/n, k_2/n)$.

Step 2. Suppose the statement is true for all $m < p$. Consider the function $f_p : \boldsymbol{\theta} \mapsto \theta_1^{k_1} \cdots \theta_p^{k_p}$ on S_p.

Case 1. Some of the k_i equal zero. Then without loss of generality (by a renumbering) we may assume that there is an integer $m < p$ for which $k_1, \ldots, k_m > 0$ and $k_{m+1}, \ldots, k_p = 0$. In this case, the map reduces to

$$(\theta_1, \ldots, \theta_p) \mapsto \theta_1^{k_1} \cdots \theta_m^{k_m}$$

which is to be maximized on S_p. This is equivalent to the maximization of the map

$$(\theta_1, \ldots, \theta_m) \mapsto \theta_1^{k_1} \cdots \theta_m^{k_m}$$

under the constraints

$$(\theta_1, \ldots, \theta_m) \in [0,1]^m \quad \text{and} \quad \theta_1 + \cdots + \theta_m \leq 1.$$

Using an argument involving compactness it is easy to see that indeed a maximum is attained in some point $(\hat{\theta}_1, \ldots, \hat{\theta}_m)$. Moreover, it is easily verified that for such a point one necessarily has

$$\hat{\theta}_1 + \cdots + \hat{\theta}_m = 1.$$

Consequently (using the induction hypothesis) we have

$$(\hat{\theta}_1, \ldots, \hat{\theta}_m) = (k_1/n, \ldots, k_m/n).$$

It follows that the function $(\theta_1, \ldots, \theta_p) \mapsto \theta_1^{k_1} \cdots \theta_p^{k_p}$ attains a maximum (only) in the point

$$(\hat{\theta}_1, \ldots, \hat{\theta}_p) = (\hat{\theta}_1, \ldots, \hat{\theta}_m, 0, \ldots, 0) = (k_1/n, \ldots, k_p/n).$$

Case 2. The case that $k_1, \ldots, k_p > 0$. Here we have to maximize the function $f_p : \boldsymbol{\theta} \mapsto \theta_1^{k_1} \cdots \theta_p^{k_p}$ under the constraints $\sum_i \theta_i = 1$ and $\theta_i \in [0,1]$ for all i. We apply the theorem of Lagrange to locate the maximum. (The existence of this maximum follows from the fact that f is continuous on the compact set $\{\boldsymbol{\theta} \in [0,1]^p : \sum_i \theta_i = 1\}$).
Define $g : \mathbb{R}^p \to \mathbb{R}$ by $g(\theta_1, \ldots, \theta_p) := \theta_1 + \cdots + \theta_p - 1$; now $S_p = \{\boldsymbol{\theta} \in [0,1]^p : g(\boldsymbol{\theta}) = 0\}$. Lagrange's theorem says that if $f|_{S_p}$ attains a maximum (or a minimum) in a point $\hat{\boldsymbol{\theta}}$, then one necessarily has

$$(\nabla f)(\hat{\boldsymbol{\theta}}) = \lambda\,(\nabla g)(\hat{\boldsymbol{\theta}}) \quad \text{for some } \lambda \in \mathbb{R}.$$

Writing this out we get:

$$k_i\, \hat{\theta}_1^{k_1} \cdots \hat{\theta}_i^{k_i - 1} \cdots \hat{\theta}_p^{k_p} = \lambda \quad \text{for all } i.$$

If $\hat{\theta}_j = 0$ for some j, then $f|_{S_p}$ attains its minimal value 0 in the point $\hat{\boldsymbol{\theta}}$. This is not what we are looking for; we therefore suppose $\hat{\theta}_i \neq 0$ for all i. Then we have

$$\hat{\theta}_1^{k_1} \cdots \hat{\theta}_p^{k_p} = \lambda \, \hat{\theta}_i / k_i \quad \text{for all } i.$$

Here we note that $\lambda \neq 0$ (because $k_i, \hat{\theta}_i \neq 0$ for all i). Setting $\mu := \hat{\theta}_1^{k_1} \cdots \hat{\theta}_p^{k_p} / \lambda$, we have $\hat{\theta}_i = \mu \, k_i$ for all i. Hence

$$1 = \sum_i \hat{\theta}_i = \sum_i \mu k_i = \mu \, n.$$

It follows that $\mu = 1/n$ and $\hat{\theta}_i = k_i/n$ for all i. □

(ii) Let $n := k_1 + \cdots + k_p$, where $k_1, \ldots, k_p \in \mathbb{N}$. Prove that the number of possibilities to partition a set of n different elements into p disjoint subsets $A_1, \ldots, A_p$, where A_i contains k_i elements, equals

$$\binom{n}{k_1, \ldots, k_p} := \frac{n!}{k_1! \cdots k_p!}.$$

Proof. First we compose A_1. This can be done in $\binom{n}{k_1}$ different ways. Then we compose A_2 which can be done in $\binom{n-k_1}{k_2}$ different ways. For A_3 there are $\binom{n-k_1-k_2}{k_3}$ possibilities; etcetera. Summarizing, we arrive at

$$\binom{n}{k_1} \binom{n-k_1}{k_2} \cdots \binom{n-k_1-\cdots-k_{p-1}}{k_p} = \binom{n}{k_1, \ldots, k_p}$$

possibilities. □

(iii) Denote the frequency vector $(k_1, \ldots, k_p)$ by $\mathbf{F} = (F_1, \ldots, F_p)$. Prove that for all $(k_1, \ldots, k_p) \in \mathbb{N}^p$

$$\mathbb{P}(\mathbf{F} = (k_1, \ldots, k_p)) = \begin{cases} \binom{n}{k_1, \ldots, k_p} \theta_1^{k_1} \cdots \theta_p^{k_p} & \text{if } \sum_i k_i = n, \\ 0 & \text{if } \sum_i k_i \neq n. \end{cases}$$

Proof. We surely have $\sum_i F_i = n$. Therefore we may restrict our attention to the case in which $k_1 + \cdots + k_p = n$. The probability that $\mathbf{X} = (X_1, \ldots, X_n)$ will exhibit an outcome

$$(\overbrace{1, \ldots, 1}^{k_1}, \overbrace{2, \ldots, 2}^{k_2}, \ldots, \overbrace{p, \ldots, p}^{k_p})$$

equals $\theta_1^{k_1} \cdots \theta_p^{k_p}$. By (ii) there are $\binom{n}{k_1, \ldots, k_p}$ possible outcomes for $\mathbf{X}$ having a frequency vector $(k_1, \ldots, k_p)$, all of them with the same probability, that is, $\theta_1^{k_1} \cdots \theta_p^{k_p}$. We conclude that

$$\mathbb{P}(\mathbf{F} = (k_1, \ldots, k_p)) = \binom{n}{k_1, \ldots, k_p} \theta_1^{k_1} \cdots \theta_p^{k_p}.$$

□

(iv) Now $\mathbf{F}$, by definition, enjoys a *multinomial* distribution with parameters n and $\boldsymbol{\theta}$. Next we define the stochastic p-vector $\mathbf{D}_k$ by

$$\mathbf{D}_k := (0, \ldots, 0, \overbrace{\ 1\ }^{X_k}, 0, \ldots, 0),$$

where the ith coordinate equals 1 if $X_k = i$. These $\mathbf{D}_1, \ldots, \mathbf{D}_n$ are statistically independent and we have $\mathbf{F} = \mathbf{D}_1 + \cdots + \mathbf{D}_n$. We define the Bernoulli distributed variables B_k^i by setting $\mathbf{D}_k = (B_k^1, \ldots, B_k^p)$. Show that $\mathbb{E}(B_k^i) = \theta_i$ and $\operatorname{cov}(B_k^i, B_k^j) = -\theta_i\theta_j$ for all $i \neq j$.

Solution. We have $B_k^i = 1$ if $X_k = i$ and otherwise $B_k^i = 0$. Hence

$$\mathbb{E}(B_k^i) = \mathbb{P}(X_k \neq i)\,0 + \mathbb{P}(X_k = i)\,1 = \mathbb{P}(X_k = i) = \theta_i.$$

If $i \neq j$, we necessarily have $B_k^i B_k^j = 0$. Consequently

$$\operatorname{cov}(B_k^i, B_k^j) = \mathbb{E}(B_k^i B_k^j) - \mathbb{E}(B_k^i)\mathbb{E}(B_k^j) = 0 - \theta_i\theta_j = -\theta_i\theta_j.$$

□

(v) Prove that for all i the variable F_i is binomially distributed and prove that $\operatorname{cov}(F_i, F_j) = -n\theta_i\theta_j$ if $i \neq j$.

Proof. The variable B_k^i is a function of $\mathbf{D}_k$. Using a generalized version of Proposition I.4.2, we conclude that $B_1^i, \ldots, B_n^i$ is an independent system of variables that are Bernoulli distributed with parameter θ_i. It follows from this that $F_i = B_1^i + \cdots + B_n^i$ is binomially distributed with parameters n and θ_i (see Exercise 26).

Furthermore, if $k \neq l$, then $\mathbf{D}_k$ and $\mathbf{D}_l$ and, a fortiori, B_k^i and B_l^j are statistically independent (all i, j). For $i \neq j$ we can therefore write

$$\begin{aligned} \operatorname{cov}(F_i, F_j) &= \operatorname{cov}(B_1^i + \cdots + B_n^i\ ,\ B_1^j + \cdots + B_n^j) \\ &= \sum_{k,l=1}^{n} \operatorname{cov}(B_k^i, B_l^j) = \sum_{k=1}^{n} \operatorname{cov}(B_k^i, B_k^j) + \sum_{k \neq l} \operatorname{cov}(B_k^i, B_l^j) \\ \text{(see (iv))} &= n \cdot (-\theta_i\theta_j) + 0 = -n\,\theta_i\theta_j. \end{aligned}$$

□

Exercise 35

Suppose the stochastic variable X is F^m_n-distributed. Prove that the variable $1/X$ is F^n_m-distributed.

Proof. Choose two independent variables U and V, where U is $\chi^2(m)$-distributed and V is $\chi^2(n)$-distributed. This is possible by §I.11 Exercise 42.

Now by construction the variables

$$Z_1 := \frac{U/m}{V/n} \quad \text{and} \quad Z_2 := \frac{V/n}{U/m}$$

are F^m_n and F^n_m-distributed respectively. The variables X and Z_1 are identically distributed, so by §I.11 Exercise 27 the variables $1/X$ and $1/Z_1 = Z_2$ are identically distributed, which proves the statement. □

Exercise 36

(i) If X and Y are essentially equal, then they are identically distributed. Prove this.

Proof. Let $(\Omega, \mathfrak{A}, \mathbb{P})$ be the underlying probability space of X and Y. Define $V := \{\omega \in \Omega : X(\omega) \neq Y(\omega)\}$. Then $V^c = (X-Y)^{-1}(0)$ and therefore $V^c, V \in \mathfrak{A}$ (see §I.11 Exercise 7 (and 1)). Furthermore, because X and Y are essentially equal we have $\mathbb{P}(V) = 0$. Now for all $A \in \mathfrak{A}$

$$X^{-1}(A) \subset Y^{-1}(A) \cup V,$$

so

$$\begin{aligned}\mathbb{P}_X(A) = \mathbb{P}\left(X^{-1}(A)\right) &\leq \mathbb{P}\left(Y^{-1}(A)\right) + \mathbb{P}(V) \\ &= \mathbb{P}\left(Y^{-1}(A)\right) = \mathbb{P}_Y(A).\end{aligned}$$

Consequently $\mathbb{P}_X(A) \leq \mathbb{P}_Y(A)$. By symmetry in our arguments, we also have $\mathbb{P}_Y(A) \leq \mathbb{P}_X(A)$. So $\mathbb{P}_X(A) = \mathbb{P}_Y(A)$ for all $A \in \mathfrak{A}$, which is the same as saying that X and Y are identically distributed. □

(ii) Give an example of two stochastic variables that are identically distributed and *not* essentially equal.

Solution. Choose two independent variables X and Y which are both Bernoulli distributed with parameter $\frac{1}{2}$. (This is possible via §I.11 Exercise 42; or throw a die twice). Now X and Y are of course identically distributed. However, they are not essentially equal:

$$\begin{aligned}\mathbb{P}(X = Y) &= \mathbb{P}(X = Y = 1) + \mathbb{P}(X = Y = 0) \\ &= \mathbb{P}(X = 1)\,\mathbb{P}(Y = 1) + \mathbb{P}(X = 0)\,\mathbb{P}(Y = 0) \\ &= (\tfrac{1}{2})^2 + (\tfrac{1}{2})^2 = \tfrac{1}{2} \neq 1.\end{aligned}$$

This shows that two variables can very well be identically distributed without being essentially equal. □

Exercise 37

(i) A sample $X_1, \ldots, X_{50}$ is drawn from a $N(\mu_X, 0.0025)$-distributed population: we obtain $\overline{X} = 4.520$. A sample $Y_1, \ldots, Y_{100}$, drawn from a $N(\mu_Y, 0.0040)$-distributed population, provides $\overline{Y} = 4.490$. Determine 97.5% confidence intervals for μ_X and μ_Y.

Solution. Proposition II.1.1 tells us that the variable $\overline{X}$ was (a priori) $N(\mu_X, 0.0025/50 = 0.000050)$-distributed. It follows that the variable $(\overline{X} - \mu_X)/\sqrt{5.0 \cdot 10^{-5}}$ was $N(0,1)$-distributed. In Table II it can be found that a priori

$$\mathbb{P}\left(\left|(\overline{X} - \mu_X)/\sqrt{5.0 \cdot 10^{-5}}\right| < 2.24\right) = 0.975.$$

Substitution of the outcome in this inequality provides us with $(4.504, 4.536)$ as a 97.5% confidence interval.

In a similar way we arrive at a 97.5% confidence interval $(4.476, 4.504)$ for μ_Y. □

(ii) Construct a 95% confidence interval for $\mu_X - \mu_Y$, using the result of (i).

Solution. With $100 \cdot (0.975)^2\% = 95.1\%$ confidence (think about this) we have

$$\mu_X - \mu_Y \in (4.504 - 4.504, 4.536 - 4.476) = (0.000, 0.060).$$ □

(iii) Compare the result of (ii) to that of §II.4 Example 1, in which we got $(0.011, 0.049)$ as a 95% confidence interval for $\mu_X - \mu_Y$.

Solution. Of course the confidence interval constructed in §II.4 is better. □

Exercise 38

(i) Let for all $n = 1, 2, \ldots$ the variable X_n be binomially distributed with parameters n and θ. Choose a statistically independent sequence $Y_1, Y_2, \ldots$ of Bernoulli distributed stochastic variables with parameter θ. Then X_n and $\sum_{i=1}^n Y_i$ are identically distributed (see Exercise 26). It follows, using Proposition II.7.2, that $\mathbb{E}(X_n) = n\theta$ and $\mathrm{var}(X_n) = n\theta(1-\theta)$. Prove that for all $x \in \mathbb{R}$

$$\lim_{n\to\infty} \mathbb{P}\left(\frac{X_n - n\theta}{\sqrt{n\theta(1-\theta)}} \leq x\right) = \Phi(x).$$

Proof. By the central limit theorem we have

$$\lim_{n\to\infty} \mathbb{P}\left(\frac{\sum_{i=1}^n Y_i - n\theta}{\sqrt{n}\sqrt{\theta(1-\theta)}} \leq x\right) = \lim_{n\to\infty} \mathbb{P}\left(\frac{(\frac{1}{n}\sum_{i=1}^n Y_i) - \theta}{\sqrt{\theta(1-\theta)}/\sqrt{n}} \leq x\right) = \Phi(x),$$

which proves the statement. □

(ii) Suppose that X is binomially distributed with parameters $n = 20$ and $\theta = 0.40$. Determine $\mathbb{P}(X \leq 6)$.

Solution. From Table I we get: $\mathbb{P}(X \leq 6) = 0.250$. □

(iii) Next we approximate X by a $N(\mu, \sigma^2)$-distributed variable Y. We take $\mu = \mathbb{E}(X) = 20 \cdot 0.40 = 8.0$ and $\sigma^2 = \text{var}(X) = 20 \cdot 0.40 \cdot 0.60 = 4.8$. Compare $\mathbb{P}(Y \leq 6)$ to $\mathbb{P}(X \leq 6)$.

Solution. We have

$$\mathbb{P}(Y \leq 6) = \Phi\left(\frac{6-8}{\sqrt{4.8}}\right) = \cdots = 1 - 0.819 = 0.181.$$

This is rather below the actual value 0.250 of $\mathbb{P}(X \leq 6)$. □

(iv) Writing $\mathbb{P}(X \leq 6) = \mathbb{P}(X < 7) \approx \mathbb{P}(Y < 7) = \mathbb{P}(Y \leq 7)$ we arrive at an approximation $\mathbb{P}(Y \leq 7)$ for $\mathbb{P}(X \leq 6)$. Compare these two probabilities.

Solution. We have

$$\mathbb{P}(Y \leq 7) = \Phi\left(\frac{7-8}{\sqrt{4.8}}\right) = \cdots = 1 - 0.677 = 0.323$$

which is rather above 0.250. □

(v) *Question.* Averaging both approximations leads to a value 0.252. Is this a fairly good approximation of $\mathbb{P}(X \leq 6)$?

Solution. Yes, it is. □

(vi) Compare $\mathbb{P}(Y \leq 6.5)$ to $\mathbb{P}(X \leq 6)$.

Solution. Computing $\mathbb{P}(Y \leq 6.5)$ we get

$$\mathbb{P}(Y \leq 6.5) = \Phi\left(\frac{6.5-8}{\sqrt{4.8}}\right) = \cdots = 1 - 0.752 = 0.248,$$

which also presents a fairly good approximation of $\mathbb{P}(X \leq 6)$. □

Exercise 39

In §II.9 the following theory is built up. Let there be given a family $\{f(\bullet, \theta) : \theta \in \Theta\}$ of probability densities, Θ being some parameter space. A sample $X_1, \ldots, X_n$ is drawn from a population from which beforehand it is known that it has a probability density out of the mentioned family.

For each outcome $(x_1, \ldots, x_n) \in \mathbb{R}^n$ we consider the function

$$\theta \mapsto L_\theta(x_1, \ldots, x_n) = \prod_{i=1}^{n} f(x_i, \theta).$$

We suppose it is (for each of the outcomes individually) maximized by exactly *one* point $\theta \in \Theta$, which we denote by $\hat{\theta}(x_1, \dots, x_n)$. This is the so-called 'maximum likelihood estimation' of θ. In this way a map $\hat{\theta}(\mathbf{X}) : \Omega \to \mathbb{R}$ arises, via

$$\hat{\theta}(\mathbf{X}) \; : \; \omega \; \mapsto \; \hat{\theta}(X_1(\omega), \dots, X_n(\omega)).$$

Prove that such a maximum likelihood estimator is always (essentially) symmetric. That is, prove that for every $\theta \in \Theta$ and permutation π of $\{1, \dots, n\}$ we have $\hat{\theta}(\mathbf{X}) = \hat{\theta}(\mathbf{X}^\pi)$, where $\mathbf{X}^\pi := (X_{\pi(1)}, \dots, X_{\pi(n)})$.

Proof. Take an arbitrary, fixed $\theta \in \Theta$ and an arbitrary permutation π of $\{1, \dots, n\}$. Then for any $\omega \in \Omega$ we have

$$\begin{aligned}
\prod_{i=1}^{n} f(X_i(\omega), \hat{\theta}(\mathbf{X})(\omega)) &= \max_{\theta \in \Theta} \left(\prod_{i=1}^{n} f(X_i(\omega), \theta) \right) \\
&= \max_{\theta \in \Theta} \left(\prod_{i=1}^{n} f(X_{\pi(i)}(\omega), \theta) \right) \\
&= \prod_{i=1}^{n} f(X_{\pi(i)}(\omega), \hat{\theta}(\mathbf{X}^\pi)(\omega)) \\
&= \prod_{i=1}^{n} f(X_i(\omega), \hat{\theta}(\mathbf{X}^\pi)(\omega)).
\end{aligned}$$

By assumption $\hat{\theta}(\mathbf{X})(\omega)$ is the only $\theta \in \Theta$ for which the maximum is taken on, so

$$\hat{\theta}(\mathbf{X})(\omega) = \hat{\theta}(\mathbf{X}^\pi)(\omega).$$

Thus $\hat{\theta}(\mathbf{X}) = \hat{\theta}(\mathbf{X}^\pi)$, which proves the statement. □

Exercise 40

Given is a population from which it is known that it enjoys a uniform distribution on the interval $(0, \theta)$, where θ is some (unknown) positive number. From this population a sample $X_1, \dots, X_n$ is drawn and a statistic T_n is defined by

$$T_n := \max(X_1, \dots, X_n).$$

(i) Prove that for all $x \in \mathbb{R}$ we have:

$$\lim_{n \to \infty} \mathbb{P}\left(\frac{n(\theta - T_n)}{\theta} \leq x \right) = \begin{cases} 1 - e^{-x} & \text{if } x \geq 0, \\ 0 & \text{elsewhere.} \end{cases}$$

Proof. The distribution function F of the uniform distribution on the interval

$(0, \theta)$ is given by

$$F(x) = \begin{cases} 0 & \text{if } x < 0, \\ x/\theta & \text{if } x \in [0, \theta], \\ 1 & \text{if } x > \theta. \end{cases}$$

As for all $x \in \mathbb{R}$ we have

$$\begin{aligned} F_{T_n}(x) &= \mathbb{P}(T_n \leq x) = \mathbb{P}(X_1 \leq x, \ldots, X_n \leq x) \\ &= \mathbb{P}(X_1 \leq x) \cdots \mathbb{P}(X_n \leq x) = F(x)^n, \end{aligned}$$

it follows that

$$F_{T_n}(x) = \begin{cases} 0 & \text{if } x < 0, \\ (x/\theta)^n & \text{if } x \in [0, \theta], \\ 1 & \text{if } x > \theta. \end{cases}$$

In general we have that

$$\begin{aligned} \mathbb{P}\left(\frac{n(\theta - T_n)}{\theta} \leq x\right) &= \mathbb{P}\left(\theta\left(1 - \frac{x}{n}\right) \leq T_n\right) \\ &= \mathbb{P}\left(\theta\left(1 - \frac{x}{n}\right) < T_n\right) = 1 - F_{T_n}\left(\theta\left(1 - \frac{x}{n}\right)\right). \end{aligned}$$

For $x < 0$ this expression equals zero. For $x \geq 0$ we get:

$$\lim_{n\to\infty} \mathbb{P}\left(\frac{n(\theta - T_n)}{\theta} \leq x\right) = \lim_{n\to\infty} \left(1 - \left(1 - \frac{x}{n}\right)^n\right) = 1 - e^{-x}.$$

□

(ii) Indicate how (i) can be used to construct confidence intervals for θ.

Solution. For large n, by (i) we know the approximate probability distribution of the variable $n(\theta - T_n)/\theta$. We see that at a prescribed confidence level of $(1 - \alpha) \times 100\%$ we can find g_1, g_2 for which $g_1 < g_2$ and, approximately,

$$\mathbb{P}\left(\frac{n(\theta - T_n)}{\theta} < g_1\right) = \tfrac{1}{2}\alpha \quad \text{and} \quad \mathbb{P}\left(\frac{n(\theta - T_n)}{\theta} > g_2\right) = \tfrac{1}{2}\alpha.$$

If T_n shows an outcome t then we arrive at the following inequality for θ:

$$g_1 < \frac{n(\theta - t)}{\theta} < g_2.$$

This is equivalent to:

$$\frac{nt}{n - g_1} < \theta < \frac{nt}{n - g_2},$$

which thus gives us a $(1 - \alpha) \times 100\%$ confidence interval for θ, given any outcome t for T_n.

□

(iii) A sample $X_1, \ldots, X_{100}$ yields an outcome 2.17 for T_{100}. Construct a 95% confidence interval for θ.

Solution. Just apply (ii). Here g_1 and g_2 are determined by

$$1 - e^{-g_1} = 0.025 \quad \text{and} \quad e^{-g_2} = 0.025,$$

which leads to $g_1 = 0.025$ and $g_2 = 3.689$. Thus the interval $(2.17, 2.25)$ emerges as a 95% confidence interval for θ. □

Exercise 41

Given is a Poisson distributed population with unknown parameter $\lambda \geq 0$. A sample $X_1, \ldots, X_n$ is drawn from this population. Is the vectorial statistic

$$\mathbf{T} = (X_1, \ldots, X_{n-1})$$

sufficient for λ?

Solution. For all $x_i, t_i \in \mathbb{N}$, setting $\mathbf{X} := (X_1, \ldots, X_n)$, one has

$$\mathbb{P}\,(\mathbf{X} = (x_1, \ldots, x_n) \mid \mathbf{T} = (t_1, \ldots, t_{n-1})) = 0$$

whenever $x_i \neq t_i$ for some $i = 1, \ldots, n-1$. In the case where $x_i = t_i$ for all $i = 1, \ldots, n-1$,

$$\begin{aligned}
&\mathbb{P}\,(\mathbf{X} = (x_1, \ldots, x_n) \mid \mathbf{T} = (t_1, \ldots, t_{n-1})) \\
&\quad = \frac{\mathbb{P}\,(\mathbf{X} = (x_1, \ldots, x_n) \text{ and } \mathbf{T} = (x_1, \ldots, x_{n-1}))}{\mathbb{P}\,(\mathbf{T} = (x_1, \ldots, x_{n-1}))} \\
&\quad = \frac{\mathbb{P}\,(\mathbf{X} = (x_1, \ldots, x_n))}{\mathbb{P}\,(\mathbf{T} = (x_1, \ldots, x_{n-1}))} = \frac{e^{-\lambda}\,\{\lambda^{x_1}/x_1!\} \cdots e^{-\lambda}\,\{\lambda^{x_n}/x_n!\}}{e^{-\lambda}\,\{\lambda^{x_1}/x_1!\} \cdots e^{-\lambda}\,\{\lambda^{x_{n-1}}/x_{n-1}!\}} \\
&\quad = e^{-\lambda}\,\frac{\lambda^{x_n}}{x_n!}.
\end{aligned}$$

We see that in this conditional probability the parameter λ plays a role. It follows that $\mathbf{T}$ is not sufficient for λ.

Intuitively this was, of course, already clear beforehand. The requirement that the population be Poisson distributed is superfluous. □

Exercise 42

A sample $X_1, \ldots, X_n$ is drawn from a Bernoulli distributed population with unknown parameter $\theta \in (0, 1)$.

(i) Use the factorization theorem to prove: $\overline{X}$ is sufficient for θ.

Proof. For the variables X_i we have:

$$\mathbb{P}(X_i = x_i) = \theta^{x_i}\,(1-\theta)^{1-x_i} \qquad (x_i = 0, 1).$$

Hence the likelihood function is given by

$$\begin{aligned} L_\theta(x_1,\dots,x_n) &= \mathbb{P}(X_1 = x_1)\cdots\mathbb{P}(X_n = x_n) \\ &= \theta^{x_1}(1-\theta)^{1-x_1}\ \cdots\ \theta^{x_n}(1-\theta)^{1-x_n} \\ &= \theta^{n\overline{x}}(1-\theta)^{n(1-\overline{x})}, \end{aligned}$$

where $x_1,\dots,x_n \in \{0,1\}$. Setting $g(x_1,\dots,x_n) := \frac{1}{n}(x_1+\dots+x_n)$, $h \equiv 1$ and

$$\varphi(\theta,y) := \theta^{ny}(1-\theta)^{n(1-y)},$$

we have $\overline{X} = g(X_1,\dots,X_n)$ and

$$L_\theta(x_1,\dots,x_n) = h(x_1,\dots,x_n)\,\varphi(\theta, g(x_1,\dots,x_n)).$$

Applying the factorization theorem (continuous version) we conclude that $\overline{X}$ is sufficient for θ. □

(ii) Starting from the definition of sufficiency, prove directly that $\overline{X}$ is a sufficient statistic for θ.

Proof. We have:

$$\begin{aligned} &\mathbb{P}\left(\mathbf{X} = (x_1,\dots,x_n)\mid \overline{X} = t\right) \\ &\qquad = \mathbb{P}(X_1 = x_1,\dots,X_n = x_n \mid X_1+\dots+X_n = nt). \end{aligned}$$

This expression equals zero if $nt \neq x_1+\dots+x_n$. Noting that $X_1+\dots+X_n$ is binomially distributed with parameters n and θ (see Exercise 26), for $nt = x_1+\dots+x_n$ we arrive at

$$\begin{aligned} \mathbb{P}\left(\mathbf{X} = \mathbf{x}\mid\overline{X} = t\right) &= \frac{\mathbb{P}\left(\mathbf{X} = \mathbf{x} \text{ and } X_1+\dots+X_n = x_1+\dots+x_n\right)}{\mathbb{P}\left(X_1+\dots+X_n = x_1+\dots+x_n\right)} \\ &= \frac{\mathbb{P}\left(\mathbf{X} = \mathbf{x}\right)}{\mathbb{P}\left(X_1+\dots+X_n = x_1+\dots+x_n\right)} \\ &= \frac{\theta^{x_1}(1-\theta)^{1-x_1}\ \cdots\ \theta^{x_n}(1-\theta)^{1-x_n}}{\binom{n}{x_1+\dots+x_n}\theta^{x_1+\dots+x_n}(1-\theta)^{n-(x_1+\dots+x_n)}} \\ &= \binom{n}{x_1+\dots+x_n}^{-1}. \end{aligned}$$

This expression does not depend on θ, so $\overline{X}$ is sufficient for the parameter θ. □

Exercise 43

Given is a gamma distributed population with parameters α and β. The numerical value of α is known, whereas β is unknown. A sample $X_1,\dots,X_n$ is drawn from this population. Is the statistic $\overline{X}$ sufficient for β?

Solution. We apply the factorization theorem (continuous version) to answer the question in the affirmative. If α is known beforehand to be equal to α_0, then the likelihood function is given by

$$L_\beta(x_1, \dots, x_n) = \frac{1}{\beta^{n\alpha_0}\,\Gamma(\alpha_0)^n}\,(x_1 \cdots x_n)^{\alpha_0 - 1}\,e^{-n\overline{x}/\beta}$$

if $x_1, \dots, x_n \geq 0$ and $L_\beta(\mathbf{x}) = 0$ otherwise. Next we set

$$g(x_1, \dots, x_n) := \frac{x_1 + \cdots + x_n}{n}, \quad h(x_1, \dots, x_n) := \frac{1}{\Gamma(\alpha_0)^n}\,(x_1 \cdots x_n)^{\alpha_0 - 1}$$

and

$$\varphi(\beta, y) := \frac{1}{\beta^{n\,\alpha_0}}\,e^{-ny/\beta}.$$

The likelihood function can be factorized as

$$L_\beta(x_1, \dots, x_n) = h(x_1, \dots, x_n)\,\varphi(\beta, g(x_1, \dots, x_n)).$$

By the factorization theorem this implies that $\overline{X}$ is sufficient for β. □

Exercise 44

It is known that a certain population enjoys a probability density f_θ given by

$$f_\theta(x) = \frac{1}{2\theta}\,1_{(-\theta,\theta)}(x),$$

where θ is some unknown positive number. A sample $X_1, \dots, X_n$ is drawn from this population in order to get information about the parameter θ.
Question. Is the statistic

$$T := \max(|X_1|, \dots, |X_n|)$$

sufficient for θ?

Solution. We apply the factorization theorem (continuous version). It is easily seen that the likelihood function can here be represented as

$$L_\theta(x_1, \dots, x_n) = \frac{1}{(2\theta)^n}\,1_{[0,\theta)}\left(\max\{|x_1|, \dots, |x_n|\}\right).$$

Next we set $g(x_1, \dots, x_n) := \max(|x_1|, \dots, |x_n|)$, $h \equiv 1$ and

$$\varphi(\theta, y) := \frac{1}{(2\theta)^n}\,1_{[0,\theta)}(y).$$

Then $T = g(X_1, \dots, X_n)$ and

$$L_\theta(x_1, \dots, x_n) = h(x_1, \dots, x_n)\,\varphi(\theta, g(x_1, \dots, x_n)).$$

By the factorization theorem this implies that T is sufficient for θ. □

Exercise 45

A certain population enjoys a probability density $f(\bullet, \theta)$, where the unknown parameter θ is an element of Θ. A sample $X_1, \dots, X_n$ is drawn from this population and its outcome is condensed as an outcome of the vectorial statistic

$$\mathbf{T} = \mathbf{g}(X_1, \dots, X_n),$$

where $\mathbf{g} : \mathbb{R}^n \to \mathbb{R}^p$ is some Borel function.
Prove that if $\psi : \mathbb{R}^p \to \mathbb{R}^q$ is a Borel function and the vectorial statistic $\psi(\mathbf{T})$ is sufficient for θ, then the statistic $\mathbf{T}$ is also sufficient for θ.

Proof. We apply the factorization theorem in both directions. If $\psi(\mathbf{T})$ is sufficient for θ, then there is a factorization of the form

$$L_\theta(x_1, \dots, x_n) = h(x_1, \dots, x_n)\ \varphi(\theta, (\psi \circ \mathbf{g})(x_1, \dots, x_n)).$$

Next, define $\tilde{\varphi} : \Theta \times \mathbb{R}^p \to [0, +\infty)$ by

$$\tilde{\varphi}(\theta, (y_1, \dots, y_p)) := \varphi(\theta, \psi(y_1, \dots, y_p)).$$

Now $\mathbf{T} = \mathbf{g}(X_1, \dots, X_n)$ and

$$L_\theta(x_1, \dots, x_n) = h(x_1, \dots, x_n)\ \tilde{\varphi}(\theta, \mathbf{g}(x_1, \dots, x_n)).$$

By the factorization theorem this implies that $\mathbf{T}$ is sufficient for θ. □

Exercise 46

Apply the inequality of Cauchy–Schwarz–Buniakowsky to the vectors

$$\mathbf{e} := (1, \dots, 1) \quad \text{and} \quad \mathbf{c} := (c_1, \dots, c_n)$$

in order to complete the proof of Theorem II.9.3.

Solution. In the proof of Theorem II.9.3 we are looking for unbiased, linear estimators of $\mu(\theta)$. These appear to be of form $c_1X_1 + \dots + c_nX_n$ with $c_1 + \dots + c_n = 1$. The variance of such an estimator is then given by $(c_1^2 + \dots + c_n^2)\,\sigma^2(\theta)$, where $\sigma^2(\theta)$ denotes the variance of the population.

Applying the inequality of Cauchy–Schwarz–Buniakowsky as suggested to the vectors $\mathbf{e}$ and $\mathbf{c}$ tells us that $|\langle \mathbf{e}, \mathbf{c} \rangle|^2 \leq \|\mathbf{e}\|^2 \|\mathbf{c}\|^2$, which in the present case gives us:

$$1 = \left| \sum_{j=1}^{n} c_j \right|^2 \leq n \sum_{j=1}^{n} c_j^2 \,.$$

So $\sum_{j=1}^n c_j^2 \geq \frac{1}{n}$ and it immediately follows that taking $c_1 = \dots = c_n = \frac{1}{n}$ yields the best choice available.

Given the fact that $\sigma^2(\theta) > 0$ it is also the unique best one, since there can only be equality in the Cauchy–Schwarz–Buniakowsky inequality if $\mathbf{c} = \lambda\mathbf{e}$ for some scalar λ. □

Exercise 47

Let $X_1, \ldots, X_n$ be a sample from a population with probability density $f(\bullet, \theta)$, where $\theta \in \Theta$. Suppose that $\mathfrak{C}$ is a linear space consisting of statistics based on $X_1, \ldots, X_n$, all of them being of finite variance. Moreover, suppose that κ is a characteristic of the population.
Prove the following well-known result. An element $T \in \mathfrak{C}_\kappa$ satisfies

$$\mathrm{var}_\theta(T) = m(\theta) \qquad \text{for all } \theta \in \Theta$$

if and only if $\mathrm{cov}_\theta(T, S) = 0$ for all $\theta \in \Theta$ and $S \in \mathfrak{C}$ that are unbiased estimators of 0.

Proof. First suppose that $\mathrm{cov}_\theta(T, S) = 0$ for all unbiased estimators S of 0 and all $\theta \in \Theta$. Now let $\tilde{T}$ be an element of $\mathfrak{C}_\kappa$. Then $T - \tilde{T}$ is an element of $\mathfrak{C}$ and it is clearly an unbiased estimator of 0. It follows that

$$\begin{aligned} \mathrm{var}_\theta(\tilde{T}) &= \mathrm{var}_\theta((\tilde{T} - T) + T) \\ &= \mathrm{cov}_\theta((\tilde{T} - T) + T, (\tilde{T} - T) + T) \\ &= \mathrm{var}_\theta(\tilde{T} - T) + 2\,\mathrm{cov}_\theta(\tilde{T} - T, T) + \mathrm{var}_\theta(T) \\ &= \mathrm{var}_\theta(\tilde{T} - T) + \mathrm{var}_\theta(T) \;\geq\; \mathrm{var}_\theta(T), \end{aligned}$$

so $\mathrm{var}_\theta(T) = \inf\{\mathrm{var}_\theta(V) : V \in \mathfrak{C}_\kappa\} = m(\theta)$ for any $\theta \in \Theta$.

As to the implication in the other direction, suppose that $T \in \mathfrak{C}_\kappa$ and $\mathrm{var}_\theta(T) = m(\theta)$ for all $\theta \in \Theta$. Now take a fixed θ and let $S \in \mathfrak{C}$ be an unbiased estimator of 0. Note that $\{\lambda S : \lambda \in \mathbb{R}\}$ is a set containing only unbiased estimators of 0, so

$$\{T + \lambda S : \lambda \in \mathbb{R}\} \subset \mathfrak{C}_\kappa$$

and consequently $\mathrm{var}_\theta(T + \lambda S) \geq \mathrm{var}_\theta(T)$ for all $\lambda \in \mathbb{R}$. As

$$\begin{aligned} \mathrm{var}_\theta(T + \lambda S) &= \mathrm{cov}_\theta(T + \lambda S, T + \lambda S) \\ &= \mathrm{var}_\theta(T) + 2\lambda\,\mathrm{cov}_\theta(S, T) + \lambda^2\,\mathrm{var}_\theta(S), \end{aligned}$$

this tells us that $2\lambda\,\mathrm{cov}_\theta(S, T) + \lambda^2\,\mathrm{var}_\theta(S) \geq 0$ for all $\lambda \in \mathbb{R}$. Calculating the discriminant of the quadratic form in λ on the left side of this inequality, this appears to be only possible in case $(2\mathrm{cov}_\theta(S, T))^2 \leq 0$, that is, if $\mathrm{cov}_\theta(S, T) = 0$. This is what we had to prove. □

Chapter 3
Hypothesis testing

3.1 Summary of Chapter III

3.1.1 The Neyman–Pearson theory (III.1)

A *statistical hypothesis* is understood to be a statement concerning the probability distribution of a population. A hypothesis is said to be *simple* if it specifies the probability distribution of the population completely; otherwise it is called *composite.* Rules are developed in order to decide whether a so-called *null hypothesis* H_0 or an *alternative hypothesis* H_1 should be considered as being true.

We say that we commit a *type I error* if in the decision procedure we reject H_0 whereas it is true. A *type II error* occurs if H_0 is accepted whereas H_1 is true. The probability of making a type I error will be denoted by α, that of making a type II error by β.

A *hypothesis test* is understood to be an ordered sequence

$$(X_1, \dots, X_n;\, H_0, H_1;\, G),$$

where $X_1, \dots, X_n$ is a sample of size n from the population, H_0 and H_1 are statistical hypotheses and $G \subset \mathbb{R}^n$ is a Borel set. The set G is called the *critical region* of the hypothesis test. The event $(X_1, \dots, X_n) \in G$ is connected to rejection of the hypothesis H_0. If $(X_1, \dots, X_n) \notin G$ then H_0 is accepted. When the hypothesis H_0 is 'simple', the probability distribution of the n-vector $(X_1, \dots, X_n)$ under H_0 will be denoted by $\mathbb{P}^{H_0}_{X_1,\dots,X_n}$.

The *level of significance* of the hypothesis test is defined by

$$\alpha := \mathbb{P}^{H_0}_{X_1,\dots,X_n}(G).$$

It presents the probability that we reject H_0 whereas it is true (i.e. the chance on a type I error). We also say that G is of *size* α.

We shall systematically assume that the population in question enjoys a (possibly discrete) probability density of which beforehand it is known that it is a member of a family $\{f(\bullet, \theta) : \theta \in \Theta\}$. The hypotheses H_0 and H_1 will always be of the following form: $H_0 : \theta \in \Theta_0$ and $H_1 : \theta \in \Theta_1$, where $\Theta_0, \Theta_1 \subset \Theta$ and $\Theta_0 \cap \Theta_1 = \emptyset$. For all $\theta_1 \in \Theta_1$ we set

$$\beta(\theta_1) := \mathbb{P}^{H_1}_{X_1,\dots,X_n}(G^c).$$

Now $\beta(\theta_1)$ presents under the alternative $H_1 : \theta = \theta_1$ the probability that a type II error will be committed. The number $1 - \beta(\theta_1)$ is called the *power* of the test in the point $\theta_1 \in \Theta_1$.

Of considerable importance is the so-called *Neyman–Pearson lemma* (Theorem III.1.1), which we now formulate. Suppose we are testing the hypothesis $H_0 : \theta = \theta_0$ versus $H_1 : \theta = \theta_1$. Then among all critical regions of size α, maximal power is realized by a (critical) region of type

$$G = \{\mathbf{x} \in \mathbb{R}^n : L_0(\mathbf{x})/L_1(\mathbf{x}) \leq c\}.$$

Here L_0 and L_1 are the *likelihood functions* under H_0 and H_1 respectively. That is to say:

$$L_0(x_1, \dots, x_n) = f(x_1, \theta_0) \cdots f(x_n, \theta_0)$$

and

$$L_1(x_1, \dots, x_n) = f(x_1, \theta_1) \cdots f(x_n, \theta_1).$$

The Neyman–Pearson theory applies to cases where both H_0 and H_1 are simple hypotheses.

Next we consider cases where H_0 and H_1 are composite hypotheses of the form $H_0 : \theta \in \Theta_0$ and $H_1 : \theta \in \Theta_1$, where $\Theta = \Theta_0 \cup \Theta_1$ and $\Theta_0 \cap \Theta_1 = \emptyset$. For these cases the *likelihood ratio* is defined by

$$\Lambda(x_1, \dots, x_n) := \frac{\sup_{\theta \in \Theta_0} L_\theta(x_1, \dots, x_n)}{\sup_{\theta \in \Theta} L_\theta(x_1, \dots, x_n)}.$$

If $(X_1, \dots, X_n)$ exhibits an outcome $(x_1, \dots, x_n)$ for which $\Lambda(x_1, \dots, x_n)$ is 'small' then H_1 is more likely to be true. The number $\Lambda(x_1, \dots, x_n)$ is always an element of the interval $[0, 1]$. Critical regions of type

$$G(\delta) = \{(x_1, \dots, x_n) : \Lambda(x_1, \dots, x_n) \leq \delta\}$$

are chosen. At a prescribed size α, they generally (but not always) present critical regions of satisfactory power $1 - \beta$.

Theorem III.1.3 states the following. Under H_0 the statistic

$$-2 \log \Lambda(X_1, \dots, X_n)$$

is for large n approximately χ^2-distributed with $\dim \Theta - \dim \Theta_0$ degrees of freedom.

3.1.2 Hypothesis tests concerning normally distributed populations (III.2)

Starting from a $N(\mu, \sigma^2)$-distributed population we wish to test hypotheses with regard to the parameters μ and/or σ. Explicit forms for the likelihood ratio are deduced and useful test statistics emerge from this.

Case 1. Suppose the numerical value of σ is known. We wish to test $H_0 : \mu = \mu_0$ against $H_1 : \mu \neq \mu_0$ (or versus $\mu < \mu_0$ or $\mu > \mu_0$). Here the likelihood ratio is a function of the test statistic

$$\frac{\overline{X} - \mu_0}{\sigma/\sqrt{n}}.$$

Under H_0 this statistic is $N(0,1)$-distributed.

Case 2. Suppose σ is not known. Again we wish to test $H_0 : \mu = \mu_0$ against $H_1 : \mu \neq \mu_0$ (or against one-sided alternatives). Here the test statistic

$$\frac{\overline{X} - \mu_0}{S/\sqrt{n}}$$

emerges. This variable enjoys under H_0 a t-distribution with $n-1$ degrees of freedom.

Case 3. Suppose μ is unknown. We wish to test $H_0 : \sigma = \sigma_0$ against $H_1 : \sigma \neq \sigma_0$ (or one-sided alternatives). Here the likelihood ratio is a function of the test statistic

$$\frac{(n-1)S^2}{\sigma_0^2},$$

which under H_0 is $\chi^2(n-1)$-distributed.

Next we consider *two* populations, one $N(\mu_X, \sigma_X^2)$-distributed and the other $N(\mu_Y, \sigma_Y^2)$-distributed, from which respectively the following independent samples are drawn: $X_1, \ldots, X_m$ and $Y_1, \ldots, Y_n$.

Case 4. Suppose μ_X and μ_Y are unknown. We wish to test $H_0 : \sigma_X = \sigma_Y$ against $H_1 : \sigma_X \neq \sigma_Y$ (or one-sided alternatives). Here we arrive at the test statistic

$$S_Y^2/S_X^2,$$

which, under H_0, is F_{m-1}^{n-1}-distributed.

Case 5. Suppose σ_X and σ_Y are known. For a certain $\Delta \in \mathbb{R}$ we wish to test $H_0 : \mu_Y - \mu_X = \Delta$ against $H_1 : \mu_Y - \mu_X \neq \Delta$ (or against one-sided alternatives). In this case the likelihood ratio leads us to the test statistic

$$\frac{\overline{Y} - \overline{X} - \Delta}{\sqrt{\sigma_X^2/m + \sigma_Y^2/n}},$$

which under H_0 is $N(0,1)$-distributed. If σ_X and/or σ_Y is unknown and $m, n \geq 30$ we can replace σ_X and/or σ_Y by S_X and/or S_Y respectively.

Case 6. We assume that σ_X and σ_Y are equal, but unknown, denoting the unknown common value of σ_X and σ_Y by σ. We wish to test $H_0 : \mu_Y - \mu_X = \Delta$

against $H_1 : \mu_Y - \mu_X \neq \Delta$ (or against one-sided alternatives). Here we arrive at the test statistic

$$\frac{\overline{Y} - \overline{X} - \Delta}{S_p\sqrt{1/m + 1/n}}, \quad \text{for} \quad S_p^2 := \frac{(m-1)S_X^2 + (n-1)S_Y^2}{m+n-2}.$$

Under H_0 this statistic is t-distributed with $m + n - 2$ degrees of freedom.

3.1.3 The χ^2-test on goodness of fit (III.3)

Given an element $\boldsymbol{\theta} = (\theta_1, \ldots, \theta_p) \in \mathbb{R}^p$ with $\theta_i \geq 0$ for all i and $\theta_1 + \cdots + \theta_p = 1$, we define the discrete probability density $f_{\boldsymbol{\theta}}$ by

$$f_{\boldsymbol{\theta}}(i) := \theta_i \qquad (i = 1, \ldots, p).$$

Next we start from a population with probability density $f_{\boldsymbol{\theta}}$. That is to say, when drawing a sample of size one, the probability of drawing the element i is θ_i. Suppose $\boldsymbol{\theta}_0 = (\theta_{01}, \ldots, \theta_{0p})$ is a fixed element. We wish to test $H_0 : \boldsymbol{\theta} = \boldsymbol{\theta}_0$ versus $H_1 : \boldsymbol{\theta} \neq \boldsymbol{\theta}_0$. To this end we draw a sample $X_1, \ldots, X_n$ of size n from the population (with replacement). Denoting the frequency of occurrence of the element i in this sequence by F_i, we compose the so-called frequency vector

$$\mathbf{F} = (F_1, \ldots, F_p).$$

Given an outcome $(k_1, \ldots, k_p)$ of $\mathbf{F}$ the likelihood function assumes the form

$$\Lambda(k_1, \ldots, k_p) = \left(\frac{n\theta_{01}}{k_1}\right)^{k_1} \cdots \left(\frac{n\theta_{0p}}{k_p}\right)^{k_p}$$

(here $k_i \in \mathbb{N}$ and $k_1 + \cdots + k_p = n$). Critical regions can be constructed based on this ratio.

Usually this kind of testing is based on the test statistic

$$\sum_{i=1}^{p} \frac{(F_i - \mathbb{E}(F_i))^2}{\mathbb{E}(F_i)},$$

which is under H_0 approximately $\chi^2(p-1)$-distributed. In the decision procedure we reject H_0 when this test statistic exhibits a large outcome.

3.1.4 The χ^2-test on independence (III.4)

A decision procedure is developed in order to test the statistical independence of two stochastic variables.

We start from two discrete variables X and Y, with range $\{1, \ldots, p\}$ and $\{1, \ldots, q\}$ respectively. A sample $(X_1, Y_1), \ldots, (X_n, Y_n)$ is drawn from a population with the distribution of (X, Y). We wish to test H_0 : 'X and Y are statistically independent' against H_1 : 'X and Y are dependent'. The probability that (X,Y) will assume the value (i,j) will be denoted by θ_{ij}. The frequency of the element (i,j) in the sample $(X_1,Y_1),\dots,(X_n,Y_n)$ is denoted by F_{ij}. Here too an explicit form for the likelihood ratio can be deduced (Theorem III.4.2). In practice we use the test statistic

$$T=\sum_{i,j}\frac{(F_{ij}-E_{ij})^2}{E_{ij}},$$

where $E_{ij}:=\frac{1}{n}F_{i\bullet}F_{\bullet j}$. Under H_0 the statistic T is for large n approximately χ^2-distributed with $(p-1)(q-1)$ degrees of freedom (Theorem III.4.3).

3.2 Exercises to Chapter III

Exercise 1

Given is a sample X_1,X_2 from a $N(0,\sigma^2)$-distributed population. We want to test $H_0:\sigma=2$ against $H_1:\sigma=1$.

(i) Construct to this test a critical region of a prescribed size α and of maximal power.

Solution. By the Neyman–Pearson lemma such a critical region is of the form $G(\delta)=\{\mathbf{x}\in\mathbb{R}^2 : L_0(\mathbf{x})/L_1(\mathbf{x})\le\delta\}$. We have

$$\begin{aligned}\frac{L_0(\mathbf{x})}{L_1(\mathbf{x})} &= \frac{\frac{1}{2\pi}\frac{1}{2^2}\exp\left[-\frac{1}{2}\left(\frac{1}{2}x_1\right)^2-\frac{1}{2}\left(\frac{1}{2}x_2\right)^2\right]}{\frac{1}{2\pi}\frac{1}{1^2}\exp\left[-\frac{1}{2}x_1^2-\frac{1}{2}x_2^2\right]}\\ &= \tfrac{1}{4}\exp\left[-\tfrac{1}{2}\left(-\tfrac{3}{4}x_1^2-\tfrac{3}{4}x_2^2\right)\right]=\tfrac{1}{4}\exp\left[\tfrac{3}{8}\left(x_1^2+x_2^2\right)\right].\end{aligned}$$

It follows that $G(\delta)$ can be written as

$$\{\mathbf{x}\in\mathbb{R}^2 : x_1^2+x_2^2\le R^2\}$$

for some $R\ge 0$. The size of $G(\delta)$ can be expressed as

$$\begin{aligned}\mathbb{P}^{H_0}_{X_1,X_2}(G(\delta)) &= \iint_{G(\delta)}\tfrac{1}{2\pi}\cdot\tfrac{1}{4}\exp\left[-\tfrac{1}{8}\left(x_1^2+x_2^2\right)\right]\,dx\\ &= \frac{1}{8\pi}\int_0^R\left(\int_{-\pi}^{\pi}e^{-\frac{1}{8}u^2}\,dv\right)u\,du\\ &= \int_0^R\tfrac{1}{4}ue^{-\frac{1}{8}u^2}\,du=\left[-e^{-\frac{1}{8}u^2}\right]_0^R=1-e^{-\frac{1}{8}R^2},\end{aligned}$$

where we passed to polar co-ordinates. Next we adjust R in such a way that the size of $G(\delta)$ equals α. Therefore we set $1-e^{-\frac{1}{8}R^2}=\alpha$: it follows that $R^2=-8\log(1-\alpha)$. Consequently

$$G(\delta)=\{\mathbf{x}\in\mathbb{R}^2 : x_1^2+x_2^2\le-8\log(1-\alpha)\}$$

is a critical region which, among all regions of size α, gives the test maximal power. □

(ii) Specify the critical region in the case that $\alpha = 0.05$.

Solution. In case $\alpha = 0.05$ (i) provides $G(\delta) = \{\mathbf{x} \in \mathbb{R}^2 \;:\; x_1^2 + x_2^2 \leq 0.41\}$. □

Exercise 2

It is known that a certain population enjoys a probability density of the form

$$f(x, \theta) = \begin{cases} \theta x^{\theta - 1} & \text{if } x \in [0, 1] \\ 0 & \text{elsewhere} \end{cases}$$

where $\theta > 0$. Basing ourselves on a sample X_1, X_2 we want to test $H_0 \;:\; \theta = 2$ against $H_1 \;:\; \theta = 3$.

(i) Construct a critical region G of size 0.05 which is of maximal power.

Solution. With probability 1 the outcome $\mathbf{x}$ of (X_1, X_2) is an element of $[0, 1] \times [0, 1]$. We therefore consider outcomes $\mathbf{x} \in [0, 1] \times [0, 1]$ only. For such $\mathbf{x}$ we have $L_0(\mathbf{x})/L_1(\mathbf{x}) = 2x_1 \cdot 2x_2/3x_1^2 \cdot 3x_2^2 = 4/(9x_1x_2)$. It follows that critical regions based on the Neyman–Pearson lemma are of type

$$G(k) = \{\mathbf{x} \in [0, 1] \times [0, 1] \;:\; x_1 x_2 \geq k\}$$

where $0 < k < 1$. Next we adjust k in such a way that $\mathbb{P}^{H_0}_{X_1, X_2}(G(k)) = \alpha = 0.05$. Then by the Neyman–Pearson lemma $G(k)$ is among all critical regions of size $\alpha = 0.05$ the one of maximal power. We have

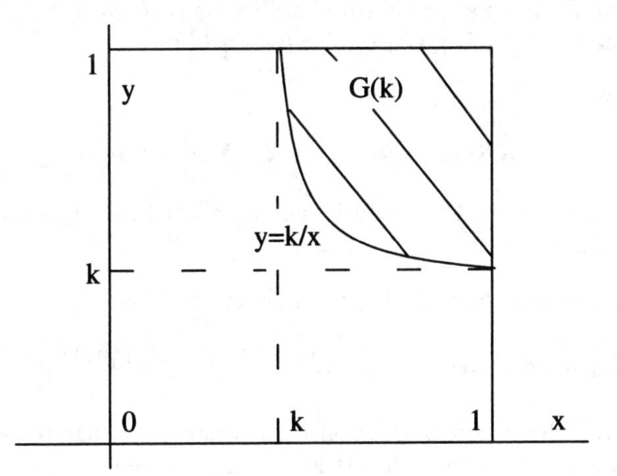

$$\begin{aligned}
\mathbb{P}^{H_0}_{X_1,X_2}(G(k)) &= \iint_{G(k)} 2x_1 2x_2\, dx_2\, dx_1 = \int_k^1 \int_{k/x_1}^1 4x_1x_2\, dx_2\, dx_1 \\
&= \int_k^1 2x_1\left(1^2 - \frac{k^2}{x_1^2}\right) dx_1 = \int_k^1 \left(2x_1 - \frac{2k^2}{x_1}\right) dx_1 \\
&= (1^2 - k^2) - 2k^2(\log 1 - \log k) = 1 - k^2 + 2k^2 \log k.
\end{aligned}$$

This is a strictly decreasing, continuous function in k. Furthermore

$$\lim_{k\downarrow 0} \mathbb{P}^{H_0}_{X_1,X_2}(G(k)) = 1 \quad \text{and} \quad \lim_{k\uparrow 1} \mathbb{P}^{H_0}_{X_1,X_2}(G(k)) = 0.$$

It follows from this that there is exactly one k such that $G(k)$ is of size 0.05. Using (for example) a pocket calculator we learn that $k \approx 0.84$. We therefore take $G = \{\mathbf{x} \in [0,1] \times [0,1] \ : \ x_1x_2 \geq 0.84\}$ as our critical region. □

(ii) Determine $1 - \beta$ in (i).

Solution.

$$\begin{aligned}
1 - \beta &= \mathbb{P}^{H_1}_{X_1,X_2}(G) = \int_k^1 \int_{k/x_1}^1 9x_1^2x_2^2\, dx_2\, dx_1 \\
&= \int_k^1 3x_1^2\left(1^3 - \frac{k^3}{x_1^3}\right) dx_1 = \left[x_1^3\right]_k^1 - 3k^3[\log x_1]_k^1 \\
&= 1 - k^3 + 3k^3 \log k = 0.10.
\end{aligned}$$

□

Exercise 3

We throw a coin; the probability that a head will be observed is θ. We wish to test $H_0 : \theta = \frac{1}{2}$ against $H_1 : \theta = \frac{3}{4}$. To this end we draw a sample X_1, X_2 where X_i $(i = 1, 2)$ is the number of throws needed before for the first time a head is observed. Now X_i is geometrically distributed with parameter θ. More specific, under H_0 the X_i have a discrete probability density given by $f_0 : n \mapsto \left(\frac{1}{2}\right)^n$ and under H_1 their density is given by $f_1 : n \mapsto \frac{3}{4}\left(\frac{1}{4}\right)^{n-1}$.

(i) Suppose

$$G(c) = \{(n_1, n_2) \in \mathbb{N} \times \mathbb{N} \ : \ 2 \leq n_1 + n_2 \leq c\}$$

is of size α. Show that $G(c)$ is, among all critical regions of size α, the one of maximal power.

Solution. Here the ratio L_0/L_1 is given by

$$L_0(n_1, n_2)/L_1(n_1, n_2) = \left(\tfrac{1}{2}\right)^{n_1+n_2} \Big/ \left(\tfrac{3}{4}\right)^2 \left(\tfrac{1}{4}\right)^{n_1+n_2-2} = \tfrac{1}{9}\, 2^{n_1+n_2}.$$

Applying the Neyman–Pearson lemma it is now easy to see that $G(c)$ is of maximal power among the critical regions of size α. □

(ii) Determine the size α of $G(c)$ if $c = 3$.

Solution. We have

$$\begin{aligned}\alpha &= \mathbb{P}^{H_0}_{X_1,X_2}(G(3)) = \mathbb{P}^{H_0}(X_1 + X_2 \leq 3) \\ &= \mathbb{P}^{H_0}(X_1 = X_2 = 1) + \mathbb{P}^{H_0}(X_1 = 1 \text{ and } X_2 = 2) \\ &\qquad + \mathbb{P}^{H_0}(X_1 = 2 \text{ and } X_2 = 1) \\ &= \tfrac{1}{2}\cdot\tfrac{1}{2} + \tfrac{1}{2}\cdot\tfrac{1}{4} + \tfrac{1}{4}\cdot\tfrac{1}{2} = \tfrac{1}{2} = 0.50.\end{aligned}$$

□

(iii) Determine $1 - \beta$ if $c = 3$.

Solution.

$$\begin{aligned}1-\beta &= \mathbb{P}^{H_1}_{X_1,X_2}(G(3)) \\ &= \mathbb{P}^{H_1}(X_1 = X_2 = 1) + \mathbb{P}^{H_1}(X_1 = 1 \text{ and } X_2 = 2) \\ &\qquad + \mathbb{P}^{H_1}(X_1 = 2 \text{ and } X_2 = 1) \\ &= f_1^2(1) + 2f_1(1)f_1(2) = \left(\tfrac{3}{4}\right)^2 + 2\cdot\tfrac{3}{4}\cdot\tfrac{3}{16} \\ &= \tfrac{18+9}{32} = \tfrac{27}{32} = 0.84.\end{aligned}$$

□

(iv) Define the critical region F by $F := \{(1,1),(2,1),(3,1),(1,3)\}$. Determine α and $1 - \beta$.

Solution. The numerical value of α is

$$\alpha = \mathbb{P}^{H_0}_{X_1,X_2}(F) = \tfrac{1}{2}\cdot\tfrac{1}{2} + \tfrac{1}{4}\cdot\tfrac{1}{2} + \tfrac{1}{8}\cdot\tfrac{1}{2} + \tfrac{1}{2}\cdot\tfrac{1}{8} = \tfrac{1}{2} = 0.50.$$

Furthermore

$$1-\beta = \mathbb{P}^{H_1}_{X_1,X_2}(F) = \tfrac{3}{4}\cdot\tfrac{3}{4} + \tfrac{3}{16}\cdot\tfrac{3}{4} + \tfrac{3}{64}\cdot\tfrac{3}{4} + \tfrac{3}{4}\cdot\tfrac{3}{64} = \tfrac{198}{256} = \tfrac{99}{128} = 0.77.$$

We see that both F and $G(3)$ are of size 0.50 and that indeed $G(3)$ generates more power than F. □

Exercise 4

We start from a population enjoying an exponential distribution with parameter $\theta > 0$. That is to say, the probability density of the population is given by

$$f(x,\theta) = \begin{cases} \frac{1}{\theta}e^{-x/\theta} & \text{if } x \geq 0, \\ 0 & \text{elsewhere.}\end{cases}$$

Through a sample X_1, X_2 we want to test $H_0 : \theta = 1$ versus $H_1 : \theta = 2$.

(i) Determine a maximum power critical region of size $\alpha = 0.10$.

Solution. Just apply the Neyman–Pearson lemma:

$$L_0(x_1, x_2)/L_1(x_1, x_2) = e^{-x_1 - x_2} / \tfrac{1}{4} e^{-\frac{1}{2}x_1 - \frac{1}{2}x_2} = 4e^{-\frac{1}{2}(x_1 + x_2)}.$$

This shows that G is necessarily of the form

$$\{\mathbf{x} \in [0, +\infty) \times [0, +\infty) \ : \ x_1 + x_2 \geq k\}.$$

Hence

$$\begin{aligned} \mathbb{P}^{H_0}_{X_1,X_2}(G) &= 1 - \int_0^k \int_0^{k-x_1} e^{-x_1 - x_2}\, dx_2\, dx_1 \\ &= 1 - \int_0^k e^{-x_1}(1 - e^{x_1 - k})\, dx_1 \\ &= 1 + [e^{-x_1}]_0^k + e^{-k} \cdot k = (k+1)\, e^{-k}. \end{aligned}$$

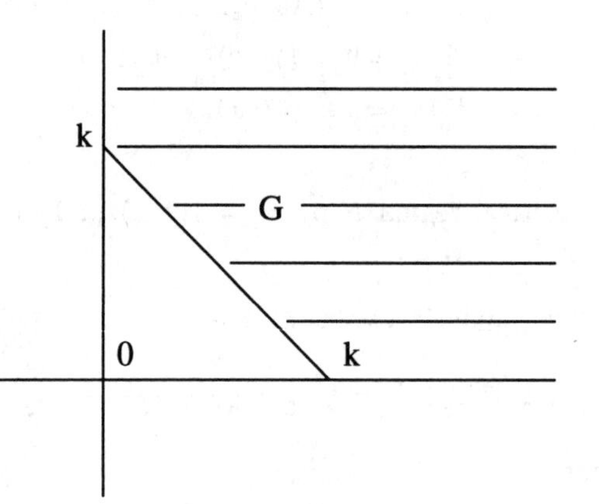

We want this expression to be equal to 0.10. Using (for example) a pocket calculator, we see that $k \approx 3.89$. So the critical region G we are looking for is given by

$$G = \{\mathbf{x} \in [0, +\infty) \times [0, +\infty) \ : \ x_1 + x_2 \geq 3.89\}.$$

□

(ii) Determine the numerical value of $1 - \beta$.

Solution.

$$\begin{aligned} 1 - \beta = \mathbb{P}^{H_1}_{X_1,X_2}(G) &= 1 - \int_0^k \int_0^{k-x_1} \tfrac{1}{4} e^{-\frac{1}{2}x_1} e^{-\frac{1}{2}x_2}\, dx_2\, dx_1 \\ &= \cdots = \tfrac{1}{2}(k+2)e^{-\frac{1}{2}k} = 0.42. \end{aligned}$$

□

(iii) Define a critical region by $F(c) := \{\mathbf{x} \in \mathbb{R}^2 : x_1, x_2 \geq c\}$. Adjust c in such a way that $F(c)$ is of size 0.10.

Solution. For all $c > 0$ the size of $F(c)$ is given by

$$\begin{aligned} \mathbb{P}(X_1 \geq c \text{ and } X_2 \geq c) &= \mathbb{P}(X_1 \geq c)^2 = \left(\int_c^{+\infty} e^{-x}\,dx\right)^2 \\ &= (e^{-c})^2 = e^{-2c} = 0.10. \end{aligned}$$

It follows that the size of $F(c)$ equals 0.10 if $c = 1.15$. The power is

$$\begin{aligned} 1 - \beta &= \mathbb{P}^{H_1}_{X_1,X_2}(F(c)) = \left(\int_c^{+\infty} \tfrac{1}{2}e^{-\frac{1}{2}x}\,dx\right)^2 \\ &= \left(\left[-e^{-\frac{1}{2}x}\right]_c^{+\infty}\right)^2 = e^{-c} = 0.32. \end{aligned}$$

□

Exercise 5

A population is Poisson distributed with parameter $\lambda > 0$. We draw a sample X_1, X_2, X_3 from this population in order to test $H_0 : \lambda = 2$ against $H_1 : \lambda = 1$. Let $G(c)$ be defined by

$$G(c) := \{(n_1, n_2, n_3) \in \mathbb{N}^3 : n_1 + n_2 + n_3 \leq c\}.$$

(i) Show that a critical region based on the Neyman–Pearson lemma is of type $G(c)$.

Solution. Such regions are of the form $\{\mathbf{n} \in \mathbb{N}^3 : L_0(\mathbf{n})/L_1(\mathbf{n}) \leq k\}$. The ratio L_0/L_1 is here given by

$$\begin{aligned} \frac{L_0(\mathbf{n})}{L_1(\mathbf{n})} &= \frac{(\lambda_1^{n_1}e^{-\lambda_1}/n_1!)\cdot(\lambda_1^{n_2}e^{-\lambda_1}/n_2!)\cdot(\lambda_1^{n_3}e^{-\lambda_1}/n_3!)}{(\lambda_2^{n_1}e^{-\lambda_2}/n_1!)\cdot(\lambda_2^{n_2}e^{-\lambda_2}/n_2!)\cdot(\lambda_2^{n_3}e^{-\lambda_2}/n_3!)} \\ &= \frac{\lambda_1^{n_1+n_2+n_3}\,e^{-3\lambda_1}}{\lambda_2^{n_1+n_2+n_3}\,e^{-3\lambda_2}} = \left(\frac{\lambda_1}{\lambda_2}\right)^{n_1+n_2+n_3} e^{3(\lambda_2-\lambda_1)}. \end{aligned}$$

Because $\lambda_1 = 2$ and $\lambda_2 = 1$ we arrive at

$$\frac{L_0(\mathbf{n})}{L_1(\mathbf{n})} = 2^{n_1+n_2+n_3}\,e^{-3}.$$

It follows from this that critical regions based on the Neyman–Pearson lemma are of type $G(c)$. □

(ii) Determine the size of $G(2)$.

Solution. The size α of $G(2)$ can be calculated as follows:

$$\begin{aligned}
\alpha &= \mathbb{P}^{H_0}_{X_1,X_2,X_3}[G(2)] \\
&= \mathbb{P}(X_1 = X_2 = X_3 = 0) + \binom{3}{1}\mathbb{P}(X_1 = X_2 = 0 \text{ and } X_3 = 1) \\
&\quad + \binom{3}{1}\mathbb{P}(X_1 = 0 \text{ and } X_2 = X_3 = 1) \\
&\quad + \binom{3}{1}\mathbb{P}(X_1 = X_2 = 0 \text{ and } X_3 = 2) \\
&= \left(\frac{2^0 e^{-2}}{0!}\right)^3 + 3\left(\frac{2^0 e^{-2}}{0!}\right)^2\left(\frac{2^1 e^{-2}}{1!}\right) + 3\left(\frac{2^0 e^{-2}}{0!}\right)\left(\frac{2^1 e^{-2}}{1!}\right)^2 \\
&\quad + 3\left(\frac{2^0 e^{-2}}{0!}\right)^2\left(\frac{2^2 e^{-2}}{2!}\right) \\
&= (1 + 6 + 12 + 6)e^{-6} = 25e^{-6} = 0.062.
\end{aligned}$$

The size of $G(2)$ can also be calculated in the following way. By §I.11 Exercise 47 the variable $X_1 + X_2 + X_3$ under H_0 enjoys a Poisson-distribution with parameter $\lambda = 6$. Hence

$$\alpha = \mathbb{P}(X_1 + X_2 + X_3 \leq 2 \mid H_0) = \sum_{k=0}^{2} \frac{6^k}{k!} e^{-6} = 25e^{-6}.$$

□

(iii) Determine the power associated with $G(2)$.

Solution. The power can be computed as follows:

$$\begin{aligned}
1 - \beta &= \mathbb{P}^{H_1}_{X_1,X_2,X_3}[G(2)] \\
&= \left(\frac{e^{-1}}{0!}\right)^3 + 3\left(\frac{e^{-1}}{0!}\right)^2\left(\frac{e^{-1}}{1!}\right) + 3\left(\frac{e^{-1}}{0!}\right)\left(\frac{e^{-1}}{1!}\right)^2 \\
&\quad + 3\left(\frac{e^{-1}}{0!}\right)^2\left(\frac{e^{-1}}{2!}\right) = \tfrac{17}{2}e^{-3} = 0.423.
\end{aligned}$$

□

(iv) *Question.* Is it possible to construct a critical region of the same size but with more power than $G(2)$?

Solution. No. The Neyman–Pearson lemma states that, given a prescribed size $\alpha = 0.062$, there is no critical region generating more power than $G(2)$ does.

□

Exercise 6

A sample $X_1, \ldots, X_n$ is drawn from a $N(0, \sigma^2)$-distributed population in order to test $H_0 : \sigma = 1$ versus $H_1 : \sigma \neq 1$.

(i) Find an explicit expression of $\Lambda(x_1, \ldots, x_n)$ in terms of $x_1, \ldots, x_n$.

Solution. The likelihood ratio $\Lambda(\mathbf{x})$ is given by $\Lambda(\mathbf{x}) = L_0(\mathbf{x})/L_1(\mathbf{x})$ where $L_0(\mathbf{x}) = (\sqrt{2\pi})^{-n} \exp[-\frac{1}{2}(x_1^2 + \cdots + x_n^2)]$ and

$$\begin{aligned} L_1(\mathbf{x}) &= \sup_{\sigma>0} L_\sigma(x_1, \ldots, x_n) \\ &= \sup_{\sigma>0} \left\{ (\sqrt{2\pi})^{-n} \sigma^{-n} \exp\left(-\frac{1}{2\sigma^2}(x_1^2 + \cdots + x_n^2) \right) \right\}. \end{aligned}$$

In order to determine the latter supremum, we maximize the function $f : (0, +\infty) \to (0, +\infty)$ defined by

$$f(\sigma) := (\sqrt{2\pi})^{-n} \sigma^{-n} \exp\left(-\frac{\|\mathbf{x}\|^2}{2\sigma^2} \right).$$

For this function we have that $f \geq 0$,

$$\lim_{\sigma \downarrow 0} f(\sigma) = 0 \quad \text{and} \quad \lim_{\sigma \to +\infty} f(\sigma) = 0.$$

From the above it follows that f attains a global maximum on $(0, +\infty)$. We examine the derivative of f in order to find the location of the maxima: at σ it equals

$$(\sqrt{2\pi})^{-n} \left(-n\sigma^{-n-1} \exp\left(-\frac{\|\mathbf{x}\|^2}{2\sigma^2} \right) + \sigma^{-n} \|\mathbf{x}\|^2 \sigma^{-3} \exp\left(-\frac{\|\mathbf{x}\|^2}{2\sigma^2} \right) \right).$$

An element σ for which f assumes a maximum necessarily satisfies:

$$\begin{aligned} f'(\sigma) = 0 \quad &\Leftrightarrow \quad \sigma^{-n-3} \|\mathbf{x}\|^2 \exp[\ldots] = n\sigma^{-n-1} \exp[\ldots] \\ &\Leftrightarrow \quad \sigma^2 = \|\mathbf{x}\|^2 / n \Leftrightarrow \sigma = \|\mathbf{x}\| / \sqrt{n}. \end{aligned}$$

Conclusion:

$$\begin{aligned} L_1(\mathbf{x}) &= f(\|\mathbf{x}\| / \sqrt{n}) \\ &= \left(\frac{\sqrt{2\pi}\, \|\mathbf{x}\|}{\sqrt{n}} \right)^{-n} \exp\left(-\frac{1}{2} \frac{\|\mathbf{x}\|^2}{\|\mathbf{x}\|^2 / n} \right) = \left(\frac{n}{2\pi e \|\mathbf{x}\|^2} \right)^{\frac{1}{2}n} \end{aligned}$$

Hence

$$\begin{aligned} \Lambda(x_1, \ldots, x_n) = L_0(\mathbf{x})/L_1(\mathbf{x}) &= \left(\frac{1}{2\pi} \right)^{\frac{1}{2}n} e^{-\frac{1}{2}\|\mathbf{x}\|^2} \left(\frac{2\pi e \|\mathbf{x}\|^2}{n} \right)^{\frac{1}{2}n} \\ &= \left(\frac{\|\mathbf{x}\|^2}{n} \right)^{\frac{1}{2}n} e^{\frac{1}{2}n - \frac{1}{2}\|\mathbf{x}\|^2}. \end{aligned}$$

□

(ii) We set $B_R = \{\mathbf{x} \in \mathbb{R}^n : \|\mathbf{x}\| < R\}$. Show that a critical region based on the likelihood ratio is of the following form: $G = \overline{B_{R_1}} \cup B_{R_2}^c$, where $R_1 < R_2$.

Solution. Critical regions based on the likelihood ratio are given by

$$G = \{\mathbf{x} \in \mathbb{R}^n \; : \; \Lambda(\mathbf{x}) \leq k\}.$$

The inequality $\Lambda(x_1, \ldots, x_n) \leq k$ is equivalent to

$$n^{-\frac{1}{2}n} e^{\frac{1}{2}n} \left(\|\mathbf{x}\|^2\right)^{\frac{1}{2}n} e^{-\frac{1}{2}\|\mathbf{x}\|^2} \leq k,$$

or

$$\left(\|\mathbf{x}\|^2\right)^n e^{-\|\mathbf{x}\|^2} \leq \left(k \left(\frac{n}{e}\right)^{\frac{1}{2}n}\right)^2 =: k_2.$$

Next we define $f : [0, +\infty) \to [0, +\infty)$ by $f(s) := s^{2n} e^{-s^2}$. Now the inequality above can be rewritten as $f(\|\mathbf{x}\|) \leq k_2$. It is easy to see that f has a graph as sketched in the figure below.

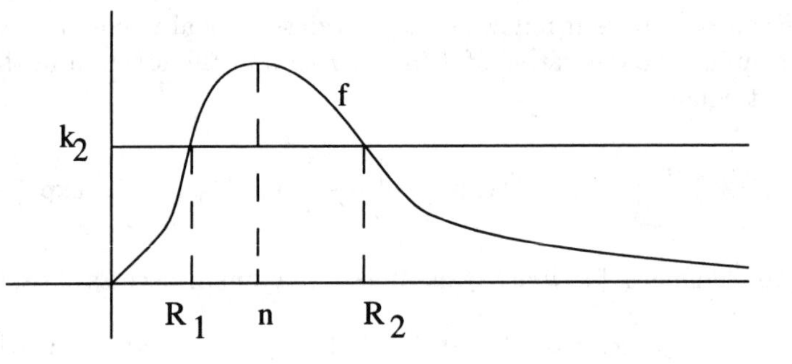

From this it follows that

$$\begin{aligned} \{\mathbf{x} \in \mathbb{R}^n \; : \; \Lambda(\mathbf{x}) \leq k\} &= \{\mathbf{x} \in \mathbb{R}^n \; : \|\mathbf{x}\| \leq R_1 \quad \text{or} \quad \|\mathbf{x}\| \geq R_2\} \\ &= \overline{B_{R_1}} \cup B_{R_2}^c. \end{aligned}$$

□

(iii) In the case where $n = 2$, determine R_1 and R_2 in such a way that both $\overline{B_{R_1}}$ and $B_{R_2}^c$ are of size 0.05. Furthermore, determine the power of the test under this critical region in the point $\sigma = 2$.

Solution. Under H_0 the variable ${X_1}^2 + {X_2}^2$ is χ^2-distributed with 2 degrees of freedom (see Proposition II.2.7). The probability equation

$$\mathbb{P}_{X_1,X_2}^{H_0}\left(\overline{B_{R_1}}\right) = \mathbb{P}[{X_1}^2 + {X_2}^2 \leq R_1^2] = 0.05$$

can now be solved by using Table IV, describing the χ^2-distribution. This gives $R_1 = 0.32$. In the same way we deduce that $R_2 = 2.45$.

The power of the test in $\sigma = 2$ (under this critical region) is given by

$$1 - \beta = \mathbb{P}\left[{X_1}^2 + {X_2}^2 \le R_1^2 \mid \sigma = 2 \right] + \mathbb{P}\left[{X_1}^2 + {X_2}^2 \ge R_2^2 \mid \sigma = 2 \right].$$

If $\sigma = 2$, the variable $\frac{1}{4}({X_1}^2 + {X_2}^2)$ is χ^2-distributed with 2 degrees of freedom. This is the same as the exponential distribution with expectation value 2, which leads to: $1 - \beta = 0.013 + 0.472 = 0.485$. □

Exercise 7

Starting from an exponentially distributed population with parameter θ, we wish to test $H_0 : \theta = 1$ against $H_1 : \theta \neq 1$.

(i) Express the likelihood ratio $\Lambda(\mathbf{x})$ in terms of $x_1, \dots, x_n$.

Solution. We are looking for $\Lambda(\mathbf{x}) = L_0(\mathbf{x})/L_1(\mathbf{x})$, where

$$L_0(\mathbf{x}) = e^{-(x_1 + \cdots + x_n)}$$

and

$$L_1(\mathbf{x}) = \sup_{\theta > 0} \left[\theta^{-n} \exp(-(x_1 + \cdots + x_n)/\theta) \right].$$

To this end we define $f : (0, +\infty) \to (0, +\infty)$ by $f(\theta) := \theta^{-n} \exp(-n\overline{x}/\theta)$. Then

$$\lim_{\theta \downarrow 0} f(\theta) = 0 \quad \text{and} \quad \lim_{\theta \to +\infty} f(\theta) = 0.$$

It follows from this that f attains a global maximum on $(0, +\infty)$. The derivative of f is given by

$$f'(\theta) = (n\overline{x}\,\theta^{-n-2} - n\theta^{-n-1}) \exp(-n\overline{x}/\theta) = n\theta^{-(n+2)}(\overline{x} - \theta) \exp(-n\overline{x}/\theta).$$

A point θ where f attains its global maximum necessarily satisfies $f'(\theta) = 0$, which is the same as saying that $\theta = \overline{x}$. So we arrive at the following expression for $L_1(\mathbf{x})$:

$$L_1(\mathbf{x}) = (\overline{x})^{-n} \exp(-n).$$

The likelihood ratio is therefore

$$\Lambda(\mathbf{x}) = \overline{x}^n e^{n - n\overline{x}} = (\overline{x} e^{1 - \overline{x}})^n.$$

□

(ii) Show that a critical region G based on the likelihood ratio is of the form $G = G_1 \cup G_2$, where $G_1 = \{\mathbf{x} \in \mathbb{R}_+^n : \overline{x} \le c_1\}$ and $G_2 = \{\mathbf{x} \in \mathbb{R}_+^n : \overline{x} \ge c_2\}$ $(c_1 < c_2)$.

Solution. This can be dealt with in the same way as Exercise 6(ii). □

(iii) Next we consider a sample X_1, X_2 and we set $c_1 = \frac{1}{2}$ and $c_2 = \frac{3}{2}$. Determine the level of significance α.

Solution. We can compute $\alpha = \alpha_1 + \alpha_2$ as follows:

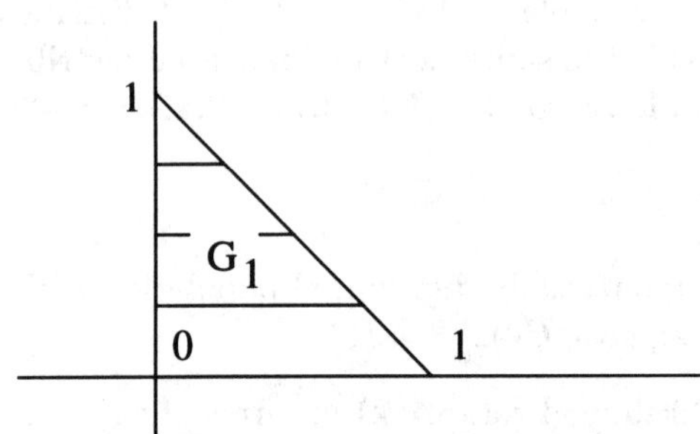

$$\begin{aligned}
\alpha_1 = \mathbb{P}^{H_0}_{X_1,X_2}(G_1) &= \int_0^1 \int_0^{1-x_1} e^{-x_1-x_2}\, dx_2\, dx_1 \\
&= \int_0^1 e^{-x_1}(1 - e^{x_1-1})\, dx_1 \\
&= (1 - e^{-1}) - e^{-1} = 1 - 2e^{-1} = 0.264.
\end{aligned}$$

In the same way we calculate

$$\begin{aligned}
\alpha_2 &= \mathbb{P}^{H_0}_{X_1,X_2}(G_2) = 1 - \mathbb{P}^{H_0}_{X_1,X_2}(G_2^c) \\
&= 1 - ((1 - e^{-3}) - 3e^{-3}) = 4e^{-3} = 0.199.
\end{aligned}$$

It follows that $\alpha = \alpha_1 + \alpha_2 = 0.463$. □

Exercise 8

Starting from a $N(\mu, \sigma^2)$-distributed population, where σ is unknown, we wish to test $H_0 : \mu = \mu_0$ against $H_1 : \mu \neq \mu_0$. Prove that the likelihood ratio is given by

$$\Lambda(x_1, \ldots, x_n) = \left[1 + \frac{1}{n-1}\left(\frac{\overline{x} - \mu_0}{s/\sqrt{n}}\right)^2\right]^{-\frac{1}{2}n}.$$

Furthermore, prove that a critical region based on the likelihood ratio Λ is of the form

$$G = \left\{\mathbf{x} \in \mathbb{R}^n \ : \ \left|\frac{\overline{x} - \mu_0}{s/\sqrt{n}}\right| \geq c\right\}.$$

Proof. Here $\Theta = \{(\mu, \sigma) \in \mathbb{R}^2 : \sigma > 0\}$, $\Theta_0 = \{(\mu_0, \sigma) \in \mathbb{R}^2 : \sigma > 0\}$ and $\Theta_1 = \{(\mu, \sigma) \in \mathbb{R}^2 \ : \mu \neq \mu_0,\ \sigma > 0\}$. The likelihood ratio can be written as

$L_0(\mathbf{x})/L_1(\mathbf{x})$, where

$$L_0(\mathbf{x}) = \sup_{\sigma>0} \left(\sqrt{2\pi}\,\sigma\right)^{-n} \exp\left(-\frac{1}{2\sigma^2}\sum_{i=1}^{n}(x_i-\mu_0)^2\right)$$

and

$$L_1(\mathbf{x}) = \sup_{\sigma>0,\mu\in\mathbb{R}} \left(\sqrt{2\pi}\,\sigma\right)^{-n} \exp\left(-\frac{1}{2\sigma^2}\sum_{i=1}^{n}(x_i-\mu)^2\right).$$

By Lemma III.2.1 we may set $\mu = \overline{x}$ in the defining supremum for $L_1(\mathbf{x})$. For any $c_1 > 0$, $c_2 < 0$ we now define $f : (0,+\infty) \to (0,+\infty)$ by $f(\sigma) := c_1\sigma^{-n}\exp[\,c_2\sigma^{-2}\,]$. We wish to maximize this function. If f attains a maximum in the point σ, then so does $\log f$ and vice versa. We consider $\log f$ instead of f, for its derivative assumes a simpler form than that of f. We have

$$(\log f)(\sigma) = \log c_1 - n\log\sigma + c_2\sigma^{-2}.$$

Hence

$$(\log f)'(\sigma) = -n\sigma^{-1} - 2c_2\sigma^{-3}.$$

This derivative vanishes in the point σ for which $\sigma^2 = -2c_2/n$. Applying this, we arrive at the following expression for $L_0(\mathbf{x})/L_1(\mathbf{x})$:

$$\frac{L_0(\mathbf{x})}{L_1(\mathbf{x})} = \frac{\left(2\pi\,\frac{1}{n}\sum_{i=1}^{n}(x_i-\mu_0)^2\right)^{-\frac{1}{2}n}\exp(-n/2)}{\left(2\pi\,\frac{1}{n}\sum_{i=1}^{n}(x_i-\overline{x})^2\right)^{-\frac{1}{2}n}\exp(-n/2)} = \left(\frac{\sum_{i=1}^{n}(x_i-\mu_0)^2}{\sum_{i=1}^{n}(x_i-\overline{x})^2}\right)^{-\frac{1}{2}n}.$$

Next we note that

$$\begin{aligned}\sum_{i=1}^{n}(x_i-\mu_0)^2 &= \sum_{i=1}^{n}\left((x_i-\overline{x})+(\overline{x}-\mu_0)\right)^2 = \cdots \\ &= \left(\sum_{i=1}^{n}(x_i-\overline{x})^2\right) + n\,(\overline{x}-\mu_0)^2.\end{aligned}$$

Therefore

$$\frac{L_0(\mathbf{x})}{L_1(\mathbf{x})} = \left(1+\frac{n\,(\overline{x}-\mu_0)^2}{\sum_{i=1}^{n}(x_i-\overline{x})^2}\right)^{-\frac{1}{2}n} = \left(1+\frac{n}{n-1}\,\frac{(\overline{x}-\mu_0)^2}{s^2}\right)^{-\frac{1}{2}n}.$$

Now for $0 < \delta < 1$ we have

$$\begin{aligned} G &= \{\mathbf{x}\in\mathbb{R}^n : \Lambda(\mathbf{x})\le\delta\} \\ &= \left\{\mathbf{x}\in\mathbb{R}^n : \left(\frac{\overline{x}-\mu_0}{s/\sqrt{n}}\right)^2 \ge (n-1)(\delta^{-2/n}-1)\right\} \\ &= \left\{\mathbf{x}\in\mathbb{R}^n : \left|\frac{\overline{x}-\mu_0}{s/\sqrt{n}}\right| \ge \sqrt{(n-1)(\delta^{-2/n}-1)}\right\} \\ &= \left\{\mathbf{x}\in\mathbb{R}^n : \left|\frac{\overline{x}-\mu_0}{s/\sqrt{n}}\right| \ge c\right\}\end{aligned}$$

for a certain c. Critical regions based on the likelihood ratio apparently are of this form. □

Exercise 9

Starting from a $N(\mu, \sigma^2)$-distributed population, where both μ and σ are unknown, we wish to test $H_0 : \sigma = \sigma_0$ against $H_1 : \sigma \neq \sigma_0$. Prove that the likelihood ratio is given by

$$\Lambda(x_1, \ldots, x_n) = e^{\frac{1}{2}n}\, n^{-\frac{1}{2}n} \left(\frac{(n-1)\, s^2}{\sigma_0^2}\right)^{\frac{1}{2}n} e^{-(n-1)\, s^2/2\sigma_0^2}$$

where $s^2 := \frac{1}{n-1} \sum_i (x_i - \overline{x})^2$. Furthermore, prove that a critical region based on this likelihood function is of the form $G = G_1 \cup G_2$ where the components G_1 and G_2 are given by

$$G_1 = \left\{\mathbf{x} \in \mathbb{R}^n : \frac{(n-1)\, s^2}{\sigma_0^2} \leq c_1\right\} \quad \text{and} \quad G_2 = \left\{\mathbf{x} \in \mathbb{R}^n : \frac{(n-1)\, s^2}{\sigma_0^2} \geq c_2\right\}.$$

Proof. Here, as in the previous exercise, we have $\Theta = \{(\mu, \sigma) \in \mathbb{R}^2 : \sigma > 0\}$. Furthermore $\Theta_0 = \{(\mu, \sigma_0) \in \mathbb{R}^2 : \mu \in \mathbb{R}\}$ and $\Theta_1 = \{(\mu, \sigma) \in \mathbb{R}^2 : \sigma \neq \sigma_0,\ \sigma > 0$ and $\mu \in \mathbb{R}\}$. The likelihood function can be expressed as $L_0(\mathbf{x})/L_1(\mathbf{x})$, where

$$L_0(\mathbf{x}) = \sup_{\mu \in \mathbb{R}} \left(\sqrt{2\pi}\, \sigma_0\right)^{-n} \exp\left(-\frac{1}{2\sigma_0^2} \sum_{i=1}^{n} (x_i - \mu)^2\right)$$

and

$$L_1(\mathbf{x}) = \sup_{\mu \in \mathbb{R},\, \sigma > 0} \left(\sqrt{2\pi}\, \sigma\right)^{-n} \exp\left(-\frac{1}{2\sigma^2} \sum_{i=1}^{n} (x_i - \mu)^2\right).$$

Let us first determine an explicit expression for $L_0(\mathbf{x})$. The expression

$$\exp\left(-\frac{1}{2\sigma_0^2} \sum_{i=1}^{n} (x_i - \mu)^2\right)$$

is (Lemma III.2.1) maximal if $\mu = \overline{x}$. It follows that

$$L_0(\mathbf{x}) = \left(\sqrt{2\pi}\, \sigma_0\right)^{-n} \exp\left(-\frac{(n-1)\, s^2}{2\sigma_0^2}\right).$$

The expression $L_1(\mathbf{x})$ can (also) only be maximal for $\mu = \overline{x}$. Therefore the only thing left is now to maximize the function

$$\sigma \mapsto \left(\sqrt{2\pi}\, \sigma\right)^{-n} \exp\left(-\frac{(n-1)\, s^2}{2\sigma^2}\right)$$

for fixed $x_1, \ldots, x_n$. It is easily deduced that the maximum in question is attained for $\sigma^2 = \frac{n-1}{n}\, s^2$. This leads to

$$L_1(\mathbf{x}) = (2\pi)^{-\frac{1}{2}n} \left(\frac{n-1}{n}\, s^2\right)^{-\frac{1}{2}n} \exp(-\tfrac{1}{2}n).$$

It follows that

$$\Lambda(\mathbf{x}) = \frac{L_0(\mathbf{x})}{L_1(\mathbf{x})} = e^{\frac{1}{2}n}\, n^{-\frac{1}{2}n} \left(\frac{(n-1)\, s^2}{\sigma_0^2}\right)^{\frac{1}{2}n} e^{-(n-1)\, s^2/2\sigma_0^2},$$

as had to be proved. As in Exercise 6 (ii) it is now easily seen that a critical region based on the likelihood function is of the form $G = G_1 \cup G_2$ with G_1 and G_2 as described. □

Exercise 10

Given are two populations: the first enjoys a $N(\mu_X, \sigma_X^2)$-distribution, the second a $N(\mu_Y, \sigma_Y^2)$-distribution. Independent samples $X_1, \ldots, X_m$ and $Y_1, \ldots, Y_n$ are drawn from these populations respectively.

(i) For fixed $\mathbf{x}$ and $\mathbf{y}$ the expression

$$L_{\mu_X, \mu_Y, \sigma_X, \sigma_Y}(x_1, \ldots, x_m, y_1, \ldots, y_n)$$

is maximal for

$$\mu_X = \overline{x},\ \mu_Y = \overline{y},\ \sigma_X^2 = \frac{1}{m}\sum_{i=1}^{m}(x_i - \overline{x})^2 \text{ and } \sigma_Y^2 = \frac{1}{n}\sum_{i=1}^{n}(y_i - \overline{y})^2.$$

This maximum equals $(2\pi\, e)^{-\frac{1}{2}(m+n)}(\sigma_X^2)^{-\frac{1}{2}m}(\sigma_Y^2)^{-\frac{1}{2}n}$. Prove this.

Proof. We have

$$L_{\mu_X, \mu_Y, \sigma_X, \sigma_Y}(\mathbf{x}, \mathbf{y}) = L_{\mu_X, \sigma_X}(\mathbf{x}) L_{\mu_Y, \sigma_Y}(\mathbf{y}).$$

Both factors on the right side are positive. Therefore, maximizing the form $L_{\mu_X, \mu_Y, \sigma_X, \sigma_Y}(\mathbf{x}, \mathbf{y})$ amounts to the same as maximizing the factors $L_{\mu_X, \sigma_X}(\mathbf{x})$ and $L_{\mu_Y, \sigma_Y}(\mathbf{y})$ separately. So by Lemma III.2.1 and Exercise 8 we learn that $\mu_X = \overline{x}$, $\mu_Y = \overline{y}$,

$$\sigma_X^2 = \frac{1}{m}\sum_{i=1}^{m}(x_i - \overline{x})^2 \quad \text{and} \quad \sigma_Y^2 = \frac{1}{n}\sum_{i=1}^{n}(y_i - \overline{y})^2$$

realize maximum values of $L_{\mu_X, \mu_Y, \sigma_X, \sigma_Y}(\mathbf{x}, \mathbf{y})$. The magnitude of this maximum can be obtained by substitution of these values. □

(ii) Under the assumption that $\sigma_X = \sigma_Y = \sigma$ the expression $L_{\mu_X,\mu_Y,\sigma_X,\sigma_Y}(\mathbf{x},\mathbf{y})$ is maximal if $\mu_X = \overline{x}$, $\mu_Y = \overline{y}$ and

$$\sigma^2 = \frac{(m-1)\,s_{\mathbf{x}}^2 + (n-1)\,s_{\mathbf{y}}^2}{m+n}$$

where $s_{\mathbf{x}}^2 = \frac{1}{m-1}\sum_{i=1}^m (x_i - \overline{x})^2$ and $s_{\mathbf{y}}^2 = \frac{1}{n-1}\sum_{i=1}^n (y_i - \overline{y})^2$. Prove this.

Proof. By Lemma III.2.1 it follows that $\mu_X = \overline{x}$ and $\mu_Y = \overline{y}$ are maximizing elements for $L_{\mu_X,\mu_Y,\sigma,\sigma}(\mathbf{x},\mathbf{y})$. Next we want to find the value of σ for which the expression

$$L_\sigma(\mathbf{x},\mathbf{y}) = \frac{1}{\left(\sqrt{2\pi}\,\sigma\right)^{m+n}} \exp\left[-\frac{1}{2\sigma^2}\left(\sum_{i=1}^m (x_i - \overline{x})^2 + \sum_{i=1}^n (y_i - \overline{y})^2\right)\right]$$

assumes a maximum (for fixed $\mathbf{x}$ and $\mathbf{y}$). This amounts to the same as finding the maximizing element σ for

$$\begin{aligned}(\log L_\sigma)(\mathbf{x},\mathbf{y}) &= -\tfrac{1}{2}(m+n)\log(2\pi) - \tfrac{1}{2}(m+n)\log\sigma^2 \\ &\quad -\tfrac{1}{2}\left(\sum_{i=1}^m (x_i - \overline{x})^2 + \sum_{i=1}^n (y_i - \overline{y})^2\right)\sigma^{-2}.\end{aligned}$$

Examining the derivative of this expression with respect to σ, we see that

$$\sigma^2 = \frac{1}{m+n}\left(\sum_{i=1}^m (x_i - \overline{x})^2 + \sum_{i=1}^n (y_i - \overline{y})^2\right)$$

realizes a maximum for $L_\sigma(\mathbf{x},\mathbf{y})$. The magnitude of this maximum is

$$\left(\frac{2\pi\,e}{m+n}\left(\sum_{i=1}^m (x_i - \overline{x})^2 + \sum_{i=1}^n (y_i - \overline{y})^2\right)\right)^{-\frac{1}{2}(m+n)}.$$

□

(iii) Testing $H_0 : \sigma_X = \sigma_Y$ against $H_1 : \sigma_X \neq \sigma_Y$ the likelihood ratio assumes the form

$$\Lambda(\mathbf{x},\mathbf{y}) = \frac{((m-1)/m)^{\frac{1}{2}m}((n-1)/n)^{\frac{1}{2}n}(s_{\mathbf{y}}^2/s_{\mathbf{x}}^2)^{\frac{1}{2}n}}{\left(\frac{m-1}{m+n} + \frac{n-1}{m+n}\cdot s_{\mathbf{y}}^2/s_{\mathbf{x}}^2\right)^{\frac{1}{2}(m+n)}}.$$

A critical region based on this ratio is of type $G = G_1 \cup G_2$, where

$$\begin{aligned} G_1 &:= \{(\mathbf{x},\mathbf{y}) \in \mathbb{R}^{m+n} : s_{\mathbf{y}}^2/s_{\mathbf{x}}^2 \leq c_1\} \quad \text{and} \\ G_2 &:= \{(\mathbf{x},\mathbf{y}) \in \mathbb{R}^{m+n} : s_{\mathbf{y}}^2/s_{\mathbf{x}}^2 \geq c_2\} \end{aligned}$$

for certain c_1, c_2 $(c_1 < c_2)$. Prove this.

Proof. Using (i) and (ii) it is easy to deduce an explicit form for $\Lambda(\mathbf{x},\mathbf{y})$:

$$\begin{aligned}\Lambda(\mathbf{x},\mathbf{y}) &= \frac{(2\pi e)^{-\frac{1}{2}(m+n)}\left(\frac{(m-1)\,s_{\mathbf{x}}^2+(n-1)\,s_{\mathbf{y}}^2}{m+n}\right)^{-\frac{1}{2}(m+n)}}{(2\pi e)^{-\frac{1}{2}(m+n)}\left(\frac{m-1}{m}\,s_{\mathbf{x}}^2\right)^{-\frac{1}{2}m}\left(\frac{n-1}{n}\,s_{\mathbf{y}}^2\right)^{-\frac{1}{2}n}} \\ &= \frac{\left(\frac{m-1}{m}\right)^{\frac{1}{2}m}\left(\frac{n-1}{n}\right)^{\frac{1}{2}n}(s_{\mathbf{x}}^2)^{\frac{1}{2}m}(s_{\mathbf{y}}^2)^{\frac{1}{2}n}}{\left(\frac{m-1}{m+n}\,s_{\mathbf{x}}^2+\frac{n-1}{m+n}\,s_{\mathbf{y}}^2\right)^{\frac{1}{2}(m+n)}} \\ &= \frac{\left(\frac{m-1}{m}\right)^{\frac{1}{2}m}\left(\frac{n-1}{n}\right)^{\frac{1}{2}n}(s_{\mathbf{y}}^2/s_{\mathbf{x}}^2)^{\frac{1}{2}n}}{\left(\frac{m-1}{m+n}+\frac{n-1}{m+n}\,(s_{\mathbf{y}}^2/s_{\mathbf{x}}^2)\right)^{\frac{1}{2}(m+n)}}.\end{aligned}$$

We see that actually $\Lambda(\mathbf{x},\mathbf{y})$ is a function is of $s_{\mathbf{y}}^2/s_{\mathbf{x}}^2$. A critical region based on the likelihood ratio is by definition of type

$$G(\delta) = \{(\mathbf{x},\mathbf{y}) \in \mathbb{R}^{m+n} \ : \ \Lambda(\mathbf{x},\mathbf{y}) \leq \delta\}.$$

Setting $\delta_1 := \left(\frac{m}{m-1}\right)^{\frac{1}{2}m}\left(\frac{n}{n-1}\right)^{\frac{1}{2}n}\left(\frac{1}{m+n}\right)^{\frac{1}{2}(m+n)}\delta$, this reads as

$$G(\delta) = \left\{(\mathbf{x},\mathbf{y}) \in \mathbb{R}^n \ : \ \frac{(s_{\mathbf{y}}^2/s_{\mathbf{x}}^2)^{\frac{1}{2}n}}{((m-1)+(n-1)(s_{\mathbf{y}}^2/s_{\mathbf{x}}^2))^{\frac{1}{2}(m+n)}} \leq \delta_1\right\}.$$

Next we notice that the graph of the function $f:(0,+\infty)\to(0,+\infty)$ defined by $f(s) := s^{\frac{1}{2}n}/((m-1)+(n-1)s)^{\frac{1}{2}(m+n)}$ is (roughly) shaped as in the picture we gave at Exercise 6 (ii). It is then immediate that (for certain $c_1 < c_2$)

$$G = \{(\mathbf{x},\mathbf{y}) \in \mathbb{R}^{m+n} \ : \ s_{\mathbf{y}}^2/s_{\mathbf{x}}^2 \leq c_1\} \cup \{(\mathbf{x},\mathbf{y}) \in \mathbb{R}^{m+n} \ : \ s_{\mathbf{y}}^2/s_{\mathbf{x}}^2 \geq c_2\}. \qquad \square$$

Exercise 11

Independent samples $X_1,\dots,X_m$ and $Y_1,\dots,Y_n$ are drawn from a $N(\mu_X,\sigma_X^2)$ and a $N(\mu_Y,\sigma_Y^2)$-distributed population respectively. Given an outcome $x_1,\dots,x_m$ of $X_1,\dots,X_m$ and $y_1,\dots,y_n$ of $Y_1,\dots,Y_n$ we write

$$\mu := \frac{(m/\sigma_X^2)\overline{x}+(n/\sigma_Y^2)\overline{y}}{m/\sigma_X^2+n/\sigma_Y^2}.$$

We assume the numerical values of σ_X and σ_Y to be known and we assume that $\mu_Y - \mu_X = \Delta$, where Δ is fixed. Prove that, for fixed $(\mathbf{x},\mathbf{y}) \in \mathbb{R}^{m+n}$, the expression $L_{\mu_X,\mu_Y}(\mathbf{x},\mathbf{y})$ is maximal for

$$\mu_X = \mu - \frac{(n/\sigma_Y^2)\Delta}{m/\sigma_X^2+n/\sigma_Y^2} \quad \text{and} \quad \mu_Y = \mu + \frac{(m/\sigma_X^2)\Delta}{m/\sigma_X^2+n/\sigma_Y^2}.$$

Proof. First we set $\alpha := m/\sigma_X^2$ and $\beta := n/\sigma_Y^2$. In this notation we have to maximize (for fixed $\mathbf{x}$ and $\mathbf{y}$) the function $(\mu_X, \mu_Y) \mapsto L_{\mu_X,\mu_Y}(\mathbf{x},\mathbf{y})$, which is defined as

$$\frac{1}{(2\pi)^{\frac{1}{2}(m+n)}\,\sigma_X^m \sigma_Y^n}\exp\left(-\frac{1}{2}\left(\frac{\alpha}{m}\sum_{i=1}^{m}(x_i-\mu_X)^2+\frac{\beta}{n}\sum_{i=1}^{n}(y_i-\mu_Y)^2\right)\right),$$

under the constraint $\mu_Y - \mu_X = \Delta$. This amounts to the same thing as minimalization of the function

$$f\,:\,(\mu_X,\mu_Y)\;\mapsto\;\frac{\alpha}{m}\sum_{i=1}^{m}(x_i-\mu_X)^2+\frac{\beta}{n}\sum_{i=1}^{n}(y_i-\mu_Y)^2$$

under the constraint that $g(\mu_X,\mu_Y) := \mu_Y - \mu_X - \Delta = 0$.

By Lagrange's theorem such a minimum (if existing) is attained in a point (μ_X, μ_Y) where

$$(\nabla f)(\mu_X,\mu_Y) = \lambda\,(\nabla g)(\mu_X,\mu_Y)$$

for some scalar λ. Working this out we arrive at

$$(2\alpha\,(\mu_X-\overline{x}),\;2\beta\,(\mu_Y-\overline{y})) = \lambda\,(-1,\;1).$$

Together with the constraint $\mu_Y - \mu_X = \Delta$ this implies that

$$\mu_X = \frac{\alpha\overline{x}+\beta(\overline{y}-\Delta)}{\alpha+\beta}\quad\text{and}\quad\mu_Y = \frac{\alpha(\overline{x}+\Delta)+\beta\,\overline{y}}{\alpha+\beta}.$$

The proof can be completed by writing out these expressions. □

Exercise 12

Under the same assumptions as in Exercise 11 we want to test $H_0 : \mu_Y - \mu_X = \Delta$ against $H_1 : \mu_Y - \mu_X \neq \Delta$. Prove that the likelihood ratio Λ is given by

$$\Lambda(\mathbf{x},\mathbf{y}) = \exp\left(-\frac{mn}{2\sigma_X^2\sigma_Y^2}\,\frac{(\overline{y}-\overline{x}-\Delta)^2}{m/\sigma_X^2+n/\sigma_Y^2}\right).$$

Furthermore, prove that a critical region based on this ratio is of type

$$G = \left\{(\mathbf{x},\mathbf{y})\in\mathbb{R}^{m+n}\;:\;\frac{|\overline{y}-\overline{x}-\Delta|}{\sqrt{\sigma_X^2/m+\sigma_Y^2/n}}\geq c\right\}.$$

Proof. As in Exercise 11 we denote $\alpha := m/\sigma_X^2$ and $\beta := n/\sigma_Y^2$. The likelihood ratio can be written as $\Lambda(\mathbf{x},\mathbf{y}) = L_0(\mathbf{x},\mathbf{y})/L_1(\mathbf{x},\mathbf{y})$, where $L_0(\mathbf{x},\mathbf{y})$ is the maximum value (for fixed $\mathbf{x}$ and $\mathbf{y}$) of the map $(\mu_X,\mu_Y)\mapsto L_{\mu_X,\mu_Y}(\mathbf{x},\mathbf{y})$, which equals

$$\frac{1}{(2\pi)^{\frac{1}{2}(m+n)}\,\sigma_X^m \sigma_Y^n}\exp\left(-\frac{1}{2}\left(\frac{\alpha}{m}\sum_{i=1}^{m}(x_i-\mu_X)^2+\frac{\beta}{n}\sum_{i=1}^{n}(y_i-\mu_Y)^2\right)\right),$$

under the constraint $\mu_Y - \mu_X = \Delta$. The denominator $L_1(\mathbf{x}, \mathbf{y})$ is the maximum value of the same map without constraints. Using Lemma III.2.1 and Exercise 11, we arrive at the following expression for the likelihood ratio:

$$\Lambda(\mathbf{x}, \mathbf{y}) = \frac{\exp\left(-\frac{1}{2}\left(\frac{\alpha}{m}\sum_{i=1}^{m}(x_i - \mu_X)^2 + \frac{\beta}{n}\sum_{i=1}^{n}(y_i - \mu_Y)^2\right)\right)}{\exp\left(-\frac{1}{2}\left(\frac{\alpha}{m}\sum_{i=1}^{m}(x_i - \overline{x})^2 + \frac{\beta}{n}\sum_{i=1}^{n}(y_i - \overline{y})^2\right)\right)} \qquad (*)$$

where

$$\mu_X = \frac{\alpha\overline{x} + \beta(\overline{y} - \Delta)}{\alpha + \beta} \quad \text{and} \quad \mu_Y = \frac{\alpha(\overline{x} + \Delta) + \beta\overline{y}}{\alpha + \beta}.$$

The expression $(*)$ can be simplified. To see this, we write (note that the cross term vanishes):

$$\begin{aligned}
\frac{\alpha}{m}\sum_{i=1}^{m}(x_i - \mu_X)^2 &= \frac{\alpha}{m}\sum_{i=1}^{m}\{(x_i - \overline{x}) + (\overline{x} - \mu_X)\}^2 \\
&= \left(\frac{\alpha}{m}\sum_{i=1}^{m}(x_i - \overline{x})^2\right) + \alpha\left(\overline{x} - \frac{\alpha\overline{x} + \beta(\overline{y} - \Delta)}{\alpha + \beta}\right)^2 \\
&= \frac{\alpha}{m}\sum_{i=1}^{m}(x_i - \overline{x})^2 + \alpha\beta^2\left(\frac{\overline{x} - \overline{y} + \Delta}{\alpha + \beta}\right)^2 \qquad (**)
\end{aligned}$$

In the same way we obtain:

$$\frac{\beta}{n}\sum_{i=1}^{n}(y_i - \mu_Y)^2 = \frac{\beta}{n}\sum_{i=1}^{n}(y_i - \overline{y})^2 + \alpha^2\beta\left(\frac{\overline{y} - \overline{x} - \Delta}{\alpha + \beta}\right)^2 \qquad (***)$$

Substitution of $(**)$ and $(***)$ into $(*)$ leads to the required form for $\Lambda(\mathbf{x}, \mathbf{y})$.

It is then more or less immediate that a critical region $\{(\mathbf{x}, \mathbf{y}) \in \mathbb{R}^{m+n} : \Lambda(\mathbf{x}, \mathbf{y}) \leq \delta\}$ is of type

$$\left\{(\mathbf{x}, \mathbf{y}) \in \mathbb{R}^{m+n} : \frac{|\overline{y} - \overline{x} - \Delta|}{\sqrt{\sigma_X^2/m + \sigma_Y^2/n}} \geq c\right\},$$

where c is a positive constant (depending on δ). □

Exercise 13

Starting from $\sigma_X = \sigma_Y = \sigma$, where σ is unknown, solve the same problem as in Exercise 11. That is to say, maximize (for fixed $\mathbf{x}$ and $\mathbf{y}$) the map

$$(\mu_X, \mu_Y, \sigma) \mapsto L_{\mu_X, \mu_Y, \sigma}(\mathbf{x}, \mathbf{y})$$

under the constraint that $\mu_Y - \mu_X = \Delta$.

Solution. $L_{\mu_X,\mu_Y,\sigma}(\mathbf{x},\mathbf{y})$ equals

$$(2\pi\sigma^2)^{-\frac{1}{2}(m+n)} \exp\left(-\frac{1}{2\sigma^2}\left(\sum_{i=1}^{m}(x_i-\mu_X)^2+\sum_{i=1}^{n}(y_i-\mu_Y)^2\right)\right).$$

This expression has to be maximized under the constraint just mentioned. This amounts to the same as maximization of the logarithm of the expression above,

$$-\tfrac{1}{2}(m+n)\log(2\pi)-(m+n)\log\sigma-\frac{1}{2\sigma^2}\left(\sum_{i=1}^{m}(x_i-\mu_X)^2+\sum_{i=1}^{n}(y_i-\mu_Y)^2\right),$$

under the given constraint. Write

$$f(\mu_X,\mu_Y,\sigma):=\log L_{\mu_X,\mu_Y,\sigma}(\mathbf{x},\mathbf{y}) \quad\text{and}\quad g(\mu_X,\mu_Y,\sigma):=\mu_Y-\mu_X-\Delta.$$

In these notations we have to maximize f under the constraint $g=0$. By Lagrange's theorem this maximum (if existing) is attained in a point $\mathbf{c}=(\mu_X,\mu_Y,\sigma)$ for which

$$(\nabla f)(\mathbf{c})=\lambda\,(\nabla g)(\mathbf{c}),$$

where λ is some scalar. Writing this out we arrive at:

$$\lambda\,(-1,\ 1,\ 0)=\left(\frac{1}{\sigma^2}\sum_{i=1}^{m}(x_i-\mu_X),\ \frac{1}{\sigma^2}\sum_{i=1}^{n}(y_i-\mu_Y),\right.$$
$$\left.-\frac{m+n}{\sigma}+\frac{1}{\sigma^3}\left[\sum_{i=1}^{m}(x_i-\mu_X)^2+\sum_{i=1}^{n}(y_i-\mu_Y)^2\right]\right)$$

It follows from this that σ^2 and μ_X,μ_Y are linked in the following way:

$$\sigma^2=\frac{1}{m+n}\left(\sum_{i=1}^{m}(x_i-\mu_X)^2+\sum_{i=1}^{n}(y_i-\mu_Y)^2\right) \qquad (*)$$

Furthermore it follows that

$$\sum_{i=1}^{m}(x_i-\mu_X) \;=\; -\sum_{i=1}^{n}(y_i-\mu_Y),$$

which is equivalent to the equality $m\overline{x}+n\overline{y}=m\mu_X+n\mu_Y$. Together with the constraint $\mu_Y-\mu_X=\Delta$, this implies that

$$\mu_X=\frac{m\overline{x}+n\overline{y}-n\Delta}{m+n} \quad\text{and}\quad \mu_Y=\frac{m\overline{x}+n\overline{y}+m\Delta}{m+n} \qquad (**)$$

Substitution of $(*)$ together with $(**)$ in $L_{\mu_X,\mu_Y,\sigma}(\mathbf{x},\mathbf{y})$ gives us the global maximum $(2\pi\,e\,\sigma^2)^{-\frac{1}{2}(m+n)}$, with σ^2 as in $(*)$. □

Exercise 14

Let $X_1, \ldots, X_m$ and $Y_1, \ldots, Y_n$ be independent samples from a $N(\mu_X, \sigma^2)$ and a $N(\mu_Y, \sigma^2)$-distributed population respectively. The numerical value of σ is unknown. We wish to test $H_0 : \mu_Y - \mu_X = \Delta$ against $H_1 : \mu_Y - \mu_X \neq \Delta$, where Δ is a fixed real number. We write

$$s_p^2 := \frac{(m-1)s_{\mathbf{x}}^2 + (n-1)s_{\mathbf{y}}^2}{m+n-2}.$$

Prove that the likelihood ratio assumes the form

$$\Lambda(\mathbf{x}, \mathbf{y}) = \left[1 + \frac{1}{m+n-2}\left(\frac{\overline{y} - \overline{x} - \Delta}{s_p\sqrt{\frac{1}{m} + \frac{1}{n}}}\right)^2\right]^{-\frac{1}{2}(m+n)}.$$

Furthermore, check that a critical region based on this ratio is of type

$$G = \left\{(\mathbf{x}, \mathbf{y}) \in \mathbb{R}^{m+n} \; : \; \frac{|\overline{y} - \overline{x} - \Delta|}{s_p\sqrt{\frac{1}{m} + \frac{1}{n}}} \geq c\right\}.$$

Proof. The likelihood ratio can be written as $\Lambda(\mathbf{x}, \mathbf{y}) = L_0(\mathbf{x}, \mathbf{y})/L_1(\mathbf{x}, \mathbf{y})$, where $L_0(\mathbf{x}, \mathbf{y})$ is the global maximum of the map

$$(\mu_X, \mu_Y, \sigma) \mapsto (2\pi\sigma^2)^{-\frac{1}{2}(m+n)} \exp\left(-\frac{1}{2\sigma^2}\left(\sum_{i=1}^{m}(x_i - \mu_X)^2 + \sum_{i=1}^{n}(y_i - \mu_Y)^2\right)\right)$$

under the constraint $\mu_Y - \mu_X = \Delta$. The denominator $L_1(\mathbf{x}, \mathbf{y})$ is the global maximum of the same map without constraints. Using Lemma III.2.1 we learn that the latter maximum is of the form

$$(2\pi\sigma^2)^{-\frac{1}{2}(m+n)} \exp\left(-\frac{1}{2\sigma^2}\left(\sum_{i=1}^{m}(x_i - \overline{x})^2 + \sum_{i=1}^{n}(y_i - \overline{y})^2\right)\right) \qquad (*)$$

The expression above can be maximized by the technique of differentiation (with respect to σ). This leads to

$$\begin{aligned}\sigma^2 &= \frac{\sum_{i=1}^{m}(x_i - \overline{x})^2 + \sum_{j=1}^{n}(y_j - \overline{y})^2}{m+n} \\ &= \frac{(m-1)s_{\mathbf{x}}^2 + (n-1)s_{\mathbf{y}}^2}{m+n} = \frac{m+n-2}{m+n}\, s_p^2.\end{aligned}$$

Substitution of this expression into $(*)$ gives

$$L_1(\mathbf{x}, \mathbf{y}) = \left(2\pi\, e\, \frac{m+n-2}{m+n}\, s_p^2\right)^{-\frac{1}{2}(m+n)}.$$

Next we want to find an explicit expression for $L_0(\mathbf{x}, \mathbf{y})$. In Exercise 13 we deduced that

$$L_0(\mathbf{x}, \mathbf{y}) = (2\pi\, e\, \sigma_0^2)^{-\frac{1}{2}(m+n)},$$

where σ_0^2 is given by

$$\sigma_0^2 = \frac{1}{m+n}\left(\sum_{i=1}^{m}\left(x_i - \frac{m\overline{x}+n\overline{y}-n\Delta}{m+n}\right)^2 + \sum_{j=1}^{n}\left(y_j - \frac{m\overline{x}+n\overline{y}+m\Delta}{m+n}\right)^2\right).$$

This expression can be simplified as follows (the cross term vanishes):

$$\begin{aligned}
&\sum_{i=1}^{m}\left(x_i - \frac{m\overline{x}+n\overline{y}-n\Delta}{m+n}\right)^2 \\
&\quad = \sum_{i=1}^{m}\left\{(x_i-\overline{x}) + \left(\overline{x} - \frac{m\overline{x}+n\overline{y}-n\Delta}{m+n}\right)\right\}^2 \\
&\quad = \left(\sum_{i=1}^{m}(x_i-\overline{x})^2\right) + 0 + mn^2\left(\frac{\overline{y}-\overline{x}-\Delta}{m+n}\right)^2.
\end{aligned}$$

In the same way

$$\begin{aligned}
&\sum_{j=1}^{n}\left(y_j - \frac{m\overline{x}+n\overline{y}+m\Delta}{m+n}\right)^2 \\
&\quad = \left(\sum_{j=1}^{n}(y_j-\overline{y})^2\right) + 0 + m^2n\left(\frac{\overline{y}-\overline{x}-\Delta}{m+n}\right)^2.
\end{aligned}$$

Combining this we see that σ_0^2 can be written as

$$\begin{aligned}
\sigma_0^2 &= \frac{1}{m+n}\left(\sum_{i=1}^{m}(x_i-\overline{x})^2 + \sum_{j=1}^{n}(y_j-\overline{y})^2\right) + \frac{m^2n+mn^2}{m+n}\left(\frac{\overline{y}-\overline{x}-\Delta}{m+n}\right)^2 \\
&= \frac{(m-1)s_{\mathbf{x}}^2+(n-1)s_{\mathbf{y}}^2}{m+n-2}\,\frac{m+n-2}{m+n} + mn\left(\frac{\overline{y}-\overline{x}-\Delta}{m+n}\right)^2 \\
&= \frac{m+n-2}{m+n}\,s_p^2 + mn\left(\frac{\overline{y}-\overline{x}-\Delta}{m+n}\right)^2.
\end{aligned}$$

Thus $L_0(\mathbf{x}, \mathbf{y})$ is equal to the following expression:

$$\left[2\pi\, e\left(\frac{m+n-2}{m+n}\,s_p^2 + mn\left(\frac{\overline{y}-\overline{x}-\Delta}{m+n}\right)^2\right)\right]^{-\frac{1}{2}(m+n)}.$$

Now we can write down the following expressions for the likelihood ratio:

$$\Lambda(\mathbf{x},\mathbf{y}) = \frac{L_0(\mathbf{x},\mathbf{y})}{L_1(\mathbf{x},\mathbf{y})} = \frac{\left(\frac{m+n-2}{m+n}\, s_p^2 + mn\left(\frac{\overline{y}-\overline{x}-\Delta}{m+n}\right)^2\right)^{-\frac{1}{2}(m+n)}}{\left(\frac{m+n-2}{m+n}\, s_p^2\right)^{-\frac{1}{2}(m+n)}}$$

$$= \left[1+\frac{1}{m+n-2}\left(\frac{\overline{y}-\overline{x}-\Delta}{s_p\sqrt{\frac{1}{m}+\frac{1}{n}}}\right)^2\right]^{-\frac{1}{2}(m+n)}.$$

Critical regions based on this ratio are of type

$$\begin{aligned} G &= \{(\mathbf{x},\mathbf{y})\in\mathbb{R}^{m+n} : \Lambda(\mathbf{x},\mathbf{y})\le\delta\} \\ &= \left\{(\mathbf{x},\mathbf{y})\in\mathbb{R}^{m+n} : \left[1+\frac{1}{m+n-2}\left(\frac{\overline{y}-\overline{x}-\Delta}{s_p\sqrt{\frac{1}{m}+\frac{1}{n}}}\right)^2\right]^{-\frac{1}{2}(m+n)} \le\delta\right\} \\ &= \left\{(\mathbf{x},\mathbf{y})\in\mathbb{R}^{m+n} : \frac{1}{m+n-2}\left(\frac{\overline{y}-\overline{x}-\Delta}{s_p\sqrt{\frac{1}{m}+\frac{1}{n}}}\right)^2 \ge \delta^{-\frac{2}{m+n}}-1\right\} \\ &= \left\{(\mathbf{x},\mathbf{y})\in\mathbb{R}^{m+n} : \frac{|\overline{y}-\overline{x}-\Delta|}{s_p\sqrt{\frac{1}{m}+\frac{1}{n}}}\ge c\right\}, \end{aligned}$$

where c is some constant depending on δ. □

Exercise 15

Develop a method to test paired differences in mean.

Solution. We are starting from a paired sample $(X_1,Y_1),\dots,(X_n,Y_n)$ and we write $D_i := Y_i - X_i$. Assume that $D_1,\dots,D_n$ is a sample from a $N(\mu,\sigma^2)$-distributed population. Under these assumptions we wish to test $H_0 : \mu=\Delta$ versus $H_1 : \mu\neq\Delta$. By Theorem III.2.3 the corresponding likelihood ratio is of the form

$$\Lambda(d_1,\dots,d_n) = \left\{1+\frac{1}{n-1}\left(\frac{\overline{d}-\Delta}{s_D/\sqrt{n}}\right)^2\right\}^{-\frac{1}{2}n},$$

where $d_i = y_i - x_i$ (all i) and s_D^2 the sample variance of the outcome $d_1,\dots,d_n$ of $D_1,\dots,D_n$. A critical region based on this ratio is of type

$$G(c) = \left\{\mathbf{d}\in\mathbb{R}^n : \left|\frac{\overline{d}-\Delta}{s_D/\sqrt{n}}\right|\ge c\right\}.$$

We know (Theorem II.4.6) that the statistic

$$\frac{\overline{D} - \Delta}{s_D/\sqrt{n}}$$

enjoys (under H_0) a t-distribution with $n-1$ degrees of freedom. This enables us to construct, for any prescribed α, a critical region $G(c)$ of size α. □

Exercise 16

Given is a population with a probability density f_θ, defined by

$$f_\theta(x) := \begin{cases} \theta\, x^{\theta-1} & \text{if } x \in (0,1), \\ 0 & \text{elsewhere.} \end{cases}$$

A sample $X_1, \ldots, X_n$ is drawn from this population; n is very large. Basing ourselves on the likelihood ratio Λ, we want to test $H_0 : \theta = \theta_0$ against $H_1 : \theta \neq \theta_0$. Adjust $\delta > 0$ in such a way that

$$G(\delta) = \{\mathbf{x} \in \mathbb{R}^n \;:\; \Lambda(\mathbf{x}) \leq \delta\}$$

is of size $\alpha = 0.05$.

Solution. Here $\Theta = (0, +\infty)$ and $\Theta_0 = \{\theta_0\}$, hence $\dim \Theta - \dim \Theta_0 = 1 - 0 = 1$. Now by Theorem III.1.2 the variable $-2 \log \Lambda(\mathbf{X})$ is approximately χ^2-distributed with 1 degree of freedom. We reject H_0 if $-2 \log \Lambda(\mathbf{X})$ exhibits a large outcome (equivalently, if $\Lambda(\mathbf{X})$ shows a small outcome). Starting from a level of significance $\alpha = 0.05$, the critical region for $-2 \log \Lambda(\mathbf{X})$ is $[3.84, +\infty)$. The critical region in $\mathbb{R}^n$, based on the likelihood ratio, is given by

$$G = \{\mathbf{x} \in \mathbb{R}^n \;:\; -2 \log \Lambda(\mathbf{x}) \geq 3.84\} = \{\mathbf{x} \in \mathbb{R}^n \;:\; \Lambda(\mathbf{x}) \leq e^{-\frac{1}{2} \cdot 3.84} = 0.15\}.$$

So for $\delta = 0.15$ the critical region $G(\delta)$ is of size $\alpha = 0.05$. □

Exercise 17

Concerning 'smoking behavior' the Dutch population is split up into four categories. Nationwide the following percentages apply:

category	1	2	3	4
percentage	46	11	27	16

To test whether these percentages also apply to Dutch students, a sample of size 500 is drawn which results in the following data:

category	1	2	3	4
frequency	206	41	155	98

Carry out a goodness of fit test to test H_0: 'the national percentages apply' against H_1: 'they don't', at a level of significance of 0.10.

Solution. Denote the ith component ($i = 1, 2, 3, 4$) of the frequency vector $\mathbf{F}$ by F_i. Under H_0 the variables F_1, F_2, F_3, F_4 are all binomially distributed with parameter $n_1 = n_2 = n_3 = n_4 = 500$ and $\theta_1 = 0.46$, $\theta_2 = 0.11$, $\theta_3 = 0.27$ and $\theta_4 = 0.16$ respectively. It follows from this that under H_0:

$$\begin{cases} \mathbb{E}(F_1) = 500 \cdot 0.46 = 230, & \mathbb{E}(F_2) = 500 \cdot 0.11 = 55, \\ \mathbb{E}(F_3) = 500 \cdot 0.27 = 135, & \mathbb{E}(F_4) = 500 \cdot 0.16 = 80. \end{cases}$$

Thus we see that the test statistic

$$T = \sum_{i=1}^{4} \frac{(F_i - \mathbb{E}(F_i))^2}{\mathbb{E}(F_i)}$$

exhibits the following outcome:

$$\frac{(206-230)^2}{230} + \frac{(41-55)^2}{55} + \frac{(155-135)^2}{135} + \frac{(98-80)^2}{80} \approx 13.1.$$

A priori T was approximately $\chi^2(3)$-distributed. Given a level of significance $\alpha = 0.10$, a critical region for T is presented by the interval $(6.25, +\infty)$. The outcome of T is an element of this interval: therefore we reject H_0. □

Exercise 18

Given is a population which is Poisson distributed with parameter $\lambda > 0$. A sample $X_1, \ldots, X_{100}$ is drawn from this population in order to test $H_0 : \lambda = 4.3$ against $H_1 : \lambda \neq 4.3$. The outcome of the sample is schematized as follows:

outcome	≤ 1	2	3	4	5	6	7	≥ 8
frequency	9	11	16	21	18	8	12	5

Carry out a goodness of fit test at a level of significance $\alpha = 0.05$.

Solution. Let F_i be the ith component ($i = 1, \ldots 8$) of the frequency vector $\mathbf{F}$. Under H_0 we have

$$\mathbb{P}(X = k) = \frac{(4.3)^k}{k!} \cdot e^{-4.3}.$$

Hence:

$$\begin{aligned}
\mathbb{E}(F_1) &= 100 \cdot e^{-4.3}(1+4.3) = 7.2,\\
\mathbb{E}(F_2) &= 100 \cdot ((4.3)^2/2!)\ e^{-4.3} = 12.5,\\
\mathbb{E}(F_3) &= 100 \cdot ((4.3)^3/3!)\ e^{-4.3} = 18.0,\\
\mathbb{E}(F_4) &= 100 \cdot ((4.3)^4/4!)\ e^{-4.3} = 19.3,\\
\mathbb{E}(F_5) &= 100 \cdot ((4.3)^5/5!)\ e^{-4.3} = 16.6,\\
\mathbb{E}(F_6) &= 100 \cdot ((4.3)^6/6!)\ e^{-4.3} = 11.9,\\
\mathbb{E}(F_7) &= 100 \cdot ((4.3)^7/7!)\ e^{-4.3} = 7.3,\\
\mathbb{E}(F_8) &= 100 - 7.2 - \cdots - 7.3 = 7.2.
\end{aligned}$$

So

$$\begin{aligned}
\sum_{i=1}^{8} \frac{(F_i - \mathbb{E}(F_i))^2}{\mathbb{E}(F_i)} \approx\ & \frac{(1.8)^2}{7.2} + \frac{(1.5)^2}{12.5} + \frac{(2.0)^2}{18.0} + \frac{(1.7)^2}{19.3}\\
& + \frac{(1.4)^2}{16.6} + \frac{(3.9)^2}{11.9} + \frac{(4.7)^2}{7.3} + \frac{(2.2)^2}{7.2} = 6.1.
\end{aligned}$$

This test statistic was (a priori) approximately $\chi^2(7)$-distributed. Thus, at a level of significance $\alpha = 0.05$, a critical region is presented by the interval $(14.1, +\infty)$. The statistic shows an outcome which is not in this interval, hence we accept H_0. □

Exercise 19

We consider a certain type of cassettes of which it is assumed that the playing time enjoys a $N(92, 1)$-distribution. A sample of size 100 is drawn concerning the playing time and the outcomes are partitioned into four categories. This leads to the following:

range	$(-\infty, 91]$	$(91, 92]$	$(92, 93]$	$(93, +\infty)$
number of outcomes	20	30	26	24

(i) Determine the expected frequencies in the four classes under H_0.

Solution. We denote the frequency of occurence in the four classes by F_1, F_2, F_3 and F_4. Suppose X is a $N(92, 1)$-distributed variable. Then

$$\begin{aligned}
\mathbb{E}(F_1) &= 100 \cdot \mathbb{P}(X \in (-\infty, 91])\\
&= 100 \cdot \mathbb{P}\left(\frac{X-92}{1} \in (-\infty, -1]\right) = 100 \cdot 0.16 = 16,\\
\mathbb{E}(F_2) &= 100 \cdot \mathbb{P}(X \in (91, 92])\\
&= 100 \cdot \mathbb{P}\left(\frac{X-92}{1} \in (-1, 0]\right) = 100(0.50 - 0.16) = 34.
\end{aligned}$$

Furthermore $\mathbb{E}(F_3) = \mathbb{E}(F_2) = 34$ and $\mathbb{E}(F_4) = \mathbb{E}(F_1) = 16$. □

(ii) At a level of significance of $\alpha = 0.05$, carry out a goodness of fit test.

Solution. The corresponding test statistic T shows the following outcome:

$$T = \sum_{i=1}^{4} \frac{(F_i - \mathbb{E}(F_i))^2}{\mathbb{E}(F_i)} = \frac{(20-16)^2}{16} + \frac{(30-34)^2}{34} + \frac{(26-34)^2}{34} + \frac{(24-16)^2}{16} = 7.35.$$

A priori the variable T is approximately $\chi^2(3)$-distributed and a critical region for T is presented by the interval $(7.81, +\infty)$. The outcome of T is not an element of this interval, so we accept H_0. □

(iii) *Question.* Is it possible that the decision procedure leads to another decision for another partition of $\mathbb{R}$?

Solution. Yes. To see this we partition $\mathbb{R}$ in a different way: $(-\infty, 91]$, $(91, 93]$ and $(93, +\infty]$. Carrying out a goodness of fit test (as we did in (ii)) we arrive at the following:

$$\mathbb{E}(F_1) = 16, \quad \mathbb{E}(F_2) = 68, \quad \mathbb{E}(F_3) = 16.$$

The corresponding test statistic T now exhibits an outcome

$$\sum_{i=1}^{3} \frac{(F_i - \mathbb{E}(F_i))^2}{\mathbb{E}(F_i)} = \frac{(20-16)^2}{16} + \frac{(56-68)^2}{68} + \frac{(24-16)^2}{16} = 7.12.$$

A priori T is (under H_0) approximately $\chi^2(2)$-distributed. At a level of significance $\alpha = 0.05$ a critical region for T is the interval $(5.99, +\infty)$. The outcome of T is an element of this interval, therefore we reject H_0. We see that our decision procedure depends on the way we partition $\mathbb{R}$. □

Exercise 20

(i) Prove that in the notation of §III.4 we have $\theta_{ij} = \theta_{i\bullet}\theta_{\bullet j}$ for all i, j if and only if there exist numbers $\alpha_1, \dots, \alpha_p \geq 0$ and $\beta_1, \dots, \beta_q \geq 0$ such that $\theta_{ij} = \alpha_i \beta_j$ for all i, j, while $\alpha_1 + \cdots + \alpha_p = 1$, $\beta_1 + \cdots + \beta_q = 1$.

Proof.

(⇒) Take $\alpha_i = \theta_{i\bullet}$ and $\beta_j = \theta_{\bullet j}$.

(⇐) Suppose there exist $\alpha_1, \dots, \alpha_p \geq 0$, $\beta_1, \dots, \beta_q \geq 0$ such that for all i, j we have $\sum_i \alpha_i = \sum_j \beta_j = 1$ and $\theta_{ij} = \alpha_i \beta_j$. Then $\theta_{i\bullet} = \sum_j \theta_{ij} = \sum_j \beta_j \alpha_i = \alpha_i$ for all i. In the same way $\theta_{\bullet j} = \beta_j$ (all j). Hence $\theta_{ij} = \theta_{i\bullet}\theta_{\bullet j}$ for all i, j. □

(ii) Given an outcome $\mathbf{k} = (k_{11}, \ldots, k_{pq})$ of the frequency vector $\mathbf{F}$, maximize the function

$$\boldsymbol{\theta} \mapsto \mathbb{P}(\mathbf{F} = \mathbf{k}) = \binom{n}{\mathbf{k}} \prod_{i,j} \theta_{ij}^{k_{ij}} \qquad (*)$$

under the constraints $\sum_j \theta_{\bullet j} = \sum_i \theta_{i\bullet} = \sum_{i,j} \theta_{ij} = 1$, $\theta_{ij} \geq 0$ and $\theta_{ij} = \theta_{i\bullet}\theta_{\bullet j}$ (all i, j).

Proof. We write $k_{i\bullet} = \sum_j k_{ij}$ and $k_{\bullet j} = \sum_i k_{ij}$. By (i) the question amounts to the same thing as maximizing the expression

$$\prod_{i,j} (\alpha_i \beta_j)^{k_{ij}} = \left(\prod_{i=1}^{p} \alpha_i^{k_{i\bullet}}\right) \left(\prod_{j=1}^{q} \beta_j^{k_{\bullet j}}\right) \qquad (**)$$

under the constraints

$$\alpha_1, \ldots, \alpha_p \geq 0, \quad \beta_1, \ldots, \beta_q \geq 0 \quad \text{and} \quad \textstyle\sum_i \alpha_i = \sum_j \beta_j = 1.$$

Using Lemma III.3.1 we see that $(**)$ is maximal for $\alpha_i = k_{i\bullet}/(\sum_j k_{j\bullet}) = k_{i\bullet}/n$ and $\beta_j = k_{\bullet j}/n$ (all i, j). Consequently $(*)$ is maximal if $\theta_{ij} = k_{i\bullet}k_{\bullet j}/n^2$. The magnitude of this maximum is

$$\sup_{\boldsymbol{\theta} \in \Theta_0} \mathbb{P}(\mathbf{F} = \mathbf{k}) = \binom{n}{\mathbf{k}} \prod_{i,j} \left(\frac{k_{i\bullet}k_{\bullet j}}{n^2}\right)^{k_{ij}}.$$

□

Exercise 21

Along the lines of the proof of Theorem III.3.3, give a heuristical proof of Theorem III.4.3.

Solution. By Theorem III.1.2 and §II.11 Exercise 34, the variable

$$-2 \log \Lambda(F_{11}, \ldots, F_{1q}, \ \ldots \ , F_{p1}, \ldots, F_{pq})$$

is for large n approximately χ^2-distributed with $\dim \Theta - \dim \Theta_0$ degrees of freedom, where

$$\Theta = \{(\theta_{11}, \ldots, \theta_{pq}) : \theta_{ij} \geq 0 \text{ for all } (i,j) \text{ and } \textstyle\sum_{i,j} \theta_{ij} = 1\}$$

and

$$\Theta_0 = \{(\theta_{11}, \ldots, \theta_{pq}) : \theta_{ij} \geq 0,\ \theta_{ij} = \theta_{i\bullet}\theta_{\bullet j} \text{ and } \textstyle\sum_i \theta_{i\bullet} = \sum_j \theta_{\bullet j} = 1 \text{ for all } i, j\}.$$

Here $\dim \Theta = pq - 1$ and using Exercise 20 (i) we see that $\dim \Theta_0 = (p-1) + (q-1) = p + q - 2$. Consequently

$$\dim \Theta - \dim \Theta_0 = pq - p - q + 1 = (p-1)(q-1).$$

In the same way as in the proof of Theorem III.3.3 we now 'derive' that

$$\sum_{i,j} \frac{(F_{ij} - E_{ij})^2}{E_{ij}} \approx -2 \log \Lambda(F_{11}, \dots, F_{pq}).$$

By Theorem III.4.2 we can write

$$\Lambda(F_{11}, \dots, F_{pq}) = \prod_{i,j} \left(\frac{F_{i\bullet} F_{\bullet j}}{n F_{ij}} \right)^{F_{ij}} = \prod_{i,j} \left(\frac{E_{ij}}{F_{ij}} \right)^{F_{ij}}.$$

Hence

$$-2 \log \Lambda(F_{11}, \dots, F_{pq}) = -2 \sum_{i,j} F_{ij} \log \left(\frac{E_{ij}}{F_{ij}} \right) \qquad (*)$$

Under H_0 we may expect that for large n

$$\frac{E_{ij}}{n} = \frac{F_{i\bullet}}{n} \frac{F_{\bullet j}}{n} \approx \theta_{i\bullet} \theta_{\bullet j} \quad \text{and} \quad \frac{F_{ij}}{n} \approx \theta_{ij} = \theta_{i\bullet} \theta_{\bullet j}.$$

Therefore

$$\frac{E_{ij}}{F_{ij}} = \frac{E_{ij}/n}{F_{ij}/n} \approx 1.$$

Exploiting the Taylor expansion of the function $x \mapsto \log(1 + x)$ around the point $x = 0$ we learn that

$$\log \left(\frac{E_{ij}}{F_{ij}} \right) = \log \left\{ 1 + \left(\frac{E_{ij}}{F_{ij}} - 1 \right) \right\} \approx \left(\frac{E_{ij}}{F_{ij}} - 1 \right) - \frac{1}{2} \left(\frac{E_{ij}}{F_{ij}} - 1 \right)^2,$$

so that

$$\begin{aligned} \sum_{i,j} F_{ij} \log \left(\frac{E_{ij}}{F_{ij}} \right) &\approx \sum_{i,j} (E_{ij} - F_{ij}) - \tfrac{1}{2} \sum_{i,j} \frac{(E_{ij} - F_{ij})^2}{F_{ij}} \\ &= 0 - \tfrac{1}{2} \sum_{i,j} \frac{(E_{ij} - F_{ij})^2}{F_{ij}} \qquad (**) \end{aligned}$$

Here we used the fact that $\sum_{i,j} E_{ij} = \sum_{i,j} F_{ij} = n$. Combining $(*)$ and $(**)$ we see that

$$\begin{aligned} -2 \log \Lambda(F_{11}, \dots, F_{pq}) &\approx (-2 \cdot -\tfrac{1}{2}) \sum_{i,j} \frac{(E_{ij} - F_{ij})^2}{F_{ij}} \\ &= \sum_{i,j} \frac{E_{ij}}{F_{ij}} \cdot \frac{(E_{ij} - F_{ij})^2}{E_{ij}} \approx \sum_{i,j} \frac{(E_{ij} - F_{ij})^2}{E_{ij}}. \end{aligned}$$

□

Exercise 22

On a civic sports-festival for students we measure the individual sporting records Y, the weight G (in kilos) and the length L (in meters) of 220 randomly chosen students. We define the variable X to be the 'Quitelet quotient'

$$X := \frac{G}{L^2},$$

and we are wondering whether the variables X and Y are statistically independent. The outcome of our experiment is presented in the following scheme:

$X \downarrow \; Y \rightarrow$	bad	average	good	very good
≤ 20	5	6	8	6
$\in (20, 25]$	16	9	61	34
$\in (25, 30]$	12	18	10	5
> 30	11	9	5	5

Carry out a χ^2-test on statistical independence on the variables X and Y at a level of significance $\alpha = 0.05$.

Solution. First we compute the outcome of the $E_{ij} = F_{i\bullet}F_{\bullet j}/220$ for all $i, j \in \{1, 2, 3, 4\}$. We have

$$F_{\bullet 1} = 25, \quad F_{\bullet 2} = 120, \quad F_{\bullet 3} = 45, \quad F_{\bullet 4} = 30,$$
$$F_{1\bullet} = 44, \quad F_{2\bullet} = 42, \quad F_{3\bullet} = 84, \quad F_{4\bullet} = 50.$$

It follows from this that

$$E_{11} = \tfrac{25\cdot 44}{220} = 5, \quad E_{12} = \tfrac{25\cdot 42}{220} = \tfrac{105}{22}, \quad E_{13} = \tfrac{25\cdot 84}{220} = \tfrac{105}{11}, \quad E_{14} = \tfrac{125}{22},$$
$$E_{21} = \tfrac{120\cdot 44}{220} = 24, \quad E_{22} = \tfrac{120\cdot 42}{220} = \tfrac{252}{11}, \quad E_{23} = \tfrac{120\cdot 84}{220} = \tfrac{504}{11}, \quad E_{24} = \tfrac{300}{11},$$
$$E_{31} = \tfrac{45\cdot 44}{220} = 9, \quad E_{32} = \tfrac{45\cdot 42}{220} = \tfrac{189}{22}, \quad E_{33} = \tfrac{45\cdot 84}{220} = \tfrac{189}{11}, \quad E_{34} = \tfrac{225}{22},$$
$$E_{41} = \tfrac{30\cdot 44}{220} = 6, \quad E_{42} = \tfrac{30\cdot 42}{220} = \tfrac{63}{11}, \quad E_{43} = \tfrac{30\cdot 84}{220} = \tfrac{126}{11}, \quad E_{44} = \tfrac{75}{11}.$$

Writing

$$M_{ij} := \frac{(F_{ij} - E_{ij})^2}{E_{ij}},$$

we can expose the outcome of the M_{ij} in the form of a matrix $\mathbf{M}$:

$$\mathbf{M} = \begin{pmatrix} 0.000 & 0.316 & 0.250 & 0.018 \\ 2.667 & 8.445 & 5.030 & 1.659 \\ 1.000 & 10.305 & 3.002 & 2.672 \\ 4.167 & 1.870 & 3.637 & 0.485 \end{pmatrix}.$$

The test statistic $T = \sum_{i,j}(F_{ij} - E_{ij})^2/E_{ij}$ exhibits the following outcome:

$$T = \sum_{i,j} M_{ij} = 45.523.$$

A priori the variable T is approximately $\chi^2(9)$-distributed (Theorem III.4.3), so a critical region for T at a level of significance $\alpha = 0.05$ is presented by the interval $(16.919, +\infty)$. The outcome of T is an element of this interval, consequently we reject H_0. $\square$

Chapter 4
Simple regression analysis

4.1 Summary of Chapter IV

4.1.1 The method of least squares (IV.1)

Presume there is a theoretical relation $y = \alpha + \beta x$ between the variables x and y, where x is a controlled and y a response variable. We make estimations a and b of α and β respectively. Once an estimation has been made, the *error* (or *residual*) corresponding to measurement no. i is defined by $e_i := y_i - (a + bx_i)$. Given a sequence $(x_1, y_1), \ldots, (x_n, y_n)$ of paired measurements for x and y, we define the *sum of squares of errors* by:

$$f(a,b) \;:=\; \sum_{i=1}^{n} e_i^2 = \sum_{i=1}^{n} \{y_i - (a + bx_i)\}^2.$$

This function f is minimized by the technique of partial differentiation. That is to say, we solve the so-called *normal equations*

$$\frac{\partial}{\partial a} f(a,b) = 0 \quad \text{and} \quad \frac{\partial}{\partial b} f(a,b) = 0$$

in order to find optimal estimations a and b for α and β. These optimal estimations are denoted by $\hat{\alpha}$ and $\hat{\beta}$, and are called the *least squares estimations* of α and β. In order to express $\hat{\alpha}$ and $\hat{\beta}$ in a friendly way in terms of the measurements $(x_1, y_1), \ldots, (x_n, y_n)$, we introduce the following notations:

$$\begin{cases} \overline{x} & := & \frac{1}{n}\sum_{i=1}^{n} x_i, \\ \overline{y} & := & \frac{1}{n}\sum_{i=1}^{n} y_i, \\ S_{\mathbf{xx}} & := & \sum_{i=1}^{n} (x_i - \overline{x})^2, \\ S_{\mathbf{yy}} & := & \sum_{i=1}^{n} (y_i - \overline{y})^2, \\ S_{\mathbf{xy}} & := & \sum_{i=1}^{n} (x_i - \overline{x})(y_i - \overline{y}). \end{cases}$$

In these notations we have: $\hat{\beta} = S_{\mathbf{xy}}/S_{\mathbf{xx}}$ and $\hat{\alpha} = \overline{y} - \hat{\beta}\,\overline{x}$.

This way of making estimations can be applied universally. For example it can be applied when we have (by theory) relations $y = \alpha$ or $y = \alpha + e^{\beta x} + \gamma x^2$. Just set up the normal equations and solve them to get optimal estimations for the parameters α, β, γ.

Now let Y_x be the stochastic variable which represents the measurements concerning y at an adjusted value x. We write $Y_i := Y_{x_i}$. A model is set up by making the following three assumptions:

1. $\mathbb{E}(Y_x) = \alpha + \beta x$,
2. $\{Y_1, \ldots, Y_n\}$ is a statistically independent system,
3. $\text{var}(Y_1) = \cdots = \text{var}(Y_n) = \sigma^2$.

Next a stochastic counterpart of the above can be formulated. Given a theoretical relation $y = \alpha + \beta x$, the least squares estimators $\hat{\alpha}$ and $\hat{\beta}$ satisfy

$$\begin{cases} \hat{\beta} & = & \sum_i (x_i - \overline{x})\, Y_i \,/ S_{\mathbf{xx}}, \\ \hat{\alpha} & = & \overline{Y} - \hat{\beta}\,\overline{x}, \\ \text{var}(\hat{\alpha}) & = & (S_{\mathbf{xx}} + n\,(\overline{x})^2)\,\sigma^2/(nS_{\mathbf{xx}}), \\ \text{var}(\hat{\beta}) & = & \sigma^2/S_{\mathbf{xx}}, \\ \text{cov}(\hat{\alpha}, \hat{\beta}) & = & -\overline{x}\sigma^2/S_{\mathbf{xx}}. \end{cases}$$

The variables $\hat{\alpha}$ and $\hat{\beta}$ are unbiased linear estimators of α and β. Among the collection of unbiased linear estimators of α and β the estimators $\hat{\alpha}$ and $\hat{\beta}$ are the ones having minimal variance; they are unique in this respect. In order to estimate $\mu_x := \alpha + \beta x$, we introduce the variable $\hat{Y}_x := \hat{\alpha} + \hat{\beta} x$. This variable presents an unbiased linear estimator of μ_x and its variance is given by

$$\left(\frac{1}{n} + \frac{(x - \overline{x})^2}{S_{\mathbf{xx}}} \right) \sigma^2.$$

4.1.2 Construction of an unbiased estimator of σ^2 (IV.2)

The stochastic variable SSE (Sum of Squares of Errors) is defined by

$$\text{SSE} := \sum_{i=1}^{n} (Y_i - \hat{Y}_i)^2.$$

Theorem IV.2.3 states that $\text{SSE}/(n-2)$ is an unbiased estimator of σ^2.

4.1.3 Normal regression analysis (IV.3)

We now sharpen the three central assumptions in §IV.1. In fact, we make the following two assumptions:

1. For all x the variable Y_x is $N(\alpha + \beta x, \sigma^2)$-distributed

2. For all sequences $x_1, \ldots, x_n$ the system $\{Y_{x_1}, \ldots, Y_{x_n}\}$ is statistically independent.

Under these assumptions we are talking about *normal regression analysis.* In normal regression analysis SSE/σ^2 is χ^2-distributed with $n-2$ degrees of freedom. Furthermore, both the statistics

$$\frac{(\hat{\alpha} - \alpha)\sqrt{n(n-2)S_{\mathbf{xx}}}}{\sqrt{\text{SSE}}\sqrt{S_{\mathbf{xx}} + n(\overline{x})^2}} \quad \text{and} \quad \frac{(\hat{\beta} - \beta)\sqrt{(n-2)S_{\mathbf{xx}}}}{\sqrt{\text{SSE}}}$$

enjoy a t-distribution with $n-2$ degrees of freedom.

4.1.4 Pearson's product-moment correlation coefficient (IV.4)

In this section a quantitative measure for the amount of linear structure in a 'cloud' of points $(x_1, y_1), \ldots, (x_n, y_n)$ is constructed. We are talking about *Pearson's product-moment correlation coefficient*, which is also called the *sample correlation coefficient.* It is defined by

$$R_P := \frac{S_{\mathbf{xy}}}{\sqrt{S_{\mathbf{xx}}}\sqrt{S_{\mathbf{yy}}}}.$$

In the following way a geometrical interpretation can be given.

Define the linear space V as the quotient of $\mathbb{R}^n$ and the one-dimensional subspace $\mathfrak{E} = \{\lambda \mathbf{e} : \lambda \in \mathbb{R}\}$, where $\mathbf{e} := (1, \ldots, 1) \in \mathbb{R}^n$. Denoting the equivalence class of $\mathbf{x} \in \mathbb{R}^n$ by $[\mathbf{x}]$, an inner product on V can be defined by

$$\langle [\mathbf{x}], [\mathbf{y}] \rangle := S_{\mathbf{xy}}.$$

It turns out that R_P can be regarded as the cosine of the angle included by the vectors $[\mathbf{x}]$ and $[\mathbf{y}]$ in V. For this reason we always have $-1 \leq R_P \leq 1$. The case $R_P = \pm 1$ occurs if and only if all the points in the scatter diagram are on one and the same straight line. The case $R_P = 0$ indicates a total absence of *linear* structure in the scatter diagram. This does not at all exclude that (possibly strong) linear structure can be observed in parts of the scatter diagram. Pearson's product-moment correlation coefficient is invariant under positive scale transformations.

4.1.5 The sum of squares of errors as a measure of the amount of linear structure (IV.5)

Again we are starting from a cloud of points $(x_1, y_1), \ldots, (x_n, y_n)$. We consider the associated outcome of the variable

$$\text{SSE} := \sum_i \left(y_i - (\hat{\alpha} + \hat{\beta} x_i)\right)^2.$$

The *Sum of Squares of Regression* is defined as

$$\text{SSR} := \sum_i \left((\hat{\alpha} + \hat{\beta}x_i) - \overline{y}\right)^2.$$

We always have $S_{\mathbf{yy}} = \text{SSE}+\text{SSR}$. We call $\text{SSR}/S_{\mathbf{yy}}$ the *fraction of explained variance* and $\text{SSE}/S_{\mathbf{yy}}$ the *fraction of non-explained variance.* We have: $\text{SSR}/S_{\mathbf{yy}} = R_P^2$, where R_P is Pearson's product-moment correlation coefficient. Usually R_P^2 is called the *coefficient of determination.*

4.2 Exercises to Chapter IV

Exercise 1

Theoretically we have the relation $y = \alpha$. Given $(x_1, y_1), \ldots, (x_n, y_n)$, a set of points, express the least squares estimation $\hat{\alpha}$ of α in terms of the x_i and y_i.

Solution. Define $f : \mathbb{R} \mapsto [0, +\infty)$ by $f(a) := \sum_{i=1}^n (y_i - a)^2$. We obtain the least squares estimation $\hat{\alpha}$ of α by minimizing this function. That is to say, we get $\hat{\alpha}$ by finding the solution of the normal equation

$$\frac{\partial}{\partial a} f(a) = 0.$$

This equation is given by $-2\sum_{i=1}^n (y_i - a) = 0$, or, $-2(n\overline{y} - na) = 0$. Hence $\hat{\alpha} = \overline{y}$. □

Exercise 2

Given is a 'cloud' of points $\{(2,3),(4,6),(5,7)\}$ and a theoretical relation $y = \alpha + \beta x^2$.

(i) Compute the value of least squares estimations $\hat{\alpha}$ and $\hat{\beta}$ of α and β.

Solution. Define $f : \mathbb{R}^2 \mapsto [0, +\infty)$ by $f(a,b) := \sum_{i=1}^3 (y_i - (a + bx_i^2))^2$. The associated normal equations are given by

$$\begin{cases} \sum_{i=1}^3 -2(y_i - (a + bx_i^2)) & = \frac{\partial}{\partial a} f(a,b) = 0, \\ \sum_{i=1}^3 -2x_i^2(y_i - (a + bx_i^2)) & = \frac{\partial}{\partial b} f(a,b) = 0. \end{cases}$$

Therefore $3\overline{y} = 3a + b\sum_{i=1}^3 x_i^2$ and $\sum_{i=1}^3 x_i^2 y_i = \sum_{i=1}^3 (a + bx_i^2)\, x_i^2$. Substitution leads to $16 = 3a + 45b$ and

$$2^2 \cdot 3 + 4^2 \cdot 6 + 5^2 \cdot 7 = 2^2\,(a + b \cdot 2^2) + 4^2\,(a + b \cdot 4^2) + 5^2\,(a + b \cdot 5^2).$$

In other words, we have $283 = 45a + 897b$ and $3a = 16 - 45b$. Solving these equations simultaneously we get $\hat{\alpha} = 2.428$ and $\hat{\beta} = 0.194$. □

(ii) Determine the numerical value of SSE.

Solution. By (i) we have

$$\begin{aligned}\text{SSE} &= f(2.428, 0.194) = \sum_{i=1}^{3} (y_i - (2.428 + 0.194x_i^2))^2 \\ &= (-0.204)^2 + (0.468)^2 + (-0.278)^2 = 0.338.\end{aligned}$$

□

Exercise 3

Prove that the determinant of the normal equations in §IV.1 is 0 if and only if the x_i are mutually equal.

Proof. We are dealing with the following system of linear equations:

$$\begin{cases} n\hat{\alpha} + n\overline{x}\hat{\beta} &= n\overline{y}, \\ n\overline{x}\hat{\alpha} + \|\mathbf{x}\|^2\hat{\beta} &= \langle \mathbf{x}, \mathbf{y} \rangle. \end{cases}$$

The determinant of this system is $n\|\mathbf{x}\|^2 - n\overline{x} \cdot n\overline{x}$, which is equal to 0 if and only if $\|\mathbf{x}\| = \sqrt{n}\,|\overline{x}|$. Setting $\mathbf{e} := (1, \ldots, 1) \in \mathbb{R}^n$, we have by the Cauchy–Schwarz inequality that $|\langle \mathbf{x}, \mathbf{e} \rangle| \leq \|\mathbf{x}\|\,\sqrt{n}$. This is the same as saying that $n\,|\overline{x}| \leq \|\mathbf{x}\|\,\sqrt{n}$, or $\sqrt{n}\,|\overline{x}| \leq \|\mathbf{x}\|$. Here equality occurs if and only if $\mathbf{x} = \lambda\mathbf{e}$ for some $\lambda \in \mathbb{R}$. Hence

$$\text{'determinant} = 0\text{'} \Leftrightarrow \|\mathbf{x}\| = \sqrt{n}\,|\overline{x}| \Leftrightarrow \mathbf{x} = \lambda\mathbf{e} \Leftrightarrow x_1 = \cdots = x_n.$$

□

Exercise 4

We are starting from a theoretical relation $y = \alpha + \beta x$ and n observed points $(x_1, y_1), \ldots, (x_n, y_n)$. Solving the normal equations we get the following expressions for $\hat{\alpha}$ and $\hat{\beta}$:

$$\hat{\beta} = \frac{n(\sum_i x_i y_i) - (\sum_i x_i)(\sum_i y_i)}{n(\sum_i x_i^2) - (\sum_i x_i)^2} \quad \text{and} \quad \hat{\alpha} = \frac{\sum_i y_i - \hat{\beta}\sum_i x_i}{n}.$$

Prove that $\hat{\beta} = S_{\mathbf{xy}}/S_{\mathbf{xx}}$ and $\hat{\alpha} = \overline{y} - \hat{\beta}\,\overline{x}$.

Proof. It is immediate that $\hat{\alpha} = \overline{y} - \hat{\beta}\,\overline{x}$. Still to be proved is the fact that $\hat{\beta} = S_{\mathbf{xy}}/S_{\mathbf{xx}}$. To see this, we write:

$$S_{\mathbf{xy}} = \textstyle\sum_i (x_i - \overline{x})(y_i - \overline{y}) = \sum_i x_i y_i - 2n\overline{x}\,\overline{y} + n\overline{x}\,\overline{y} = \sum_i x_i y_i - n\overline{x}\,\overline{y} \quad \text{and}$$

$$S_{\mathbf{xx}} = \textstyle\sum_i (x_i - \overline{x})^2 = \sum_i x_i^2 - 2\overline{x} \cdot n\overline{x} + n\overline{x}\,\overline{y} = \sum_i x_i^2 - n\overline{x}^2.$$

Now the rest is trivial.

□

Exercise 5

We start from the theoretical relation $y = Rx$, where x is a controlled and y a response variable. In order to get an impression of the magnitude of R we are dealing with the estimator $\hat{R} = \sum_i x_i Y_i / \sum_i x_i^2$ for R.
Prove that

(i) $\hat{R}$ is an estimator that is linear in $Y_1, \dots, Y_n$,

(ii) $\hat{R}$ is an unbiased estimator of R,

(iii) $\mathrm{var}(\hat{R}) = \sigma^2 / \sum_i x_i^2$,

(iv) Among the unbiased linear estimators of R the variable $\hat{R}$ is the one of minimal variance.

Proof.

(i) The variable $\hat{R}$ is linear in $Y_1, \dots, Y_n$ as we can write

$$\hat{R} = \sum_{i=1}^{n} \frac{x_i}{\|\mathbf{x}\|^2} Y_i.$$

(ii) The variable $\hat{R}$ is an unbiased estimator of R because

$$\mathbb{E}(\hat{R}) = \sum_{i=1}^{n} \frac{x_i}{\|\mathbf{x}\|^2} \mathbb{E}(Y_i) = \sum_{i=1}^{n} R \frac{x_i^2}{\|\mathbf{x}\|^2} = R.$$

(iii) The variance of $\hat{R}$ equals $\mathrm{var}(\hat{R}) = \sigma^2 / \sum_i x_i^2$ as

$$\mathrm{var}(\hat{R}) = \mathrm{var}\left(\sum_{i=1}^{n} \frac{x_i}{\|\mathbf{x}\|^2} Y_i\right) = \sum_{i=1}^{n} \frac{x_i^2}{\|\mathbf{x}\|^4} \mathrm{var}(Y_i) = \frac{\sigma^2 \|\mathbf{x}\|^2}{\|\mathbf{x}\|^4} = \frac{\sigma^2}{\|\mathbf{x}\|^2}.$$

(iv) To prove property (iv), let $\tilde{R}$ be any linear unbiased estimator of R, say $\tilde{R} = \sum_i c_i Y_i$. By unbiasedness we have $\mathbb{E}(\tilde{R}) = R$: this implies that $\sum_i c_i (R x_i) = R \sum_i c_i x_i = R$, that is: $\sum_i c_i x_i = 1$. Defining $f : \mathbb{R}^n \to \mathbb{R}$ by

$$f(c_1, \dots, c_n) := c_1 x_1 + \cdots + c_n x_n - 1,$$

we can also express this by setting $f(c_1, \dots, c_n) = 0$. We have

$$\mathrm{var}(\tilde{R}) = \textstyle\sum_i c_i^2 \sigma^2 = \sigma^2 \sum_i c_i^2;$$

it thus appears we have to maximize $g : \mathbb{R}^n \to \mathbb{R}$ defined by $g(c_1, \dots, c_n) := \sum_i c_i^2$ under the constraint $f(c_1, \dots, c_n) = 0$. If such a maximum is realized in a point $\mathbf{c} \in \mathbb{R}^n$, then (by Lagrange's theorem) we necessarily have

$$(\nabla g)(\mathbf{c}) = \mu (\nabla f)(\mathbf{c}) \quad \text{for some } \mu \in \mathbb{R}.$$

In this case we have $(\nabla f)(\mathbf{c}) = (x_1, \ldots, x_n)$ and $(\nabla g)(\mathbf{c}) = (2c_1, \ldots, 2c_n)$, hence $c_i = (\mu/2)\, x_i$ for all i. Substituting this in the constraint $f(\mathbf{c}) = 0$ we get $(\mu/2) \sum_i x_i^2 = 1$, or $\mu = 2/\|\mathbf{x}\|^2$. Then we have $c_i = x_i/\|\mathbf{x}\|^2$ for all i, and these coefficients are exactly the ones belonging to $\hat{R}$.

So far we have proved that *if* there exists an unbiased linear estimator of minimal variance, then it is $\hat{R}$. The existence problem (frequently ignored in applied mathematics) can be solved as in the proof of Proposition IV.1.3. □

Exercise 6

We start from the same model as in Exercise 5. However, now we use the following estimator of R:

$$\tilde{R} := \frac{1}{n} \sum_{i=1}^{n} Y_i/x_i.$$

(i) Prove that $\tilde{R}$ is an unbiased estimator of R.

Proof. Whatever the value of R may be, we have $\mathbb{E}(Y_i) = Rx_i$. Hence

$$\mathbb{E}(\tilde{R}) = \frac{1}{n} \sum_{i=1}^{n} \frac{1}{x_i} \mathbb{E}(Y_i) = R.$$

□

(ii) Express the variance of $\tilde{R}$ in terms of $n, \sigma, x_1, \ldots, x_n$.

Solution.

$$\operatorname{var}(\tilde{R}) = \frac{1}{n^2} \sum_{i=1}^{n} \frac{1}{x_i^2} \operatorname{var}(Y_i) = \frac{\sigma^2}{n^2} \sum_{i=1}^{n} \frac{1}{x_i^2}.$$

□

(iii) Prove that $\operatorname{var}(\tilde{R}) \geq \operatorname{var}(\hat{R})$ and that equality occurs if and only if $x_1^2 = \cdots = x_n^2$.

Proof. Write $\mathbf{a} := (x_1, \ldots, x_n)$ and $\mathbf{b} := (1/x_1, \ldots, 1/x_n)$; then by the Cauchy–Schwarz inequality we have $\langle \mathbf{a}, \mathbf{b} \rangle^2 \leq \|\mathbf{a}\|^2 \|\mathbf{b}\|^2$. In this particular case this means that $n^2 \leq \|\mathbf{x}\|^2 (\sum_i 1/x_i^2)$: equality occurs if and only if $\mathbf{a} = \lambda \mathbf{b}$ for some scalar λ. The latter occurs if and only if $x_1^2 = \cdots = x_n^2$.

We have proved that for all $\mathbf{x}$

$$\left(\frac{1}{n^2} \sum_{i=1}^{n} \frac{1}{x_i^2} \right) \sigma^2 \geq \frac{1}{\|\mathbf{x}\|^2} \sigma^2,$$

where equality occurs if and only if $x_1^2 = \cdots = x_n^2$. This is exactly what we wanted to prove. □

(iv) *Question.* Which estimator do you think is better?

Solution. In most cases we have $\mathrm{var}(\tilde{R}) > \mathrm{var}(\hat{R})$. Therefore estimations made by $\hat{R}$ can be expected to be more accurate than ones made by $\tilde{R}$. For this reason we prefer $\hat{R}$. □

Exercise 7

Prove that, under the central assumptions 1, 2 and 3 of §IV.1, we have:

$$\mathrm{var}(\hat{\alpha}) = \frac{(S_{\mathbf{xx}} + n(\overline{x})^2)\sigma^2}{nS_{\mathbf{xx}}}.$$

Proof. Under the assumptions just mentioned we can write (see also Proposition IV.1.2):

$$\begin{aligned}
\mathrm{var}(\hat{\alpha}) &= \mathrm{var}(\overline{Y} - \hat{\beta}\,\overline{x}) = \mathrm{var}\left(\sum_{i=1}^{n}\left(\frac{1}{n} - \overline{x}\left(\frac{x_i - \overline{x}}{S_{\mathbf{xx}}}\right)\right)Y_i\right) \\
&= \sum_{i=1}^{n}\left(\frac{1}{n} - \overline{x}\left(\frac{x_i - \overline{x}}{S_{\mathbf{xx}}}\right)\right)^2 \mathrm{var}(Y_i) = \sigma^2\sum_{i=1}^{n}\left(\frac{S_{\mathbf{xx}} - n\overline{x}(x_i - \overline{x})}{nS_{\mathbf{xx}}}\right)^2 \\
&= \frac{\sigma^2}{n^2S_{\mathbf{xx}}^2}\sum_{i=1}^{n}\left(S_{\mathbf{xx}}^2 - 2n\overline{x}S_{\mathbf{xx}}(x_i - \overline{x}) + n^2\overline{x}^2(x_i - \overline{x})^2\right) \\
&= \frac{\sigma^2}{n^2S_{\mathbf{xx}}^2}\left\{nS_{\mathbf{xx}}^2 - 2n\overline{x}S_{\mathbf{xx}}\sum_{i=1}^{n}(x_i - \overline{x}) + n^2\overline{x}^2\sum_{i=1}^{n}(x_i - \overline{x})^2\right\} \\
&= \frac{\sigma^2}{n^2S_{\mathbf{xx}}^2}\left\{nS_{\mathbf{xx}}^2 + n^2\overline{x}^2S_{\mathbf{xx}}\right\} = \frac{(S_{\mathbf{xx}} + n(\overline{x})^2)\sigma^2}{nS_{\mathbf{xx}}}.
\end{aligned}$$

□

Exercise 8

Prove that in the case of normal regression analysis the variable

$$\frac{(\hat{Y}_x - \mu_x)\sqrt{n(n-2)S_{\mathbf{xx}}}}{\sqrt{\mathrm{SSE}}\sqrt{S_{\mathbf{xx}} + n(x - \overline{x})^2}}$$

is t-distributed with $n-2$ degrees of freedom.

Proof. In normal regression analysis $\hat{Y}_x$ is normally distributed, being a non-trivial combination of independent normally distributed variables. By Proposition IV.1.4 the variable $\hat{Y}_x$ is an unbiased estimator of μ_x and furthermore

$$\mathrm{var}(\hat{Y}_x) = \left(\frac{1}{n} + \frac{(x - \overline{x})^2}{S_{\mathbf{xx}}}\right)\sigma^2.$$

It follows from this that

$$\frac{\hat{Y}_x - \mu_x}{\sigma\sqrt{\frac{1}{n} + \frac{(x-\overline{x})^2}{S_{\mathbf{xx}}}}} = \frac{(\hat{Y}_x - \mu_x)\sqrt{nS_{\mathbf{xx}}}}{\sigma\sqrt{S_{\mathbf{xx}} + n(x-\overline{x})^2}} \qquad (*)$$

is $N(0,1)$-distributed.

By Theorem IV.3.2 the variable SSE/σ^2 is χ^2-distributed with $n-2$ degrees of freedom. Applying Theorem IV.3.2 (i) together with Proposition I.4.2, we see that SSE/σ^2 and the statistic in $(*)$ are statistically independent.

By Proposition II.3.1 the variable

$$\sqrt{n-2}\left(\frac{(\hat{Y}_x - \mu_x)\sqrt{nS_{\mathbf{xx}}}}{\sigma\sqrt{S_{\mathbf{xx}} + n(x-\overline{x})^2}}\right) \Big/ \sqrt{\mathrm{SSE}/\sigma^2}$$

is apparently t-distributed with $n-2$ degrees of freedom. □

Exercise 9

Given is a set of points $(1,1),(2,1),(2,3),(3,2),(4,3)$.

(i) Compute the numerical value of $\hat{\alpha}$ and $\hat{\beta}$.

Solution. Firstly we have $\overline{x} = \frac{1}{5}(1+2+2+3+4) = \frac{12}{5}$. Furthermore

$$S_{\mathbf{xx}} = \sum_{i=1}^{5}(x_i - \overline{x})^2 = \tfrac{1}{25}(49+4+4+9+64) = \tfrac{130}{25} = \tfrac{26}{5}.$$

Hence

$$\begin{aligned}\hat{\beta} &= \left(\sum_{i=1}^{5}(x_i - \overline{x})Y_i\right) \Big/ S_{\mathbf{xx}} \\ &= \tfrac{5}{26}\left(-\tfrac{7}{5}\cdot 1 - \tfrac{2}{5}\cdot 1 - \tfrac{2}{5}\cdot 3 + \tfrac{3}{5}\cdot 2 + \tfrac{8}{5}\cdot 3\right) = \tfrac{15}{26}.\end{aligned}$$

The value of $\hat{\alpha}$ is given by $\hat{\alpha} = \overline{Y} - \hat{\beta}\overline{x} = \frac{1}{5}(1+1+3+2+3) - \frac{15}{26}\cdot\frac{12}{5} = 2 - \frac{18}{13} = \frac{8}{13}$. Note that $\overline{Y} = 2$. □

(ii) Determine the error for $x = 3$.

Solution. The error for $x = 3 = x_4$ is $e_{x_4} = 2 - (\frac{8}{13} + \frac{15}{26}\cdot 3) = -\frac{9}{26}$. □

(iii) Compute the coefficient of determination, that is, the fraction of explained variance $\mathrm{SSR}/S_{\mathbf{yy}}$.

Solution. First we note that $S_{\mathbf{yy}} = \sum_{i=1}^{5}(y_i - \overline{y})^2 = 1+1+1+0+1 = 4$. Moreover

$$\begin{aligned}\mathrm{SSR} &= \sum_{i=1}^{5}(\hat{y}_i - \overline{y})^2 = \textstyle\sum_{i=1}^{5}\left(\frac{8}{13} + \frac{15}{26}x_i - 2\right)^2 \\ &= \left(\tfrac{-21}{26}\right)^2 + 2\cdot\left(\tfrac{-6}{26}\right)^2 + \left(\tfrac{9}{26}\right)^2 + \left(\tfrac{24}{26}\right)^2 = \tfrac{45}{26},\end{aligned}$$

so the fraction explained variance is given by $\mathrm{SSR}/S_{\mathbf{yy}} = \frac{45}{26}/4 = \frac{45}{104}$. □

(iv) From now on we assume that normal regression analysis can be applied. Make a point estimation of σ^2.

Solution. SSE$/(n-2)$ is (Theorem IV.2.3) an unbiased estimator of σ^2 which in our case provides us with the value SSE$/3 = \frac{1}{3}(S_{\mathbf{yy}} - \text{SSR}) = \frac{1}{3}(4 - \frac{45}{26}) = \frac{59}{78} = 0.7564$ as a point estimation for σ^2. □

(v) Construct a 90% confidence interval for σ.

Solution. By Theorem IV.3.2 the statistic SSE$/\sigma^2$ is χ^2-distributed with 3 degrees of freedom. Using Table IV, describing the χ^2-distribution, we find that there is (a priori) a 90% probability that SSE$/\sigma^2$ will assume a value in the interval (0.352, 7.815). Substituting the value SSE $= \frac{59}{26}$, we get $\sigma^2 \in (0.29, 6.45)$. Thus (0.54, 2.54) appears as a 90% confidence interval for σ. □

(vi) Construct a 90% confidence interval for μ_x in the point $x = 5$.

Solution. In Exercise 8 we have seen that the statistic

$$\frac{(\hat{Y}_x - \mu_x)\sqrt{n(n-2)S_{\mathbf{xx}}}}{\sqrt{\text{SSE}}\sqrt{S_{\mathbf{xx}} + n(x - \overline{x})^2}}$$

is t-distributed with $n - 2$ degrees of freedom. Consequently, this statistic will (a priori) assume an outcome in the interval $(-2.353, 2.353)$ with a probability of 90%. We compute:

$$\begin{aligned}\sqrt{\text{SSE}}\sqrt{S_{\mathbf{xx}} + n(x - \overline{x})^2} &= \sqrt{\tfrac{59}{26}}\sqrt{\tfrac{26}{5} + 5\left(5 - \tfrac{12}{5}\right)^2} \\ &= \sqrt{\tfrac{59}{26}}\sqrt{\tfrac{26+13^2}{5}} = \sqrt{\tfrac{59}{26}}\sqrt{39} = \sqrt{\tfrac{3\cdot 59}{2}}.\end{aligned}$$

So with a probability of 90% we have

$$-2.353 < \sqrt{\tfrac{2}{3\cdot 59}}\sqrt{5\cdot 3\cdot \tfrac{26}{5}}\left(\hat{Y}_x - \mu_x\right) < 2.353 \qquad (*)$$

The variable $\hat{Y}_x$ shows an outcome given by $\hat{Y}_x = \hat{\alpha} + \hat{\beta}x = \frac{8}{13} + \frac{15}{26}\cdot 5 = \frac{91}{26} = \frac{7}{2}$. Substituting this outcome in $(*)$ we get

$$-2.353 < \tfrac{2}{59}\sqrt{13\cdot 59}\left(\tfrac{7}{2} - \mu_x\right) < 2.353.$$

We thus obtain for μ_x the inequality $1.0 < \mu_x < 6.0$. In this way the interval (1.0, 6.0) emerges as a 90% confidence interval for μ_x. □

Exercise 10

Prove Lemma IV.4.1.

Proof. By definition, $S_{\mathbf{xy}} = \sum_i (x_i - \overline{x})(y_i - \overline{y})$. It is therefore immediate that:

(i) $S_{\mathbf{yx}} = \sum_i (y_i - \overline{y})(x_i - \overline{x}) = S_{\mathbf{xy}}$,

(ii) $S_{\mathbf{xx}} = \sum_i (x_i - \overline{x})^2 \geq 0$,

(iii) $S_{\mathbf{xx}} = 0 \Leftrightarrow x_i = \overline{x}$ for all $i \Leftrightarrow \mathbf{x} = \overline{x}\mathbf{e}$,

(iv) for all $\lambda, \mu \in \mathbb{R}$ we have

$$S_{\mathbf{x}+\lambda\mathbf{e},\mathbf{y}+\mu\mathbf{e}} = \textstyle\sum_i ((x_i + \lambda) - (\overline{x + \lambda e}))((y_i + \mu) - (\overline{y + \mu e}))$$

and

$$\overline{x + \lambda e} = \frac{1}{n} \textstyle\sum_i (x_i + \lambda) = \overline{x} + \lambda \quad \text{and} \quad \overline{y + \mu e} = \overline{y} + \mu.$$

Therefore

$$S_{\mathbf{x}+\lambda\mathbf{e},\mathbf{y}+\mu\mathbf{e}} = \textstyle\sum_i (x_i - \overline{x})(y_i - \overline{y}) = S_{\mathbf{xy}}.$$

□

Exercise 11

We assume that we are in a case where normal regression analysis can be applied.

(i) *Question.* Are $\overline{Y}$ and $\hat{\beta}$ statistically independent?

Answer. Yes. Namely, let $\mathfrak{V} := \{\sum_{i=1}^{n} c_i Y_i \ : \ c_i \in \mathbb{R}\}$. Then of course $\overline{Y} \in \mathfrak{V}$ and (Proposition IV.1.2) also $\hat{\beta} \in \mathfrak{V}$. To prove statistical independence of two elements in $\mathfrak{V}$ it is (Theorem I.6.7) sufficient to verify that their covariance is zero. For the variables $\overline{Y}$ and $\hat{\beta}$ we have:

$$\begin{aligned}
\operatorname{cov}(\overline{Y}, \hat{\beta}) &= \frac{1}{n} \sum_i \operatorname{cov}(Y_i, \hat{\beta}) = \frac{1}{n} \sum_i \operatorname{cov}\left(Y_i, \sum_j \left(\frac{x_j - \overline{x}}{S_{\mathbf{xx}}}\right) Y_j\right) \\
&= \frac{1}{n} \sum_{i,j} \left(\frac{x_j - \overline{x}}{S_{\mathbf{xx}}}\right) \operatorname{cov}(Y_i, Y_j) = \frac{1}{n} \sum_i \left(\frac{x_i - \overline{x}}{S_{\mathbf{xx}}}\right) \operatorname{var}(Y_i) \\
&= \frac{\sigma^2}{n S_{\mathbf{xx}}} \sum_i (x_i - \overline{x}) = 0.
\end{aligned}$$

□

(ii) *Question.* Are $\overline{Y}$ and $\hat{\alpha}$ statistically independent?

Answer. No. It is easily verified that $\operatorname{cov}(\overline{Y}, \hat{\alpha}) \neq 0$, namely:

$$\begin{aligned}
\operatorname{cov}(\overline{Y}, \hat{\alpha}) &= \operatorname{cov}(\overline{Y}, \overline{Y} - \hat{\beta}\overline{x}) = \operatorname{var}(\overline{Y}) - \overline{x} \operatorname{cov}(\overline{Y}, \hat{\beta}) \\
&= \operatorname{var}(\overline{Y}) = \frac{1}{n^2} \cdot n\sigma^2 = \frac{\sigma^2}{n} \neq 0.
\end{aligned}$$

By Proposition I.5.7 this implies that $\overline{Y}$ and $\hat{\alpha}$ are not statistically independent.

□

Exercise 12

Prove that in the case of normal regression analysis we have

$$\operatorname{var}\left(\frac{\text{SSE}}{n-2}\right) = \frac{2\sigma^4}{n-2}.$$

Proof. By Theorem IV.3.2 the variable SSE/σ^2 is χ^2-distributed with $n-2$ degrees of freedom. Therefore, applying Theorem II.2.8, we see that

$$\frac{1}{\sigma^4}\operatorname{var}(\text{SSE}) = \operatorname{var}\left(\frac{\text{SSE}}{\sigma^2}\right) = 2(n-2).$$

Hence

$$\operatorname{var}\left(\frac{\text{SSE}}{n-2}\right) = \frac{1}{(n-2)^2}\,2(n-2)\sigma^4 = \frac{2\sigma^4}{n-2}. \qquad \square$$

Exercise 13

Let $(x_1, y_1), \ldots, (x_n, y_n)$ be an arbitrary cloud of points. Set $\Omega := \{1, \ldots, n\}$ and let $\mathfrak{A}$ be the collection of all subsets of Ω. For all $A \in \mathfrak{A}$ we define $\mathbb{P}(A) := (\#A)/n$. Now $(\Omega, \mathfrak{A}, \mathbb{P})$ is a probability space on which we define the variables $X, Y : \Omega \to \mathbb{R}$ by

$$X(i) := x_i \quad \text{and} \quad Y(i) := y_i.$$

Show that Pearson's product-moment correlation coefficient R_P is the same as the correlation coefficient $\rho(X,Y)$ of the variables X and Y (see §I.5).

Solution. We have $\mathbb{E}(X) = \sum_i x_i \mathbb{P}(X = x_i) = \overline{x}$ and $\mathbb{E}(Y) = \overline{y}$. Furthermore $\mathbb{E}(X^2) = \sum_i x_i^2 \mathbb{P}(X = x_i) = \|\mathbf{x}\|^2/n$ and $\mathbb{E}(Y^2) = \|\mathbf{y}\|^2/n$. It follows that $\operatorname{var}(X) = (\|\mathbf{x}\|^2/n) - \overline{x}^2$ and $\operatorname{var}(Y) = (\|\mathbf{y}\|^2/n) - \overline{y}^2$. Finally $\mathbb{E}(XY) = \sum_{i,j} x_i y_j \mathbb{P}(X = x_i \text{ and } Y = y_j) = (\sum_i x_i y_i)/n$; hence $\operatorname{cov}(X,Y) = (\sum_i x_i y_i)/n - \overline{x}\,\overline{y}$. We learn that

$$\begin{aligned} \rho(X,Y) = \frac{\operatorname{cov}(X,Y)}{\sqrt{\operatorname{var}(X)}\sqrt{\operatorname{var}(Y)}} &= \frac{\sum_i x_i y_i - n\overline{x}\,\overline{y}}{n\sqrt{\frac{1}{n}\|\mathbf{x}\|^2 - \overline{x}^2}\sqrt{\frac{1}{n}\|\mathbf{y}\|^2 - \overline{y}^2}} \\ &= \frac{\sum_i x_i y_i - n\overline{x}\,\overline{y}}{\sqrt{\|\mathbf{x}\|^2 - n\overline{x}^2}\sqrt{\|\mathbf{y}\|^2 - n\overline{y}^2}}. \end{aligned}$$

On the other hand we have

$$R_P = \frac{S_{\mathbf{xy}}}{\sqrt{S_{\mathbf{xx}}}\sqrt{S_{\mathbf{yy}}}},$$

where $S_{\mathbf{xy}} = \sum_i (x_i - \overline{x})(y_i - \overline{y}) = \sum_i x_i y_i - n\overline{x}\,\overline{y}$, $S_{\mathbf{xx}} = \|\mathbf{x}\|^2 - n\overline{x}^2$ and $S_{\mathbf{yy}} = \|\mathbf{y}\|^2 - n\overline{y}^2$. Altogether we see that $\rho(X,Y) = R_P$. $\square$

Exercise 14

(i) Find the maximum likelihood estimator of σ^2, under the assumptions of normal regression analysis.

Solution. Here we have to maximize the function

$$f_0 : (\alpha, \beta, \sigma) \mapsto \frac{1}{(2\pi)^{\frac{1}{2}n}\, \sigma^n} \exp\left[-\frac{1}{2} \frac{\sum_i (y_i - (\alpha + \beta x_i))^2}{\sigma^2} \right]$$

for $\alpha, \beta \in \mathbb{R}$ and $\sigma > 0$. By construction $(\hat{\alpha}, \hat{\beta})$ is the best choice for (α, β), so we have to maximize the function

$$f : \sigma \mapsto \sigma^{-n} k_1 \exp\{-k_2 \sigma^{-2}\}$$

on the domain $(0, +\infty)$, where

$$k_1 = (2\pi)^{-\frac{1}{2}n} > 0 \quad \text{and} \quad k_2 = \tfrac{1}{2} \textstyle\sum_i (y_i - (\hat{\alpha} + \hat{\beta} x_i))^2 \geq 0.$$

If $k_2 = 0$ then the function $\sigma \mapsto \sigma^{-n} k_1$ appears; in that case there is no maximum. If $k_2 > 0$, then it is easier to maximize the function $\log f$, that is: the function

$$g : \sigma \mapsto \log k_1 - n \log \sigma - k_2 \sigma^{-2}.$$

For g we have $g'(\sigma) = -n/\sigma + 2k_2/\sigma^3$; this expression equals 0 if $2k_2/\sigma^3 = n/\sigma$. It follows that for $\sigma^2 = 2k_2/n = \frac{1}{n} \sum_i (y_i - (\hat{\alpha} + \hat{\beta} x_i))^2$ there is a maximum (check this). Consequently the maximum likelihood estimator for σ^2 is given by

$$\frac{1}{n} \sum_i (y_i - (\hat{\alpha} + \hat{\beta} x_i))^2 = \frac{\text{SSE}}{n}.$$

□

(ii) *Question.* Is the maximum likelihood estimator of σ^2 also unbiased?

Answer. No. By Theorem IV.2.3 the statistic $\text{SSE}/(n-2)$ is an unbiased estimator of σ^2. It follows from this that the maximum likelihood estimator given in (i) is not. □

Exercise 15

Under the assumptions of normal regression analysis the variable Y_x is $N(\alpha + \beta x, \sigma^2)$-distributed. Moreover, given a fixed sequence $x_1, \ldots, x_n$ the variables $Y_{x_1}, \ldots, Y_{x_n}$ constitute a statistically independent system.

Now consider the case of $m + n$ response variables, $Y_1 = Y_{x_1}, \ldots, Y_{m+n} = Y_{x_{m+n}}$, where $x_1 = \cdots = x_m = 0$ and $x_{m+1} = \cdots = x_{m+n} = 1$.

(i) Prove that SSE $= (m+n-2)S_p^2$ where S_p^2 is the pooled variance of the samples $Y_1, \ldots, Y_m$ and $Y_{m+1}, \ldots, Y_{m+n}$.

Proof. As

$$\mathrm{SSE} = \sum_i (Y_i - \hat{Y}_i)^2 = \sum_i (Y_i - (\hat{\alpha} + \hat{\beta} x_i))^2$$

we first deduce closed expressions for $\hat{\alpha}$ and $\hat{\beta}$. We denote

$$V_1 := Y_1, \ \ldots, V_m := Y_m \quad \text{and} \quad W_1 := Y_{m+1}, \ \ldots, W_n := Y_{m+n}.$$

Then

$$\overline{V} = \frac{1}{m} \sum_{i=1}^{m} Y_i \quad \text{and} \quad \overline{W} = \frac{1}{n} \sum_{i=m+1}^{m+n} Y_i.$$

Next, we note that $(m+n-2)\, S_p^2 = (m-1) S_V^2 + (n-1) S_W^2$ where S_V^2 and S_W^2 denote the sample variances:

$$S_V^2 = \frac{1}{m-1} \sum_{i=1}^{m} (V_i - \overline{V})^2 \quad \text{and} \quad S_W^2 = \frac{1}{n-1} \sum_{j=1}^{n} (W_j - \overline{W})^2.$$

It is clear that $\overline{x} = \frac{n}{m+n}$. Moreover,

$$S_{\mathbf{xx}} = \sum_{i=1}^{m+n} (x_i - \overline{x})^2 = m(0 - \tfrac{n}{m+n})^2 + n(1 - \tfrac{n}{m+n})^2 = \tfrac{mn}{m+n}.$$

From this (see Proposition IV.1.2) we get that

$$\begin{aligned}
\hat{\beta} &= \sum_{i=1}^{m+n} (x_i - \overline{x}) Y_i / S_{\mathbf{xx}} \\
&= \frac{m+n}{mn} \left[\sum_{i=1}^{m} \left(-\frac{n}{m+n} \right) Y_i + \sum_{i=m+1}^{m+n} \left(1 - \frac{n}{m+n} \right) Y_i \right] \\
&= \frac{m+n}{mn} \left[-\frac{mn}{m+n} \overline{V} + \frac{mn}{m+n} \overline{W} \right] = \overline{W} - \overline{V}.
\end{aligned}$$

Because $\overline{Y} = \frac{m}{m+n} \overline{V} + \frac{n}{m+n} \overline{W}$ (check this) we may write

$$\hat{\alpha} = \overline{Y} - \hat{\beta} \overline{x} = \frac{m}{m+n} \overline{V} + \frac{n}{m+n} \overline{W} - \frac{n}{m+n} (\overline{W} - \overline{V}) = \overline{V}.$$

Now we are able to compute SSE:

$$\begin{aligned}
\mathrm{SSE} &= \sum_{i=1}^{m} (V_i - \overline{V})^2 + \sum_{i=m+1}^{m+n} (Y_i - (\overline{V} + (\overline{W} - \overline{V})))^2 \\
&= \sum_{i=1}^{m} (V_i - \overline{V})^2 + \sum_{j=1}^{n} (W_j - \overline{W})^2 = (m-1) S_V^2 + (n-1) S_W^2.
\end{aligned}$$

It follows that $\mathrm{SSE} = (m+n-2)S_p^2$. □

(ii) Under the conditions mentioned above, Theorem IV.3.3 states that the variable

$$\frac{(\hat{\beta}-\beta)\sqrt{(m+n-2)S_{\mathbf{xx}}}}{\sqrt{\mathrm{SSE}}} \qquad (*)$$

is t-distributed with $m+n-2$ degrees of freedom.

Moreover we have Theorem II.4.4: suppose $V_1, \ldots, V_m$ and $W_1, \ldots, W_n$ are independent samples from populations that are $N(\mu_V, \sigma^2)$ and $N(\mu_W, \sigma^2)$-distributed respectively. Then the statistic

$$\frac{(\overline{V}-\overline{W})-(\mu_V-\mu_W)}{S_p\sqrt{\frac{1}{m}+\frac{1}{n}}}$$

is t-distributed with $m+n-2$ degrees of freedom.

Deduce Theorem II.4.4 from Theorem IV.3.3.

Solution. We keep the notations introduced in (i). As before we denote

$$Y_1 := V_1, \ \ldots, \ Y_m := V_m, \ Y_{m+1} := W_1, \ \ldots, \ Y_{m+n} := W_n.$$

In these notations the systems

$$Y_1, \ldots, Y_m \quad \text{and} \quad Y_{m+1}, \ldots, Y_{m+n}$$

constitute two independent samples from a $N(\alpha, \sigma^2)$ and a $N(\alpha+\beta, \sigma^2)$-distributed population respectively, where $\alpha := \mu_V$ and $\beta := \mu_W - \mu_V$. Furthermore, the variable in $(*)$ assumes (see (i)) the form:

$$\frac{\{(\overline{W}-\overline{V})-(\mu_W-\mu_V)\}\ \sqrt{(m+n-2)\cdot\frac{mn}{m+n}}}{\sqrt{m+n-2}\ S_p}.$$

In turn this equals

$$\frac{(\overline{W}-\overline{V})-(\mu_W-\mu_V)}{S_p\ \sqrt{\frac{1}{m}+\frac{1}{n}}} = -\frac{(\overline{V}-\overline{W})-(\mu_V-\mu_W)}{S_p\sqrt{\frac{1}{m}+\frac{1}{n}}},$$

which is thus t-distributed with $m+n-2$ degrees of freedom. As the t-distribution is symmetric around the origin, the proof is complete. □

Exercise 16

A function of type

$$f : (a,b) \ \mapsto \ pa^2 + 2qab + rb^2 + sa + tb + u$$

where p, q, r, s, t, u are constants, is called a *quadratic form* in a and b. It is called *degenerate* if $pr - q^2 = 0$.

(i) Concerning the quadratic form $(a,b) \mapsto \sum_i \{y_i - (a + bx_i)\}^2$ in §IV.1 there is a 'missing link' in the argument. What is it ?

Solution. In §IV.1 it is taken for granted that this quadratic form attains a minimal value. When applying Lagrange's theorem in a proper way this should be proved beforehand. □

(ii) For f as described above there exists an orthogonal linear map $\mathbf{Q} : \mathbb{R}^2 \to \mathbb{R}^2$ and constants $c, d, e, \lambda_1, \lambda_2$ such that

$$f(\mathbf{Q}(x,y)) = \lambda_1(x-c)^2 + \lambda_2(y-d)^2 + e.$$

Prove this.

Proof. There exists a symmetric linear operator $\mathbf{A} : \mathbb{R}^2 \to \mathbb{R}^2$ for which $f(a,b) = \langle (a,b), \mathbf{A}(a,b)\rangle + \langle (a,b),(s,t)\rangle + u$, namely the operator corresponding to the matrix

$$[\mathbf{A}] := \begin{pmatrix} p & q \\ q & r \end{pmatrix}.$$

Elementary linear algebra guarantees the existence of an orthogonal linear map $\mathbf{Q} : \mathbb{R}^2 \to \mathbb{R}^2$ and a linear map $\mathbf{\Lambda}$ such that $[\mathbf{\Lambda}] = \mathrm{diag}(\lambda_1, \lambda_2)$ (where $\lambda_1, \lambda_2 \in \mathbb{R}$) for which $\mathbf{A} = \mathbf{Q}\mathbf{\Lambda}\mathbf{Q}^*$. Thus

$$\begin{aligned} f(\mathbf{Q}(x,y)) &= \langle \mathbf{Q}(x,y), \mathbf{Q}\mathbf{\Lambda}\mathbf{Q}^*\mathbf{Q}(x,y)\rangle + \langle \mathbf{Q}(x,y),(s,t)\rangle + u \\ &= \langle \mathbf{Q}^*\mathbf{Q}(x,y), \mathbf{\Lambda}(x,y)\rangle + \langle (x,y), \mathbf{Q}^*(s,t)\rangle + u \\ &= \langle (x,y), \mathbf{\Lambda}(x,y)\rangle + \langle (x,y), \mathbf{Q}^*(s,t)\rangle + u \\ &= \lambda_1 x^2 + \lambda_2 y^2 + [\mathbf{Q}^*(s,t)]_1 x + [\mathbf{Q}^*(s,t)]_2 y + u. \end{aligned}$$

The required form for f is easily deduced from this. □

(iii) If $f \geq 0$ then f attains a global minimum in at least one point (a,b). Prove this.

Proof. By (ii) we have an orthogonal linear operator $\mathbf{Q}$ such that

$$f(\mathbf{Q}(x,y)) = \lambda_1(x-c)^2 + \lambda_2(y-d)^2 + e.$$

From the fact that $f \geq 0$ it follows that $\lambda_1, \lambda_2 \geq 0$ (check this). Therefore

$$f(\mathbf{Q}(x,y)) \geq e \quad \text{for all } (x,y).$$

We now have $f \geq e$ as for all (a,b)

$$f(a,b) = f(\mathbf{Q}\,[\mathbf{Q}^*(a,b)]) \geq e.$$

As $f(\mathbf{Q}(c,d)) = e$ the statement is correct. □

(iv) Show that f is degenerate if and only if the determinant of the system of linear equations

$$\frac{\partial}{\partial a} f(a,b) = 0 \quad \text{and} \quad \frac{\partial}{\partial b} f(a,b) = 0$$

is zero.

Solution. The normal equations just mentioned lead us to the system

$$\begin{cases} 2pa + 2qb + s &= 0, \\ 2qa + 2rb + t &= 0. \end{cases}$$

In matrix notation:
$$\begin{pmatrix} 2p & 2q \\ 2q & 2r \end{pmatrix} \begin{pmatrix} a \\ b \end{pmatrix} = \begin{pmatrix} -s \\ -t \end{pmatrix}$$

The corresponding determinant equals zero if and only if $2p \cdot 2r - 2q \cdot 2q = 0$, that is, if and only if f is degenerate. □

(v) We assume that among $x_1, \dots, x_n$ there are at least two distinct values. Prove that there is a unique point (a,b) where the map $f : (a,b) \mapsto \sum_i \{y_i - (a + bx_i)\}^2$ attains a global minimum.

Proof. It is clear that f is a quadratic form in a and b for which $f \geq 0$. By (iii) there exists at least one point in $\mathbb{R}^2$ where f attains a global minimum. Using Exercise 3 together with (iv) we see that such a minimizing point is necessarily unique. □

Exercise 17

There is a missing link in the proof of Proposition IV.1.3. Trace it and repair it.

Solution. The situation is as follows. For fixed $\sigma^2 > 0$ and $\mathbf{x} = (x_1, \dots, x_n) \in \mathbb{R}^n$ (the controlled variables) we have to minimize the function

$$f : (c_1, \dots, c_n) \mapsto \sigma^2 \sum_i c_i^2 = \sigma^2 \|\mathbf{c}\|^2.$$

This has to be done under the constraints that

$$g_1(c_1, \dots, c_n) := \sum_i c_i = \langle \mathbf{c}, \mathbf{e} \rangle = 0$$

and

$$g_2(c_1, \dots, c_n) := \sum_i c_i x_i = \langle \mathbf{c}, \mathbf{x} \rangle = 1$$

(where $\mathbf{e} = (1, \dots, 1) \in \mathbb{R}^n$).

In order to apply Lagrange's theorem in a proper way, we need to know *a priori* that a minimal value is actually attained: this is the missing link.

Next we try to repair it. First we verify the existence of a $\mathbf{c} \in \mathbb{R}^n$ such that $g_1(\mathbf{c}) = 0$ and $g_2(\mathbf{c}) = 1$. If $x_1 = \cdots = x_n$ this is clearly impossible (we would get $\sum_i c_i = 0$ and $\sum_i c_i = 1$ as constraints). Suppose $x_i \neq x_j$ for some pair (i, j). We are looking for a $\mathbf{c} \in \mathbb{R}^n$ for which $\mathbf{Ac} = (0, 1)$, where $\mathbf{A} : \mathbb{R}^n \to \mathbb{R}^2$ is the linear operator belonging to the matrix

$$[\mathbf{A}] = \begin{pmatrix} 1 & \cdots & 1 \\ x_1 & \cdots & x_n \end{pmatrix}.$$

The rank of $\mathbf{A}$ is 2 , by which we may draw the conclusion that a convenient $\mathbf{c}$ really exists.

We take such a $\mathbf{c}_0 \in \mathbb{R}^n$ and set $K := f(\mathbf{c}_0) = \sigma^2 \|\mathbf{c}_0\|^2$. At this stage it doesn't make sense to consider $\mathbf{c} \notin \{\mathbf{y} \in \mathbb{R}^n \ : \ \|\mathbf{y}\|^2 \leq K/\sigma^2\}$. So the domain in which we are interested becomes $A_1 \cap A_2 \cap A_3$, where

$$\begin{aligned} A_1 &:= \{\mathbf{c} \in \mathbb{R}^n \ : \ \|\mathbf{c}\| \leq \sqrt{K}/\sigma\}, \\ A_2 &:= \{\mathbf{c} \in \mathbb{R}^n \ : \ \langle \mathbf{c}, \mathbf{e} \rangle = 0\} = g_1^{-1}(0), \\ A_3 &:= \{\mathbf{c} \in \mathbb{R}^n \ : \ \langle \mathbf{c}, \mathbf{x} \rangle = 1\} = g_2^{-1}(1). \end{aligned}$$

Here g_1 and g_2 are linear operators: they are therefore continuous. It follows that A_1, A_2 and A_3 are all closed in $\mathbb{R}^n$, and so is $A_1 \cap A_2 \cap A_3$.

As f is continuous and $A_1 \cap A_2 \cap A_3$ compact (bounded and closed), f attains a minimal value on $(A_1 \cap) A_2 \cap A_3$. This completes the proof. □

Exercise 18

In this exercise we return to the scientist in the introduction of §IV.1: see Exercise 5. In our model we additionally assume that for all i the variable Y_i enjoys a $N(Rx_i, \sigma^2)$-distribution.

(i) Determine the probability distribution of the estimator $\hat{R}$.

Solution. We know (see Exercise 5) that

$$\mathbb{E}(\hat{R}) = R \quad \text{and} \quad \text{var}(\hat{R}) = \sigma^2/\|\mathbf{x}\|^2.$$

Moreover, being a non-trivial linear combination of independent normally distributed variables, $\hat{R}$ is itself normally distributed (combine the Theorems I.6.1 and I.6.6). Hence $\hat{R}$ must be $N(R, \sigma^2/\|\mathbf{x}\|^2)$-distributed. □

(ii) We define the variable T by

$$T := \sum_{i=1}^{n} (Y_i - \hat{R}x_i)^2.$$

Prove that $\hat{R}$ and T are statistically independent.

Proof. Define $\mathfrak{V}$ as the linear space spanned by the variables $Y_1, \ldots, Y_n$. Then, of course,

$$\hat{R} \in \mathfrak{V} \quad \text{and} \quad Y_j - \hat{R}x_j \in \mathfrak{V} \quad \text{for all } j.$$

Furthermore we have:

$$\begin{aligned} \operatorname{cov}(\hat{R}, Y_j - \hat{R}x_j) &= \operatorname{cov}(\hat{R}, Y_j) - x_j \operatorname{cov}(\hat{R}, \hat{R}) \\ &= \frac{1}{\|\mathbf{x}\|^2} \sum_i x_i \operatorname{cov}(Y_i, Y_j) - x_j \operatorname{var}(\hat{R}) \\ &= \frac{x_j\, \sigma^2}{\|\mathbf{x}\|^2} - \frac{x_j\, \sigma^2}{\|\mathbf{x}\|^2} = 0. \end{aligned}$$

Applying Theorem I.6.7 we see that the variables

$$\hat{R} \quad \text{and} \quad (Y_1 - \hat{R}x_1, \ldots, Y_n - \hat{R}x_n)$$

are statistically independent. Next we apply Proposition I.4.2 and conclude that $\hat{R}$ and T are independent. □

(iii) Determine the probability distribution of the variable T/σ^2.

Solution. Write

$$\begin{aligned} \frac{1}{\sigma^2} \sum_{i=1}^n (Y_i - Rx_i)^2 &= \frac{1}{\sigma^2} \sum_{i=1}^n \left\{ (Y_i - \hat{R}x_i) + (\hat{R}x_i - Rx_i) \right\}^2 \\ &= \frac{1}{\sigma^2} \sum_{i=1}^n (Y_i - \hat{R}x_i)^2 + \frac{2}{\sigma^2} (\hat{R} - R) \sum_{i=1}^n (Y_i - \hat{R}x_i)\, x_i \\ &\quad + \frac{1}{\sigma^2} (\hat{R} - R)^2 \sum_{i=1}^n x_i^2 \\ &= T/\sigma^2 \ + \ 0 \ + \ Z^2. \end{aligned}$$

Here Z is the $N(0,1)$-distributed variable

$$\frac{\hat{R} - R}{\sqrt{\sigma^2 / \sum_i x_i^2}}$$

which is, by (ii), independent of T/σ^2. Next, using Proposition II.2.10, we conclude that T/σ^2 enjoys a $\chi^2(n-1)$-distribution. □

(iv) Determine the maximum likelihood estimator of the (vectorial) parameter (R, σ^2).

Solution. The likelihood function of our model is given by

$$L_{R,\sigma^2}(y_1, \dots, y_n) = \frac{1}{(2\pi)^{n/2}\,\sigma^n} \exp\left[-\frac{1}{2\sigma^2}\sum_{i=1}^{n}(y_i - Rx_i)^2\right].$$

Note that the parameter space is here given by $\Theta = \mathbb{R} \times (0, +\infty)$. Maximization of $L_{R,\sigma^2}(\mathbf{y})$ as a function of (R, σ^2) certainly requires the minimization of $\sum_i (y_i - Rx_i)^2$ against R. This leads us to the fact that $\hat{R} = \sum_i x_i Y_i / \sum_i x_i^2$ is indeed the maximum likelihood estimator for R. Maximization against the variable σ^2 is now achieved as in Exercise 14: we see that

$$\hat{\sigma}^2 = \frac{1}{n}\sum_{i=1}^{n}(Y_i - \hat{R}x_i)^2,$$

which shows that $\frac{1}{n}T$ is the maximum likelihood estimator for σ^2. □

(v) Is, for fixed $x_1, \dots, x_n$, the vectorial statistic $(\hat{R}, \frac{1}{n}T)$ sufficient for (R, σ^2)?

Solution. Yes, it is. To see this we apply the factorization theorem (Theorem II.10.3). Define the functions $g_1, g_2 : \mathbb{R}^n \to \mathbb{R}$ by

$$g_1(\mathbf{y}) := \frac{\sum_i x_i y_i}{\sum_i x_i^2} \quad \text{and} \quad g_2(\mathbf{y}) := \frac{1}{n}\sum_{i=1}^{n}(y_i - g_1(\mathbf{y})x_i)^2.$$

Setting $\mathbf{g}(\mathbf{y}) := (g_1(\mathbf{y}), g_2(\mathbf{y}))$ we evidently have

$$\mathbf{g}(\mathbf{Y}) = (\hat{R}, \tfrac{1}{n}T).$$

Furthermore,

$$\begin{aligned}
\sum_{i=1}^{n}(y_i - Rx_i)^2 &= \sum_{i=1}^{n}\{(y_i - g_1(\mathbf{y})x_i) + (g_1(\mathbf{y})x_i - Rx_i)\}^2 \\
&= \sum_{i=1}^{n}(y_i - g_1(\mathbf{y})x_i)^2 + 2\sum_{i=1}^{n}(y_i - g_1(\mathbf{y})x_i)(g_1(\mathbf{y}) - R)x_i \\
&\qquad + \sum_{i=1}^{n}(g_1(\mathbf{y})x_i - Rx_i)^2 \\
&= ng_2(\mathbf{y}) + 0 + (g_1(\mathbf{y}) - R)^2\,\|\mathbf{x}\|^2.
\end{aligned}$$

Hence we can write the likelihood function as

$$L_{R,\sigma^2}(\mathbf{y}) = \frac{1}{(2\pi\sigma^2)^{n/2}} \exp\left[-\frac{1}{2\sigma^2}\left\{ng_2(\mathbf{y}) + (g_1(\mathbf{y}) - R)^2\,\|\mathbf{x}\|^2\right\}\right].$$

Next we define the function $\varphi : \mathbb{R} \times (0, +\infty) \times \mathbb{R}^2 \to \mathbb{R}$ by

$$\varphi(s, t, u, v) := \frac{1}{(2\pi\,t)^{n/2}} \exp\left[-\frac{1}{2t}\left\{nv + (u - s)^2\,\|\mathbf{x}\|^2\right\}\right].$$

Setting $h \equiv 1$ we can factorize the likelihood function as follows:

$$L_{R,\sigma^2}(\mathbf{y}) = h(\mathbf{y})\, \varphi(R, \sigma^2, \mathbf{g}(\mathbf{y})).$$

By the factorization theorem this implies that the statistic $\mathbf{g}(\mathbf{Y}) = (\hat{R}, \frac{1}{n}T)$ is sufficient for the parameter $(R, \sigma^2) \in \Theta = \mathbb{R} \times (0, +\infty)$. □

Exercise 19

The crux in the proof of Proposition IV.1.3 was the minimization of the expression $c_1^2 + \cdots + c_n^2$ under the two constraints

$$c_1 + \cdots + c_n = 0 \quad \text{and} \quad c_1x_1 + \cdots + c_nx_n = 1.$$

Find the minimizing $c_1, \ldots, c_n$ by applying the Cauchy–Schwarz–Buniakowsky inequality to the vectors

$$(c_1, \ldots, c_n) \quad \text{and} \quad (x_1 - \overline{x}, \ldots, x_n - \overline{x}).$$

Solution. Applying this inequality to the vectors just mentioned, we see that we necessarily have

$$\sum_{j=1}^{n} c_j(x_j - \overline{x}) \leq \left(\sum_{j=1}^{n} c_j^2\right)^{1/2} S_{\mathbf{xx}}^{1/2}.$$

As under our conditions on $\mathbf{c}$ always imply $\sum_{j=1}^{n} c_j(x_j - \overline{x}) = \sum_{j=1}^{n} c_jx_j - \overline{x}\sum_{j=1}^{n} c_j = 1 - 0 = 1$, we see that for all such $\mathbf{c}$

$$\sum_{j=1}^{n} c_j^2 \geq S_{\mathbf{xx}}^{-1}.$$

Equality can be reached if and only if $c_j = \lambda(x_j - \overline{x})$ for some constant λ, all $j = 1, \ldots, n$. It is now easy to see that $\lambda = S_{\mathbf{xx}}^{-1}$ gives us precisely what we need: consequently the unique minimizing vector $\mathbf{c}$ is given by $S_{\mathbf{xx}}^{-1}(x_1 - \overline{x}, \ldots, x_n - \overline{x})$.

□

Exercise 20

Determine the value of R_P belonging to the cloud of points

$$(-n, n),\ (-(n-1), n-1),\ \ldots,\ (0,0),\ \ldots,\ (n,n).$$

Also make a scatter diagram.

Solution. The scatter diagram is left to the reader. As to R_P we have that

$$S_{\mathbf{xy}} = \left(\sum_{j=-n}^{n} j\,|j|\right) - \frac{1}{2n+1}\left(\sum_{j=-n}^{n} j\right)\left(\sum_{j=-n}^{n} |j|\right) = 0,$$

so $R_P = 0$. □

Chapter 5
Normal analysis of variance

5.1 Summary of Chapter V

5.1.1 One-way analysis of variance (V.1)

We start from a *factor* which is adjustable at p levels. Associated with these levels are p (possibly different) populations. The mean of population no. i is denoted by μ_i. The *grand population mean* is defined by $\mu := \frac{1}{p}\sum_i \mu_i$. For each i the number $\alpha_i := \mu_i - \mu$ is called the *effect at level no. i.* In order to study the p populations in question we draw for each i a sample $X_{i1}, \ldots, X_{in}$ from population no. i. It is assumed that the system $\{X_{ij} : i = 1, \ldots, p,\ j = 1, \ldots, n\}$ is statistically independent. Moreover, we presume that for all i, j the variable X_{ij} enjoys a $N(\mu_i, \sigma^2)$-distribution. Under these assumptions we speak of *one-way normal analysis of variance* (one-way, because there is only one factor playing a role).

The *level sample means* are defined as $\overline{X}_{i\bullet} := \frac{1}{n}\sum_j X_{ij}$ and the *grand sample mean* as $\overline{X}_{\bullet\bullet} := \frac{1}{np}\sum_{i,j} X_{ij}$. These statistics are unbiased estimators of μ_i and μ respectively. Furthermore the Mean of Squared Errors (MSE) is defined by

$$\text{MSE} := \frac{1}{p(n-1)}\sum_{i,j}(X_{ij} - \overline{X}_{i\bullet})^2.$$

This variable presents an unbiased estimator of σ^2. The Mean of Squared Level deviations (MSL) is defined by

$$\text{MSL} := \frac{n}{p-1}\sum_{i}(\overline{X}_{i\bullet} - \overline{X}_{\bullet\bullet})^2.$$

We want to test

$$H_0 : \alpha_i = 0 \text{ for all } i \quad \text{versus} \quad H_1 : \alpha_i \neq 0 \text{ for some } i.$$

In our test we agree the outcome of the test statistic $T :=$ MSL/MSE to be conclusive in the decision procedure. Under H_0 the variable T enjoys a $F^{p-1}_{p(n-1)}$-distribution (Theorem IV.1.7). The hypothesis H_0 is rejected if T exhibits a large outcome.

5.1.2 Two-way analysis of variance (V.2)

In this section we are dealing with *two* factors, A and B, which are adjustable at p and q levels respectively. Adjusting factor A at level no. i and, simultaneously, factor B at level no. j, the associated population is denoted by 'population (i,j)'. The mean of this population is denoted by μ_{ij}. Furthermore we write

$$\mu := \frac{1}{pq}\sum_{i,j}\mu_{ij}, \quad \mu_{i\bullet} := \frac{1}{q}\sum_{j=1}^{q}\mu_{ij} \quad \text{and} \quad \mu_{\bullet j} := \frac{1}{p}\sum_{i=1}^{p}\mu_{ij}.$$

We call $\mu_{i\bullet}$ the *partial level mean* corresponding to factor A at level i ($\mu_{\bullet j}$ is named in the same way).

There is a unique decomposition of μ_{ij} of the form

$$\mu_{ij} = \mu + \alpha_i + \beta_j + \gamma_{ij},$$

where $\sum_i \alpha_i = 0$, $\sum_j \beta_j = 0$ and $\sum_i \gamma_{ik} = \sum_j \gamma_{lj} = 0$ for all k, l. In this decomposition we necessarily have $\alpha_i = \mu_{i\bullet} - \mu$ (the *effect of factor A at level* i), $\beta_j = \mu_{\bullet j} - \mu$ (the *effect of factor B at level* j) and $\gamma_{ij} = \mu_{ij} - \mu - \alpha_i - \beta_j$ (the *interaction between factors A and B at the simultaneous level* (i,j)).

Next we draw for each (i,j) a sample $X_{ij1}, \ldots, X_{ijn}$ from population (i,j). We assume that the system $\{X_{ijk} : i = 1, \ldots, p, \; j = 1, \ldots, q \text{ and } k = 1, \ldots, n\}$ is statistically independent. Moreover, we assume that population (i,j) enjoys a $N(\mu_{ij}, \sigma^2)$-distribution (all i,j). Under these assumptions we speak of *two-way normal analysis of variance.*

In a self-explaining notation the variables $\overline{X}_{\bullet\bullet\bullet}$, $\overline{X}_{i\bullet\bullet}$, $\overline{X}_{\bullet j\bullet}$ and $\overline{X}_{ij\bullet}$ are introduced. Moreover, the following 'sums of squares' are defined:

$$\begin{cases} \text{SS(A)} & := & qn\sum_i \left(\overline{X}_{i\bullet\bullet} - \overline{X}_{\bullet\bullet\bullet}\right)^2, \\ \text{SS(B)} & := & pn\sum_j \left(\overline{X}_{\bullet j\bullet} - \overline{X}_{\bullet\bullet\bullet}\right)^2, \\ \text{SS(AB)} & := & n\sum_{i,j} \left(\overline{X}_{ij\bullet} - \overline{X}_{i\bullet\bullet} - \overline{X}_{\bullet j\bullet} + \overline{X}_{\bullet\bullet\bullet}\right)^2, \\ \text{SSE} & := & \sum_{i,j,k} \left(X_{ijk} - \overline{X}_{ij\bullet}\right)^2, \\ \text{SST} & := & \sum_{i,j,k} \left(X_{ijk} - \overline{X}_{\bullet\bullet\bullet}\right)^2. \end{cases}$$

We have SST = SSE + SS(A) + SS(B) + SS(AB). The following statistics are introduced:

$$\begin{cases} \text{MSA} & := & \text{SS(A)}/(p-1), \\ \text{MSB} & := & \text{SS(B)}/(q-1), \\ \text{MS(AB)} & := & \text{SS(AB)}/(p-1)(q-1), \\ \text{MSE} & := & \text{SSE}/pq(n-1). \end{cases}$$

Next we carry out tests, in which the null hypothesis is of form:

- $H_0 : \alpha_i = 0$ for all i. The test statistic is given by $T_A :=$ MSA/MSE. Under H_0 the variable T_A is $F^{p-1}_{pq(n-1)}$-distributed. The hypothesis H_0 is rejected if T_A exhibits a large outcome.

- $H_0 : \beta_j = 0$ for all j. The test statistic is given by $T_B :=$ MSB/MSE. Under H_0 the variable T_B is $F^{q-1}_{pq(n-1)}$-distributed. The hypothesis H_0 is rejected if T_B exhibits a large outcome.

- $H_0 : \gamma_{ij} = 0$ for all i, j. The test statistic is $T_{AB} :=$ MS(AB)/MSE. Under H_0 the variable T_{AB} is $F^{(p-1)(q-1)}_{pq(n-1)}$-distributed. The hypothesis H_0 is rejected if T_{AB} exhibits a large outcome.

5.2 Exercises to Chapter V

Exercise 1

Under the assumptions of one-way normal analysis of variance we have:

$$\operatorname{cov}\left(\overline{X}_{\bullet\bullet}, \overline{X}_{i\bullet}\right) = \frac{\sigma^2}{pn}.$$

Prove this.

Proof. First of all we write

$$\operatorname{cov}\left(\overline{X}_{j\bullet}, \overline{X}_{i\bullet}\right) = \operatorname{cov}\left(\frac{1}{n}\sum_{k=1}^{n} X_{jk}, \frac{1}{n}\sum_{l=1}^{n} X_{il}\right) = \frac{1}{n^2}\sum_{k,l} \operatorname{cov}\left(X_{jk}, X_{il}\right).$$

The latter sum vanishes if $i \neq j$, because then the pair of variables X_{jk} and X_{il} is statistically independent (all k, l). For $i = j$ we get

$$\frac{1}{n^2}\sum_{k,l} \operatorname{cov}\left(X_{jk}, X_{il}\right) = \frac{1}{n^2}\sum_{k=1}^{n} \operatorname{var}(X_{ik}) = \frac{n\sigma^2}{n^2} = \frac{\sigma^2}{n}.$$

Therefore

$$\begin{aligned}
\operatorname{cov}\left(\overline{X}_{\bullet\bullet}, \overline{X}_{i\bullet}\right) &= \frac{1}{p}\sum_{j=1}^{p} \operatorname{cov}\left(\overline{X}_{j\bullet}, \overline{X}_{i\bullet}\right) \\
&= \frac{1}{p}\operatorname{cov}\left(\overline{X}_{i\bullet}, \overline{X}_{i\bullet}\right) = \frac{1}{p}\cdot\frac{\sigma^2}{n} = \frac{\sigma^2}{pn}.
\end{aligned}$$

□

Exercise 2

We start from the underlying assumptions of normal analysis of variance.

(i) *Question.* What is, under H_0, the distribution of the variable

$$\frac{1}{\sigma^2}\sum_{i,j}(X_{ij}-\overline{X}_{\bullet\bullet})^2 \ ?$$

Solution. Under H_0 the X_{ij} are all $N(\mu,\sigma^2)$-distributed. Therefore the X_{ij} can be looked upon as one sample, of size pn, from a $N(\mu,\sigma^2)$-distributed population. However, then (by Theorem II.2.12) the variable $(np-1)S^2/\sigma^2 = \frac{1}{\sigma^2}\sum_{i,j}(X_{ij}-\overline{X}_{\bullet\bullet})^2$ enjoys a $\chi^2(np-1)$-distribution. □

(ii) Verify that

$$\sum_{i,j}(X_{ij}-\overline{X}_{\bullet\bullet})^2 = \sum_{i,j}(X_{ij}-\overline{X}_{i\bullet})^2 + n\sum_{i}(\overline{X}_{i\bullet}-\overline{X}_{\bullet\bullet})^2.$$

Solution.

$$\begin{aligned}
&\sum_{i,j}(X_{ij}-\overline{X}_{\bullet\bullet})^2 \\
&\quad= \sum_{i,j}\left((X_{ij}-\overline{X}_{i\bullet})+(\overline{X}_{i\bullet}-\overline{X}_{\bullet\bullet})\right)^2 \\
&\quad= \sum_{i,j}(X_{ij}-\overline{X}_{i\bullet})^2 + 2\sum_{i,j}(X_{ij}-\overline{X}_{i\bullet})(\overline{X}_{i\bullet}-\overline{X}_{\bullet\bullet}) \\
&\qquad\qquad + \sum_{i,j}(\overline{X}_{i\bullet}-\overline{X}_{\bullet\bullet})^2 \\
&\quad= \sum_{i,j}(X_{ij}-\overline{X}_{i\bullet})^2 + 2\sum_{i}\left\{(\overline{X}_{i\bullet}-\overline{X}_{\bullet\bullet})\sum_{j}(X_{ij}-\overline{X}_{i\bullet})\right\} \\
&\qquad\qquad + n\sum_{i}(\overline{X}_{i\bullet}-\overline{X}_{\bullet\bullet})^2 \\
&\quad= \sum_{i,j}(X_{ij}-\overline{X}_{i\bullet})^2 + n\sum_{i}(\overline{X}_{i\bullet}-\overline{X}_{\bullet\bullet})^2.
\end{aligned}$$

□

(iii) Prove that $\sum_{i,j}(X_{ij}-\overline{X}_{i\bullet})^2$ and $\sum_i(\overline{X}_{i\bullet}-\overline{X}_{\bullet\bullet})^2$ are statistically independent.

Proof. See the proof of Theorem V.1.7. □

(iv) Define

$$Y := \frac{n}{\sigma^2}\sum_{i}(\overline{X}_{i\bullet}-\overline{X}_{\bullet\bullet})^2.$$

Prove that, under H_0, the variable Y is $\chi^2(p-1)$-distributed.

Proof. Set

$$X := \frac{1}{\sigma^2}\sum_{i,j}\left(X_{ij}-\overline{X}_{i\bullet}\right)^2 \quad\text{and}\quad V := \frac{1}{\sigma^2}\sum_{i,j}\left(X_{ij}-\overline{X}_{\bullet\bullet}\right)^2.$$

Assume that H_0 is valid. Then by (ii) we have $V = X + Y$ and by (iii) X and Y are statistically independent. According to Proposition V.1.1 the variable X is $\chi^2(pn-p)$-distributed, whereas (by (i)) V enjoys a $\chi^2(pn-1)$-distribution. Applying Proposition II.2.10 we conclude that Y is χ^2-distributed with $(pn-1)-(pn-p) = p-1$ degrees of freedom. □

Exercise 3

Consider a case of normal one-way analysis of variance in which the factor involved can be adjusted at two levels. We want to test $H_0 : \mu_1 = \mu_2$ against $H_1 : \mu_1 \neq \mu_2$, basing our decision procedure on the outcome of two samples $X_{11}, \ldots, X_{1n}$ and $X_{21}, \ldots, X_{2n}$. Now there are in a natural way two test statistics in sight. Namely, aside from the test statistic $T :=$ MSL/MSE which enjoys under H_0 (Theorem V.1.7) an F-distribution, we have the test statistic

$$\frac{\overline{X}_{1\bullet}-\overline{X}_{2\bullet}}{S_p\sqrt{\frac{1}{n}+\frac{1}{n}}},$$

which is t-distributed under H_0 (Theorem II.4.5).

(i) Show that $\text{MSE} = S_p^2$.

Solution. On the one hand

$$\begin{aligned} S_p^2 &= \frac{(n-1)S_1^2+(n-1)S_2^2}{2n-2} \\ &= \frac{\sum_{i=1}^n\left(X_{1i}-\overline{X}_{1\bullet}\right)^2+\sum_{i=1}^n\left(X_{2i}-\overline{X}_{2\bullet}\right)^2}{2n-2}. \end{aligned}$$

On the other hand we have

$$\begin{aligned} \text{MSE} &= \frac{1}{p(n-1)}\sum_{i,j}\left(X_{ij}-\overline{X}_{i\bullet}\right)^2 \\ &= \frac{\sum_{j=1}^n\left(X_{1j}-\overline{X}_{1\bullet}\right)^2+\sum_{j=1}^n\left(X_{2j}-\overline{X}_{2\bullet}\right)^2}{2(n-1)}, \end{aligned}$$

which shows that $\text{MSE} = S_p^2$. □

(ii) Prove that

$$\left(\frac{\overline{X}_{1\bullet}-\overline{X}_{2\bullet}}{S_p\sqrt{\frac{1}{n}+\frac{1}{n}}}\right)^2 = \frac{\text{MSL}}{\text{MSE}}.$$

Proof. Using (i) it suffices to prove that

$$\text{MSL} = \frac{\left(\overline{X}_{1\bullet} - \overline{X}_{2\bullet}\right)^2}{2/n}.$$

This is easy, for we can write

$$\begin{aligned}
\text{MSL} &= \tfrac{n}{2-1}\left(\left(\overline{X}_{1\bullet} - \overline{X}_{\bullet\bullet}\right)^2 + \left(\overline{X}_{2\bullet} - \overline{X}_{\bullet\bullet}\right)^2\right) \\
&= n\left(\left(\tfrac{1}{2}\overline{X}_{1\bullet} - \tfrac{1}{2}\overline{X}_{2\bullet}\right)^2 + \left(\tfrac{1}{2}\overline{X}_{2\bullet} - \tfrac{1}{2}\overline{X}_{1\bullet}\right)^2\right) \\
&= n \cdot 2 \cdot \tfrac{1}{4}\left(\overline{X}_{1\bullet} - \overline{X}_{2\bullet}\right)^2 = \tfrac{1}{2}n\left(\overline{X}_{1\bullet} - \overline{X}_{2\bullet}\right)^2.
\end{aligned}$$

□

(iii) *Question.* Can these two test statistics (at the same level of significance) generate different decisions?

Solution. No. Set

$$Y := \frac{\overline{X}_{1\bullet} - \overline{X}_{2\bullet}}{S_p\sqrt{\frac{1}{n} + \frac{1}{n}}}.$$

Suppose that Y exhibits an outcome y; then the outcome of T is y^2 (see (ii)). The P-value belonging to Y is $\mathbb{P}(|Y| \geq |y|)$; the P-value associated with T is $\mathbb{P}\left(T \geq y^2\right)$. Of course we have

$$\mathbb{P}(|Y| \geq |y|) = \mathbb{P}\left(Y^2 \geq y^2\right) = \mathbb{P}\left(T \geq y^2\right).$$

This shows that both P-values are the same. Hence the corresponding decision procedures are identical. □

Exercise 4

We start from two-way normal analysis of variance. Thus we are dealing with a statistically independent system of pqn variables X_{ijk}, where X_{ijk} is $N(\mu_{ij}, \sigma^2)$-distributed (all i, j, k). Prove that the quantities $\overline{X}_{1\bullet\bullet} - \overline{X}_{\bullet\bullet\bullet}$ and $\overline{X}_{11\bullet} - \overline{X}_{1\bullet\bullet} - \overline{X}_{\bullet 1\bullet} + \overline{X}_{\bullet\bullet\bullet}$ are statistically independent.

Proof. As we did before we set $\mathfrak{V} := \{\sum_{i,j,k} c_{ijk}X_{ijk} : c_{ijk} \in \mathbb{R}\}$. The two quantities mentioned above are both members of $\mathfrak{V}$. For this reason we can apply Theorem I.6.7 in order to prove statistical independence. This theorem says that, under the imposed conditions, it suffices to prove that the covariance between the variables in question is zero. This can be verified as follows:

$$\begin{aligned}
&\operatorname{cov}\left(\overline{X}_{1\bullet\bullet} - \overline{X}_{\bullet\bullet\bullet}\,,\,\overline{X}_{11\bullet} - \overline{X}_{1\bullet\bullet} - \overline{X}_{\bullet 1\bullet} + \overline{X}_{\bullet\bullet\bullet}\right) = \\
&\operatorname{cov}\left(\overline{X}_{1\bullet\bullet}, \overline{X}_{11\bullet}\right) - \operatorname{cov}\left(\overline{X}_{1\bullet\bullet}, \overline{X}_{1\bullet\bullet}\right) - \operatorname{cov}\left(\overline{X}_{1\bullet\bullet}, \overline{X}_{\bullet 1\bullet}\right) + 2\operatorname{cov}\left(\overline{X}_{1\bullet\bullet}, \overline{X}_{\bullet\bullet\bullet}\right) \\
&\quad -\operatorname{cov}\left(\overline{X}_{\bullet\bullet\bullet}, \overline{X}_{11\bullet}\right) + \operatorname{cov}\left(\overline{X}_{\bullet\bullet\bullet}, \overline{X}_{\bullet 1\bullet}\right) - \operatorname{cov}\left(\overline{X}_{\bullet\bullet\bullet}, \overline{X}_{\bullet\bullet\bullet}\right).
\end{aligned}$$

All these covariances must be calculated. We illustrate how this can be done by picking out the sixth term on the right side of the equality above:

$$\begin{aligned}
\operatorname{cov}\left(\overline{X}_{\bullet\bullet\bullet}, \overline{X}_{\bullet 1 \bullet}\right) &= \frac{1}{p^2qn^2} \sum_{ijklm} \operatorname{cov}\left(X_{ijk}, X_{l1m}\right) \\
&= \frac{1}{p^2qn^2} \sum_{iklm} \operatorname{cov}\left(X_{i1k}, X_{l1m}\right) \\
&\doteq \frac{1}{p^2qn^2} \sum_{i,k} \operatorname{cov}\left(X_{i1k}, X_{i1k}\right) = \frac{pn\,\sigma^2}{p^2qn^2} = \frac{\sigma^2}{pqn}.
\end{aligned}$$

Proceeding in this way we come to

$$\operatorname{cov}(\ldots, \ldots) = \frac{\sigma^2}{qn} - \frac{\sigma^2}{qn} - \frac{\sigma^2}{pqn} + 2 \cdot \frac{\sigma^2}{pqn} - \frac{\sigma^2}{pqn} + \frac{\sigma^2}{pqn} - \frac{\sigma^2}{pqn} = 0,$$

thus accomplishing the proof. □

Exercise 5

In two-way analysis of variance we define $T_{\bullet\bullet\bullet} := \sum_{i,j,k} X_{ijk}$ and $T_{i\bullet\bullet} := \sum_{j,k} X_{ijk}$. Show that

$$\mathrm{SS(A)} = \frac{1}{qn} \sum_i (T_{i\bullet\bullet})^2 - \frac{1}{pqn} (T_{\bullet\bullet\bullet})^2.$$

Proof. Using the fact that $T_{\bullet\bullet\bullet} = pqn\,\overline{X}_{\bullet\bullet\bullet}$ and $T_{i\bullet\bullet} = qn\,\overline{X}_{i\bullet\bullet}$ we can write

$$\begin{aligned}
\mathrm{SS(A)} &= qn \sum_i \left(\overline{X}_{i\bullet\bullet} - \overline{X}_{\bullet\bullet\bullet}\right)^2 \\
&= qn \sum_i \left(\frac{1}{qn} T_{i\bullet\bullet} - \frac{1}{pqn} T_{\bullet\bullet\bullet}\right)^2 = \frac{1}{qn} \sum_i \left(T_{i\bullet\bullet} - \frac{1}{p} T_{\bullet\bullet\bullet}\right)^2 \\
&= \frac{1}{qn} \sum_i (T_{i\bullet\bullet})^2 - \frac{2}{pqn} \sum_i (T_{\bullet\bullet\bullet})(T_{i\bullet\bullet}) + \frac{1}{p^2qn} \sum_i (T_{\bullet\bullet\bullet})^2 \\
&= \frac{1}{qn} \sum_i (T_{i\bullet\bullet})^2 - \frac{2}{pqn} (T_{\bullet\bullet\bullet})^2 + \frac{1}{pqn} (T_{\bullet\bullet\bullet})^2 \\
&= \frac{1}{qn} \sum_i (T_{i\bullet\bullet})^2 - \frac{1}{pqn} (T_{\bullet\bullet\bullet})^2.
\end{aligned}$$

□

Exercise 6

There are four commercial travellers. With commercial traveller no. i we associate a sample $X_{i1}, \ldots, X_{i5}$, presenting the number of car-kilometers covered by the person in question. We presume a case of normal one-way analysis of variance. The outcome of the sample is summarized in the following scheme:

traveller	week 1	week 2	week 3	week 4	week 5
1	1533	1657	1845	1732	1784
2	1891	1923	2073	1873	1972
3	1347	1462	1417	1582	1345
4	1907	2137	1849	1994	2098

We prescribe a level of significance $\alpha = 0.10$. Basing ourselves on this scheme, we are asked to test

$$H_0 : \mu_1 = \mu_2 = \mu_3 = \mu_4 \quad \text{versus} \quad H_1 : \mu_i \neq \mu_j \text{ for some } (i,j).$$

Solution. First we compute the outcome of the test statistic $T = \text{MSL}/\text{MSE}$. The variable enjoys (a priori) under H_0 an $F^{4-1}_{4(5-1)}$-distribution, that is to say an F^3_{16}-distribution. We examine to which extent the outcome of T is likely under H_0.

We compute the outcome of T as follows. Firstly,

$$\text{MSL} = \frac{5}{4-1}\sum_{i=1}^{4}\left(\overline{X}_{i\bullet} - \overline{X}_{\bullet\bullet}\right)^2$$

and

$$\text{MSE} = \frac{1}{4(5-1)}\sum_{i,j}\left(X_{ij} - \overline{X}_{i\bullet}\right)^2.$$

We have

$$\overline{X}_{1\bullet} = 1710.2,\ \overline{X}_{2\bullet} = 1946.4,\ \overline{X}_{3\bullet} = 1430.6,\ \overline{X}_{4\bullet} = 1997,\ \overline{X}_{\bullet\bullet} = 1771.05,$$

which shows that $\text{MSL} = \frac{5}{3}\left((-60.85)^2 + (175.35)^2 + (-340.45)^2 + (225.95)^2\right) = 335683.25$. Concerning MSE we compute:

$$\begin{cases} \sum_j \left(X_{1j} - \overline{X}_{1\bullet}\right)^2 = 58322.8, & \sum_j \left(X_{2j} - \overline{X}_{2\bullet}\right)^2 = 25687.2, \\ \sum_j \left(X_{3j} - \overline{X}_{3\bullet}\right)^2 = 38409.2, & \sum_j \left(X_{4j} - \overline{X}_{4\bullet}\right)^2 = 59814. \end{cases}$$

Hence

$$\text{MSE} = \tfrac{1}{16}\left(58322.8 + \cdots + 59814\right) = 11389.58$$

and

$$T = 335683.25/11389.58 = 29.47.$$

Now in the table describing the F-distribution we read off that $\mathbb{P}(T \geq 5.29) = 0.01$. The actual outcome of T is 29.47; therefore we reject H_0. □

Exercise 7

We consider an experiment in which two-way normal analysis of variance can be applied. Factor A can be adjusted at $p = 4$ levels and factor B at $q = 6$ levels. At every simultaneous level we draw a sample of size $n = 10$. This provides: SST $= 10$, SSE $= 4$, SS(A) $= 3$ and SS(B) $= 2$. Test at a significance level of $\alpha = 0.05$ whether there is interaction between the two factors.

Solution. In the notations of §V.2 the null hypothesis can be formulated as:

$$H_0 \, : \, \gamma_{ij} = 0 \quad \text{for all } i, j.$$

The test statistic $T_{AB} = \text{MS(AB)}/\text{MSE}$ enjoys under H_0 (a priori) a $F^{(4-1)(6-1)}_{4\cdot 6\cdot(10-1)}$-distribution, that is, a F^{15}_{216}-distribution. The variable MSE shows the following outcome:

$$\text{MSE} = \text{SSE}/(4 \cdot 6 \cdot (10-1)) = \tfrac{4}{216} = \tfrac{1}{54}.$$

Moreover, using Proposition V.2.4 we have

$$\text{SS(AB)} = \text{SST} - \text{SSE} - \text{SS(A)} - \text{SS(B)} \; = \; 10 - 4 - 3 - 2 = 1$$

and $\text{MS(AB)} = \text{SS(AB)}/(4-1)(6-1) = 1/15$, so that

$$T_{AB} = \tfrac{1}{15} \Big/ \tfrac{1}{54} = \tfrac{54}{15} = \tfrac{18}{5} = 3.60.$$

In Table V, describing the F-distribution, we find that $\mathbb{P}\,(T_{AB} \geq 2.04) = 0.01$, where we assume that T_{AB} is F^{15}_{∞}-distributed. It follows that $\mathbb{P}\,(T_{AB} \geq 3.60) < 0.05$; we reject H_0. □

Exercise 8

We are starting from a set of independent samples $X_{i1}, \ldots, X_{in_i}$, where $i = 1, 2, \ldots, p$. For all possible i, j the variable X_{ij} is $N(\mu_i, \sigma^2)$-distributed. We write $N := \sum_i n_i$. The mean of sample no. i is denoted by $\overline{X}_{i\bullet}$.

(i) *Question.* What is the probability distribution of the variable

$$\frac{1}{\sigma^2} \sum_{i=1}^{p} \sum_{j=1}^{n_i} (X_{ij} - \overline{X}_{i\bullet})^2 \; ?$$

Solution. According to Theorem II.2.12 the variable $\frac{1}{\sigma^2} \sum_{j=1}^{n_i} (X_{ij} - \overline{X}_{i\bullet})^2$ is $\chi^2(n_i - 1)$-distributed (for all i). Furthermore, these variables constitute a statistically independent system. Applying Proposition II.2.9 we conclude that

$$\frac{1}{\sigma^2} \sum_{i=1}^{p} \sum_{j=1}^{n_i} (X_{ij} - \overline{X}_{i\bullet})^2$$

enjoys a χ^2-distribution with $\sum_i (n_i - 1) = N - p$ degrees of freedom. □

(ii) Determine the expectation value of

$$\frac{1}{\sigma^2}\sum_{i=1}^{p} n_i(\overline{X}_{i\bullet}-\overline{X}_{\bullet\bullet})^2.$$

Solution. Splitting up the variable $X_{ij}-\overline{X}_{\bullet\bullet}$ as $X_{ij}-\overline{X}_{\bullet\bullet}=(X_{ij}-\overline{X}_{i\bullet})+(\overline{X}_{i\bullet}-\overline{X}_{\bullet\bullet})$ we deduce that

$$\frac{1}{\sigma^2}\sum_{i=1}^{p}\sum_{j=1}^{n_i}(X_{ij}-\overline{X}_{\bullet\bullet})^2 = \frac{1}{\sigma^2}\sum_{i=1}^{p} n_i(\overline{X}_{i\bullet}-\overline{X}_{\bullet\bullet})^2+\frac{1}{\sigma^2}\sum_{i=1}^{p}\sum_{j=1}^{n_i}(X_{ij}-\overline{X}_{i\bullet})^2. \qquad (*)$$

By Proposition II.1.3 we have

$$\mathbb{E}\left[\frac{1}{\sigma^2}\sum_{j=1}^{n_i}(X_{ij}-\overline{X}_{i\bullet})^2\right]=n_i-1.$$

Hence

$$\mathbb{E}\left[\frac{1}{\sigma^2}\sum_{i=1}^{p}\sum_{j=1}^{n_i}(X_{ij}-\overline{X}_{i\bullet})^2\right]=\sum_{i=1}^{p}(n_i-1)=N-p. \qquad (**)$$

Next we apply Lemma V.1.3 to the variables in question. By this lemma, setting $\mu:=(n_1\mu_1+\cdots+n_p\mu_p)/N$, we may write

$$\mathbb{E}\left[(X_{ij}-\overline{X}_{\bullet\bullet})^2\right]=(\mu_i-\mu)^2+(N-1)\sigma^2/N.$$

Consequently

$$\begin{aligned}\mathbb{E}\left[\frac{1}{\sigma^2}\sum_{i=1}^{p}\sum_{j=1}^{n_i}(X_{ij}-\overline{X}_{\bullet\bullet})^2\right] &= \frac{1}{\sigma^2}\sum_{i=1}^{p}\sum_{j=1}^{n_i}\left[(\mu_i-\mu)^2+(N-1)\sigma^2/N\right]\\ &= \frac{1}{\sigma^2}\sum_{i=1}^{p}\left[n_i(\mu_i-\mu)^2+n_i(N-1)\sigma^2/N\right]\\ &= (N-1)+\frac{1}{\sigma^2}\sum_{i=1}^{p} n_i(\mu_i-\mu)^2. \qquad (***)\end{aligned}$$

It follows from (*), (**) and (***) that

$$\mathbb{E}\left[\frac{1}{\sigma^2}\sum_{i=1}^{p} n_i(\overline{X}_{i\bullet}-\overline{X}_{\bullet\bullet})^2\right]$$

$$=\left\{(N-1)+\frac{1}{\sigma^2}\sum_{i=1}^{p} n_i(\mu_i-\mu)^2\right\}-(N-p)$$

$$=(p-1)+\frac{1}{\sigma^2}\sum_{i=1}^{p} n_i(\mu_i-\mu)^2.$$

□

(iii) *Question.* What is, under the null hypothesis that $\mu_1 = \cdots = \mu_p$, the probability distribution of $\frac{1}{\sigma^2}\sum_{i=1}^{p} n_i(\overline{X}_{i\bullet}-\overline{X}_{\bullet\bullet})^2$?

Solution. Firstly (see (ii)) we can write

$$\frac{1}{\sigma^2}\sum_{i=1}^{p}\sum_{j=1}^{n_i}(X_{ij}-\overline{X}_{\bullet\bullet})^2$$

$$=\frac{1}{\sigma^2}\sum_{i=1}^{p} n_i(\overline{X}_{i\bullet}-\overline{X}_{\bullet\bullet})^2+\frac{1}{\sigma^2}\sum_{i=1}^{p}\sum_{j=1}^{n_i}(X_{ij}-\overline{X}_{i\bullet})^2.$$

From (i) we know that the right hand term on the right hand side of this equality is $\chi^2(N-p)$-distributed. Moreover, under H_0 the variables $X_{11}, X_{12}, \ldots, X_{pn_p}$ constitute a sample from a $N(\mu,\sigma^2)$-distributed population. Therefore, according to Theorem II.2.12, the term

$$\frac{1}{\sigma^2}\sum_{i=1}^{p}\sum_{j=1}^{n_i}(X_{ij}-\overline{X}_{\bullet\bullet})^2$$

is $\chi^2(N-1)$-distributed. Looking back at the proof of Theorem V.1.7 we see that (under H_0) the stochastic vectors

$$(\overline{X}_{1\bullet}-\overline{X}_{\bullet\bullet},\ldots,\overline{X}_{p\bullet}-\overline{X}_{\bullet\bullet}) \quad\text{and}\quad (X_{11}-\overline{X}_{1\bullet},\ldots,X_{pn}-\overline{X}_{p\bullet})$$

are statistically independent. It follows from this that

$$\frac{1}{\sigma^2}\sum_{i=1}^{p} n_i(\overline{X}_{i\bullet}-\overline{X}_{\bullet\bullet})^2 \quad\text{and}\quad \frac{1}{\sigma^2}\sum_{i=1}^{p}\sum_{j=1}^{n_i}(X_{ij}-\overline{X}_{i\bullet})^2$$

are independent. Finally we apply Proposition II.2.10 and conclude that

$$\frac{1}{\sigma^2}\sum_{i=1}^{p} n_i(\overline{X}_{i\bullet}-\overline{X}_{\bullet\bullet})^2$$

is χ^2-distributed with $(N-1)-(N-p)=p-1$ degrees of freedom. □

Chapter 6
Non-parametric methods

6.1 Summary of Chapter VI

In this chapter hypothesis tests are discussed about populations that are not necessarily normally distributed. In fact, hardly any condition will be imposed on the probability distribution of the populations in question.

6.1.1 The sign test, Wilcoxon's signed-rank test (VI.1)

We first discuss the *sign test.* Here we are starting from a population having a continuous distribution function with an unambiguous median m. We wish to test the hypothesis $H_0 : m = m_0$ versus alternatives of the form $H_1 : m \neq m_0$, $H_1 : m > m_0$ or $H_1 : m < m_0$. As always, we base our decision procedure on a sample $X_1, \ldots, X_n$ from the population being considered. We define the variable T^+ as the number of outcomes larger than m_0 in the sequence $X_1, \ldots, X_n$. The variable T^- is defined as being the frequency of outcomes smaller than m_0. Under H_0 both T^+ and T^- enjoy a binomial distribution with parameters n and $\theta = \frac{1}{2}$. Hence, under H_0 these variables have an expectation value of $\frac{1}{2}n$ and a variance of $\frac{1}{4}n$. If T^+ exhibits an outcome much larger or smaller than $\frac{1}{2}n$ we reject H_0. A suitable critical region for T^+ can be found by using Table I. For large n the variable T^+ is approximately normally distributed.

A refinement of the sign test is *Wilcoxon's signed-rank test.* Here we suppose that the probability distribution of the population is symmetric around the median m. Now for all i we define the variable D_i by $D_i := X_i - m_0$. The number R_i is understood to be the rank number of $|D_i|$ in the sequence $|D_1|, \ldots, |D_n|$, when ordered from small to large. Next the variables T^+ and T^- are defined as

$$T^+ := \sum_{D_i > 0} R_i \quad \text{and} \quad T^- := \sum_{D_i < 0} R_i.$$

With probability 1 we have $T^+ + T^- = \frac{1}{2}n(n+1)$. For small n the probability distribution of T^+ and T^- (under H_0) is tabulated in Table VI. For large n these variables are approximately normally distributed. Using Theorem VI.1.5 the parameters for such a normal distribution can suitably be adjusted.

6.1.2 Wilcoxon's rank-sum test (VI.2)

Here we are starting from two populations, $\mathfrak{X}$ and $\mathfrak{Y}$, of which beforehand it is known that their probability distribution functions are translated versions of each other. That is to say: if $\mathfrak{X}$ has a distribution function F, then $\mathfrak{Y}$ has a distribution function of type $x \mapsto F(x - \theta)$. Basing ourselves on samples we want to test the hypothesis $H_0 : \theta = 0$ against alternatives of type $H_1 : \theta \neq 0$, $H_1 : \theta > 0$ or $H_1 : \theta < 0$. More specific, let $X_1, \ldots, X_m$ and $Y_1, \ldots, Y_n$ be statistically independent samples from population $\mathfrak{X}$ and $\mathfrak{Y}$ respectively. We denote the rank number of X_i (of Y_j) in the sequence $\{X_1, \ldots, X_m, Y_1, \ldots, Y_n\}$, when ordered from small to large, by $\mathfrak{r}(X_i)$ (by $\mathfrak{r}(Y_j)$). The *Wilcoxon rank-sums* are now defined by

$$T_X := \sum_{i=1}^{m} \mathfrak{r}(X_i) \quad \text{and} \quad T_Y := \sum_{j=1}^{n} \mathfrak{r}(Y_j).$$

For T_X and T_Y we have:

$$T_X + T_Y = \tfrac{1}{2}(m+n)(m+n+1).$$

For small m and n the probability distribution (under H_0) of T_X (T_Y) is tabulated in Table VII. For large m and n the variable T_X is approximately $N(\mu, \sigma^2)$-distributed with

$$\mu = \tfrac{1}{2}m(m+n+1) \quad \text{and} \quad \sigma^2 = \tfrac{1}{12}mn(m+n+1).$$

Similar considerations apply to T_Y. The null hypothesis is rejected if T_X exhibits an outcome which is much larger or much smaller than $\frac{1}{2}m(m+n+1)$. This decision procedure is called *Wilcoxon's rank-sum test* or the *Mann–Whitney test.*

6.1.3 The runs test (VI.3)

We are considering random sequences which consist of symbols A and B. The frequency of occurence of symbol A is denoted by n_1 and that of B by n_2. The total number of chains in such a sequence is denoted by T, the total number of A-chains by T_1 and the total number of B-chains by T_2. We want to test the null hypothesis

H_0 : every arrangement consisting of n_1 symbols A and n_2 symbols B enjoys the same probability of occurence.

against alternatives of type

H_1 : not H_0 (two-sided),

H_1 : there is a tendency toward clustering,

H_1 : there is a tendency toward mixing.

Theorem VI.3.2 describes, under H_0, the probability distribution of the variable T. Critical regions for T can be constructed by using Table VIII. For large n_1 and n_2 the variable T is approximately normally distributed (Theorem VI.3.3) with parameters

$$\mathbb{E}(T) = 1 + \frac{2n_1n_2}{(n_1+n_2)} \quad \text{and} \quad \operatorname{var}(T) = \frac{2n_1n_2(2n_1n_2 - n_1 - n_2)}{(n_1+n_2)^2(n_1+n_2-1)}.$$

We are talking about the *runs test* .

6.1.4 Rank correlation tests (VI.4)

A *bivariate sample* (a vectorial sample of dimension 2) of size n is understood to be a statistically independent system $(X_1, Y_1), \ldots, (X_n, Y_n)$ of stochastic 2-vectors, all of them with the same probability distribution (the 'distribution of the population').

As a measure for the correlation between the X and Y *Spearman's rank correlation coefficient* R_S is introduced. This is in fact Pearson's product-moment correlation coefficient (see §IV.4) belonging to the 'cloud' of points

$$(r(X_1), r(Y_1)), \ \ldots, \ (r(X_n), r(Y_n)),$$

where $r(Z_i)$ denotes the rank (from small to large) of Z_i in the sequence $Z_1, \ldots,$ Z_n. This R_S can be expressed as

$$R_S = \frac{[12\sum_i r(X_i)r(Y_i)] - 3n(n+1)^2}{(n-1)\,n\,(n+1)}.$$

We wish to test the null hypothesis

$$H_0 \ : \ \text{there is no rank correlation}$$

against suitable alternatives, using R_S as a test statistic. For small n the probability distribution of R_S (under H_0) is tabulated (see Table IX). For large n the statistic R_S is (under H_0) approximately $N(0, \frac{1}{n-1})$-distributed (Theorem VI.4.3).

Another measure for the linear association between the two variables X and Y is provided by *Kendall's coefficient of concordance* (also: 'Kendall's tau'). Given the mentioned two-dimensional sample it is defined in the following way. First we determine the permutation $\pi : \{1, \ldots, n\} \to \{1, \ldots, n\}$ for which

$$X_{\pi(1)} < X_{\pi(2)} < \cdots < X_{\pi(n)}.$$

For each ordered pair (p, q) of integers we set $\varepsilon(p,q) := 1$ if $p < q$ and $\varepsilon(p,q) :=$ -1 if $p > q$. We call this the *concordance* of p and q. Kendall's coefficient of concordance R_K is now defined as

$$R_K := \frac{2}{n(n-1)} \sum_{i<j} \varepsilon\left(r(Y_{\pi(i)}), r(Y_{\pi(j)})\right).$$

In this way always $-1 \leq R_K \leq +1$ and $R_K = \pm 1$ if there is perfect 'ranking concordance' between the outcomes of the X_i and Y_i. When testing the same null hypothesis as before, in Table X we can find critical values for R_K.

6.1.5 The Kruskal–Wallis test (VI.5)

In this section a non-parametric analogue of one-way normal analysis of variance is developed. We are starting from p populations of which the distribution functions are mutually shifted. More specific, we assume that the distribution function of population no. i is given by

$$x \mapsto F(x - \theta_i),$$

where F is some continuous distribution function (not depending on i). We wish to test $H_0 : \theta_1 = \cdots = \theta_p = 0$ against H_1 : '$\theta_i \neq 0$ for at least one i'. The decision procedure is based on p statistically independent samples $X_{11}, \ldots, X_{1n_1}$, $X_{21}, \ldots, X_{2n_2}, \ldots, X_{p1}, \ldots, X_{pn_p}$ from population $1, 2, \ldots, p$ respectively. Ordering the sequence $X_{11}, \ldots, X_{1n_1}, \ldots, X_{p1}, \ldots, X_{pn_p}$ from small to large, the number $\mathfrak{r}(X_{ij})$ is the rank number of X_{ij} in this sequence. We define

$$S_i := \sum_{j=1}^{n_i} \mathfrak{r}(X_{ij}) \qquad (i = 1, \ldots, p).$$

Next, the *Kruskal–Wallis test statistic* is defined by

$$T := \frac{12}{N(N+1)} \sum_{i=1}^{p} n_i \left(\frac{S_i}{n_i} - \frac{N+1}{2} \right)^2,$$

where $N := n_1 + \cdots + n_p$. For small $p, n_1, \ldots, n_{p-1}, n_p$ the probability distribution of T (under H_0) is tabulated. For $n_1, \ldots, n_p \geq 5$ the variable T is (under H_0) approximately $\chi^2(p-1)$-distributed. In the Kruskal–Wallis test the null hypothesis is rejected if T exhibits a large outcome.

6.1.6 Friedman's test (VI.6)

A non-parametric analogue of two-way normal analysis of variance is developed. To begin with, we assume that there are two factors A and B which can be adjusted at p and q levels respectively. In this way pq populations come into our field of view. Population (i, j) is understood to be that belonging to the simultaneous level (i, j). We assume that the distribution function of population (i, j) is of the form

$$x \mapsto F(x - \theta_{ij}),$$

where F is some continuous distribution function (not depending on (i,j)). We wish to test the hypothesis

$$H_0 : \theta_{i\bullet} = 0 \text{ for all } i \quad \text{against} \quad H_1 : \theta_{i\bullet} \neq 0 \text{ for some } i.$$

To this we draw for each (i,j) a sample $X_{ij1}, \ldots, X_{ijn}$ from population (i,j). We assume that these samples are statistically independent. Ordering the sequence

$$X_{1j1}, \ldots, X_{1jn}, \ \ldots \ , X_{pj1}, \ldots, X_{pjn}$$

from small to large for all j, the number R_{ijk} is defined as the rank number of X_{ijk}. Furthermore we define $R_{i\bullet\bullet} := \frac{1}{qn}\sum_{j,k} R_{ijk}$. As a test statistic we use *Friedman's statistic*

$$T = \frac{12q}{p(pn+1)} \sum_{i=1}^{p} \left(R_{i\bullet\bullet} - \tfrac{1}{2}(pn+1)\right)^2,$$

which is, under H_0, for large q and n approximately $\chi^2(p-1)$-distributed. A large outcome of T is pushing us toward rejection of the null hypothesis.

6.2 Exercises to Chapter VI

Exercise 1

Suppose $X_1, \ldots, X_n$ is a sample from a population which has a continuous distribution function F. For an arbitrary $m_0 \in \mathbb{R}$ we set $D_i := X_i - m_0$. Prove that for all $i \neq j$ we have

$$\mathbb{P}(D_i = 0) = 0 \quad \text{and} \quad \mathbb{P}(|D_i| = |D_j|) = 0.$$

Solution. As F is continuous we have $\mathbb{P}(X_i = a) = 0$ for all $a \in \mathbb{R}$ and i. Because of this for all i we have

$$\mathbb{P}(D_i = 0) = \mathbb{P}(X_i = m_0) = 0;$$

this proves the first part of the statement. Furthermore we notice that for all $i \neq j$ we have

$$\begin{aligned}
\mathbb{P}(|D_i| = |D_j|) &= \mathbb{P}[(X_i, X_j) \in A] = \mathbb{P}_{(X_i,X_j)}(A) \\
&= (\mathbb{P}_{X_i} \otimes \mathbb{P}_{X_j})(A) \overset{\text{(Fubini)}}{=} \int\!\!\int 1_A(x,y)\, d\mathbb{P}_{X_j}(y)\, d\mathbb{P}_{X_i}(x) \\
&= \int \left[\mathbb{P}_{X_j}(\{x\}) + \mathbb{P}_{X_j}(\{-x + 2m_0\})\right] d\mathbb{P}_{X_i}(x) = 0
\end{aligned}$$

with A as sketched in the figure.

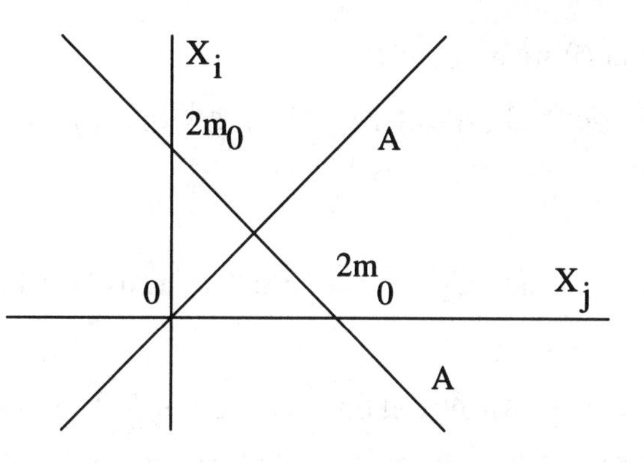

Exercise 2

Let T^+ be the test statistic in Wilcoxon's signed-rank test.

(i) Setting $N := \frac{1}{2}n(n+1)$, prove that for all $k \in \{0, \dots, N\}$ under H_0 we have $\mathbb{P}(T^+ = k) = \mathbb{P}(T^+ = N - k)$.

Proof. By Proposition VI.1.3 we always have $T^+ + T^- = N$. Both the variables T^+ and T^- can (under H_0) be decomposed as a sum

$$1\,B_1 + 2\,B_2 + \cdots + n\,B_n,$$

where the $B_1, \dots, B_n$ constitute a statistically independent system of variables, all enjoying a Bernoulli distribution with parameter $\theta = \frac{1}{2}$. It follows that T^+ and T^- are identically distributed. Consequently

$$\mathbb{P}(T^+ = k) = \mathbb{P}(T^- = N - k) = \mathbb{P}(T^+ = N - k)$$

for all possible k. □

(ii) Compute for $n = 5$ the numerical value of $\mathbb{P}(T^+ = 12)$.

Solution. We start from a population which has a continuous distribution function F and a strict median m_0. Let $X_1, \dots, X_5$ be a sample from this population. We set $D_i := X_i - m_0$ and we assign to every X_i the rank number of $|D_i|$ in the sequence $|D_1|, \dots, |D_5|$, when ordered from small to large. Now T^+ is the sum of the rank numbers belonging to outcomes X_i which are above m_0. The variable T^+ can be decomposed as

$$T^+ = 1B_1 + 2B_2 + \cdots + 5B_5,$$

where the $B_1, \dots, B_5$ constitute a statistically independent system of variables enjoying a Bernoulli distribution with parameter $\theta = \frac{1}{2}$. Therefore

$$\begin{aligned}
\mathbb{P}(T^+ = 3) &= \mathbb{P}(B_1 = B_2 = 1 \text{ and } B_3 = B_4 = B_5 = 0) \\
&\quad + \mathbb{P}(B_3 = 1 \text{ and } B_1 = B_2 = B_4 = B_5 = 0) \\
&= 2 \cdot (\tfrac{1}{2})^5 = 0.0625.
\end{aligned}$$

As we saw in (i) we may write

$$\mathbb{P}(T^+ = 12) = \mathbb{P}(T^+ = \tfrac{1}{2} \cdot 5 \cdot 6 - 12 = 3) = 0.0625.$$

□

Exercise 3

(i) Prove that the equality $\sum_{k=1}^{n} k = \frac{1}{2}n(n+1)$ holds for all $n \geq 1$.

Proof.

1. For $n = 1$ the equality above reads as $1 = \frac{1}{2} \cdot 1 \cdot 2$, which is correct.
2. Suppose that for some n the statement is correct; then

$$\begin{aligned}\sum_{k=1}^{n+1} k &= (n+1) + \sum_{k=1}^{n} k \\ &= (n+1) + \tfrac{1}{2}n(n+1) = \tfrac{1}{2}(n+1)(n+2),\end{aligned}$$

which shows that it is also correct for $n+1$.

3. Now apply the principle of induction. □

(ii) Prove that the equality $\sum_{k=1}^{n} k^2 = \frac{1}{6}n(n+1)(2n+1)$ holds for all $n \geq 1$.

Proof.

1. We have $1^2 = \frac{1}{6} \cdot 1 \cdot 2 \cdot 3$, which shows that the equality in question is correct for $n = 1$.
2. Suppose that the statement is correct for some n. Then

$$\begin{aligned}\sum_{k=1}^{n+1} k^2 &= (n+1)^2 + \sum_{k=1}^{n} k^2 = (n+1)^2 + \tfrac{1}{6}n(n+1)(2n+1) \\ &= \tfrac{1}{6}(n+1)(2n^2+7n+6) = \tfrac{1}{6}(n+1)(n+2)(2n+3).\end{aligned}$$

This shows that it is necessarily also correct for $n+1$.

3. Apply the principle of induction. □

(iii) Prove that $\sum_{k=1}^{n} \left(k - \frac{1}{2}(n+1)\right)^2 = \frac{1}{12}(n-1)\,n\,(n+1)$ for all $n \geq 1$.

Proof.

$$\begin{aligned}\sum_{k=1}^{n} \left(k - \tfrac{1}{2}(n+1)\right)^2 &= \sum_{k=1}^{n} \left(k^2 - (n+1)k + \tfrac{1}{4}(n+1)^2\right) \\ &= \tfrac{1}{6}n(n+1)(2n+1) - \tfrac{1}{2}n(n+1)^2 + \tfrac{1}{4}n(n+1)^2 \\ &= \tfrac{1}{12}n(n+1) \cdot (2(2n+1) - 3(n+1)) \\ &= \tfrac{1}{12}(n-1)\,n\,(n+1).\end{aligned}$$

□

Exercise 4

Let $X_1, \dots, X_m$ and $Y_1, \dots, Y_n$ be independent samples from two populations which both have some continuous distribution function. Prove that for all i, j we have $\mathbb{P}(X_i = Y_j) = 0$ and that for all $i \neq j$ we have both $\mathbb{P}(X_i = X_j) = 0$ and $\mathbb{P}(Y_i = Y_j) = 0$.

Proof. Quite analogous to that of Proposition VI.1.2: see Exercise 1. □

Exercise 5

Develop a strategy to deal with ties in the outcome of a sample.

Solution. If there are ties in the outcome of a sample we can choose the so-called *mid-rank strategy.* To illustrate this method, suppose we are dealing with an outcome $x_1, \dots, x_n$ of the sample $X_1, \dots, X_n$. Moreover, suppose that $x_2 = x_4 = x_5 \neq x_i$ for $i \in \{1, 3, 6, \dots, n\}$. If the rank numbers k, $k+1$ and $k+2$ are involved, then we set

$$r(X_2) = r(X_4) = r(X_5) := \frac{k + (k+1) + (k+2)}{3} = k + 1.$$

In Wilcoxon's signed-rank test the measurements for which $D_i = 0$ are usually skipped. □

Exercise 6

A sample $X_1, \dots, X_{10}$ is drawn from a population with an unambiguous median. The outcome is given by

$$(X_1, \dots, X_{10}) = (5.1,\ 8.0,\ 6.4,\ 16.0,\ 6.1,\ 10.4,\ 9.9,\ 7.2,\ 12.3,\ 0.1).$$

(i) Apply the sign test in order to test $H_0 : m = 5.3$ versus $H_1 : m \neq 5.3$ at a level of significance $\alpha = 0.10$.

Solution. Let T^+ be the frequency of occurrence of outcomes larger than 5.3. Under H_0 the variable T^+ is binomially distributed with parameters $n = 10$ and $\theta = \frac{1}{2}$. We observe an outcome $T^+ = 8$. Using Table I we see that, at the prescribed level of significance, a critical region for T^+ is given by $\{0, 1\} \cup \{9, 10\}$. We do not reject H_0. □

(ii) Use Wilcoxon's signed-rank test to test the hypothesis $H_0 : m = 5.3$ against $H_1 : m \neq 5.3$ at a level of significance $\alpha = 0.10$.

Solution. Setting $D_i := X_i - 5.3$ we can write the given outcome as

$$(D_1, \dots, D_{10}) = (-0.2,\ 2.7,\ 1.1,\ 10.7,\ 0.8,\ 5.1,\ 4.6,\ 1.9,\ 7.0,\ -5.2).$$

The corresponding rank numbers are given by

$$(r(X_1), \dots, r(X_{10})) = (1,\ 5,\ 3,\ 10,\ 2,\ 7,\ 6,\ 4,\ 9,\ 8)$$

so that $T^- = 1 + 8 = 9$ and $T^+ = \frac{1}{2} \cdot 10 \cdot 11 - 9 = 46$. Using Table VI we see that the critical region for T^+ (or, equivalently, T^-) is $\{0, 1, \ldots, 10\} \cup \{45, \ldots, 55\}$. Consequently in this decision procedure H_0 is rejected. □

Exercise 7

In a self-explanatory notation the following configuration is observed:

$$AD^4A^3D^3ADADA^2.$$

Test, at a level of significance $\alpha = 0.05$, whether there is a tendency toward mixing.

Solution. Taking $n = 9$ and $m = 8$, we consult Table VIII. In this table it can be read off that the (one-sided) critical region for T is $\{14, \ldots, 17\}$. The variable T in this experiment exhibits an outcome $T = 9$, which is not an element of this critical region. We therefore accept H_0 in this decision procedure. □

Exercise 8

(i) Let $A, B \in \{1, 2, \ldots\}$ and let $l \in \{-B, \ldots, A\}$. Prove that

$$\sum_{j=\max(0,-l)}^{\min(B,A-l)} \binom{A}{j+l}\binom{B}{j} = \binom{A+B}{l+B}.$$

Proof. For all $s \in (0, +\infty)$ the following equality holds:

$$(1+s)^A\left(1+\frac{1}{s}\right)^B = \frac{(1+s)^{A+B}}{s^B}. \tag{$*$}$$

Now on the one hand we have by Newton's binomium

$$\begin{aligned}(1+s)^A\left(1+\frac{1}{s}\right)^B &= \left(\sum_{i=0}^{A}\binom{A}{i}s^i\right)\left(\sum_{j=0}^{B}\binom{B}{j}s^{-j}\right)\\ &= \sum_{i=0}^{A}\sum_{j=0}^{B}\binom{A}{i}\binom{B}{j}s^{i-j}. \end{aligned} \tag{$**$}$$

On the other hand we have

$$\frac{(1+s)^{A+B}}{s^B} = \sum_{k=0}^{A+B}\binom{A+B}{k}s^{k-B}. \tag{$***$}$$

By $(*)$ the expressions $(**)$ and $(***)$ are identical. Equating the corresponding coefficients of s^l we obtain

$$\binom{A+B}{l+B} = \sum_{j=\max(0,-l)}^{\min(B,A-l)}\binom{A}{j+l}\binom{B}{j}$$

for each $l \in \{-B, \ldots, A\}$. □

(ii) Again let $A, B \in \{1, 2, \dots\}$ and let $l \in \{-B+1, \dots, A\}$. Prove that

$$\sum_{j=\max(0,-l)}^{\min(B,A-l)} j \binom{A}{j+l}\binom{B}{j} = A\binom{A+B-1}{l+B-1} - l\binom{A+B}{l+B}.$$

Proof. For all $s, t \in (0, +\infty)$ we have

$$(1+s)^A\left(1+\frac{1}{t}\right)^B = \frac{(1+s)^A(1+t)^B}{t^B}. \qquad (*)$$

Applying (again) Newton's binomium we get

$$\begin{aligned}(1+s)^A\left(1+\frac{1}{t}\right)^B &= \left(\sum_{i=0}^{A}\binom{A}{i}s^i\right)\left(\sum_{j=0}^{B}\binom{B}{j}t^{-j}\right)\\ &= \sum_{i=0}^{A}\sum_{j=0}^{B}\binom{A}{i}\binom{B}{j}s^i t^{-j}.\end{aligned}$$

Differentiation of $(*)$ with respect to s shows that

$$A(1+s)^{A-1}(1+t)^B t^{-B} = \sum_{i=0}^{A}\sum_{j=0}^{B}\binom{A}{i}\binom{B}{j}\, i\, s^{i-1}t^{-j}.$$

Setting $s = t$, this reads as

$$A(1+s)^{A+B-1}s^{-B} = \sum_{i=0}^{A}\sum_{j=0}^{B}\binom{A}{i}\binom{B}{j}\, i\, s^{i-j-1}.$$

Hence

$$A\sum_{k=0}^{A+B-1}\binom{A+B-1}{k}s^{k-B+1} = \sum_{i=0}^{A}\sum_{j=0}^{B}\binom{A}{i}\binom{B}{j}\, i\, s^{i-j}$$

for all $s \in (0, +\infty)$. Equating the corresponding coefficients of s^l, for all $l \in \{-B+1, \dots, A\}$ we obtain

$$A\binom{A+B-1}{l+B-1} = \sum_{j=\max(0,-l)}^{\min(B,A-l)} (j+l)\binom{A}{j+l}\binom{B}{j}.$$

By (i) we see that

$$\sum_{j=\max(0,-l)}^{\min(B,A-l)} l\binom{A}{j+l}\binom{B}{j} = l\binom{A+B}{l+B}.$$

The required identity now follows immediately. $\square$

(iii) Prove that in the runs test

$$\mathbb{E}(T) = \frac{2n_1n_2}{n_1+n_2} + 1.$$

Proof. Without loss of generality we may assume that $n_2 \le n_1$. We first consider the case where $n_2 < n_1$. Then it is easily verified that T will assume an outcome in the set of integers $\{2, \ldots, 2n_2+1\}$. Now

$$\mathbb{E}(T) = \sum_{k \text{ even}} k \cdot \mathbb{P}(T=k) + \sum_{k \text{ odd}} k \cdot \mathbb{P}(T=k).$$

According to Theorem VI.3.2 we can write

$$\binom{n_1+n_2}{n_1} \sum_{k \text{ even}} k \cdot \mathbb{P}(T=k) = \sum_{m=1}^{n_2} (2m)\, 2 \binom{n_1-1}{m-1}\binom{n_2-1}{m-1}$$

$$= 4 \sum_{m=0}^{n_2-1} m \binom{n_1-1}{m}\binom{n_2-1}{m} + 4 \sum_{m=0}^{n_2-1} 1 \binom{n_1-1}{m}\binom{n_2-1}{m}$$

(using (i) and (ii) by setting $A = n_1 - 1,\ B = n_2 - 1$ and $l = 0$)

$$= 4(n_1-1)\binom{n_1+n_2-3}{n_2-2} + 4\binom{n_1+n_2-2}{n_2-1}. \qquad (*)$$

Furthermore, again using Theorem VI.3.2,

$$\binom{n_1+n_2}{n_1} \sum_{k \text{ odd}} k \cdot \mathbb{P}(T=k)$$

$$= \sum_{m=1}^{n_2} (2m+1) \binom{n_2-1}{m-1}\binom{n_1-1}{m}$$

$$+ \sum_{m=1}^{n_2-1} (2m+1) \binom{n_2-1}{m}\binom{n_1-1}{m-1}.$$

An outcome $2n_2+1$ for T is possible only if the sequence starts with an element from the sequence which contains n_1 symbols.

$$\sum_{m=1}^{n_2} (2m+1) \binom{n_2-1}{m-1}\binom{n_1-1}{m}$$

(setting $A = n_2 - 1,\ B = n_1 - 1$ and $l = -1$ in (ii))

$$= 2\sum_{m=1}^{A+1} m \binom{A}{m+l}\binom{B}{m} + \sum_{m=1}^{A+1} \binom{A}{m+l}\binom{B}{m}$$

$$= 2A\binom{A+B-1}{l+B-1} - 2l\binom{A+B}{l+B} + \binom{A+B}{l+B}$$

$$= 2(n_2-1)\binom{n_1+n_2-3}{n_1-3} + 3\binom{n_1+n_2-2}{n_1-2} \qquad (**)$$

and

$$\sum_{m=1}^{n_2-1} (2m+1) \binom{n_2-1}{m} \binom{n_1-1}{m-1}$$

(setting $A = n_1 - 1,\ B = n_2 - 1$ and $l = -1$ in (ii))

$$= \sum_{m=1}^{B} (2m+1) \binom{A}{m+l} \binom{B}{m}$$

$$= 2\left(A \binom{A+B-1}{l+B-1} - l \binom{A+B}{l+B}\right) + \binom{A+B}{l+B}$$

$$= 2(n_1-1) \binom{n_1+n_2-3}{n_2-3} + 3 \binom{n_1+n_2-2}{n_2-2}. \qquad (***)$$

Note that this makes sense only if $n_1, n_2 \geq 3$.

In order to get the expectation value of T, the terms in $(*)$, $(**)$ and $(***)$ should first be added. We shall verify that this results in the following expression:

$$\binom{n_1+n_2}{n_1}\left(\frac{2n_1n_2}{n_1+n_2}+1\right) = \frac{(n_1+n_2)!}{n_1!n_2!}\,\frac{2n_1n_2+n_1+n_2}{n_1+n_2} = \frac{(2n_1n_2+n_1+n_2)(n_1+n_2-1)!}{n_1!\,n_2!},$$

which proofs that indeed $\mathbb{E}(T) = \frac{2n_1n_2}{n_1+n_2} + 1$.

For typographical reasons we multiply things with $n_1!\,n_2!$. Thus we obtain

$$\begin{aligned}
n_1!\,n_2!\ \Big\{&4(n_1-1)\binom{n_1+n_2-3}{n_2-2} + 4\binom{n_1+n_2-2}{n_2-1} \\
&+ 2(n_2-1)\binom{n_1+n_2-3}{n_1-3} + 3\binom{n_1+n_2-2}{n_1-2} \\
&+ 2(n_1-1)\binom{n_1+n_2-3}{n_2-3} + 3\binom{n_1+n_2-2}{n_2-2}\Big\} \\
= \ & 4(n_1-1)\,(n_1+n_2-3)!\ n_2(n_2-1)n_1\ + 4\,(n_1+n_2-2)!\ n_2n_1 \\
&+ 2(n_2-1)\,(n_1+n_2-3)!\ n_1(n_1-1)(n_1-2) \\
&+ 3\,(n_1+n_2-2)!\ n_1(n_1-1) \\
&+ 2(n_1-1)\ (n_1+n_2-3)!\ n_2(n_2-1)(n_2-2) \\
&+ 3\,(n_1+n_2-2)!\ n_2(n_2-1) \\
= \ & (n_1+n_2-3)!\ (n_2-1)(n_1-1) \\
&\{4n_1n_2 + 2n_1(n_1-2) + 2n_2(n_2-2)\} \\
&+ (n_1+n_2-2)!\,\{4n_1n_2 + 3n_1(n_1-1) + 3n_2(n_2-1)\}.
\end{aligned}$$

Observing that $4n_1n_2+2n_1(n_1-2)+2n_2(n_2-2) = 2(n_1+n_2-2)(n_1+n_2)$, this can be rewritten as

$$(n_1+n_2-2)!\,\{\ldots\},$$

where the term between braces is given by

$$\begin{aligned}
2(n_2-1)(n_1-1)\,(n_1+n_2) &+ (4n_1n_2+3n_1(n_1-1)+3n_2(n_2-1)) \\
&= (2n_1^2n_2 - 2n_1^2 + 2n_1n_2^2 - 2n_1n_2 - 2n_1n_2 + 2n_1 \\
&\quad - 2n_2^2 + 2n_2) + (4n_1n_2 + 3n_1^2 - 3n_1 + 3n_2^2 - 3n_2) \\
&= 2n_1^2n_2 + 2n_1n_2^2 + n_1^2 - n_1 + n_2^2 - n_2 \\
&= (n_1+n_2-1)(2n_1n_2+n_1+n_2).
\end{aligned}$$

Thus we get the desired expression $(n_1+n_2-1)!\,(2n_1n_2+n_1+n_2)$.

Cases where $n_1, n_2 \in \{1,2\}$ should be treated ad hoc. The case $n_1 = n_2$ can be handled analogously. (J.D. Gibbons & S. Chakraborti give a proof which is more economical and far more elegant in their book 'Non-parametric Statistical Inference'.) □

Exercise 9

A bivariate (two-dimensional) population enjoys a probability distribution given by

$$\mathbb{P}((-1,1)) = \mathbb{P}((0,0)) = \mathbb{P}((1,1)) = \tfrac{1}{3}.$$

(i) Show that the correlation coefficient of this population is zero.

Solution. Denoting this correlation coefficient by $\rho(X,Y)$, we can write (see Definition I.5.2):

$$\begin{aligned}
&\rho(X,Y)\,\sigma_X\sigma_Y \\
&\quad= \mathrm{cov}(X,Y) = \mathbb{E}(XY) - \mathbb{E}(X)\,\mathbb{E}(Y) \\
&\quad= \sum_{k=-1}^{1} k\,\mathbb{P}(XY=k) - \left(\sum_{k=-1}^{1} k\,\mathbb{P}(X=k)\right)\left(\sum_{k=0}^{1} k\,\mathbb{P}(Y=k)\right) \\
&\quad= -\tfrac{1}{3} + 0 + \tfrac{1}{3} - (-\tfrac{1}{3} + 0 + \tfrac{1}{3})\,(0 + \tfrac{2}{3}) = 0.
\end{aligned}$$

Because $\sigma_X, \sigma_Y > 0$ this implies that $\rho(X,Y) = 0$. □

(ii) A sample $(X_1,Y_1), (X_2,Y_2)$ is drawn from this population. What is the probability that we will observe a tie in the variables X or Y ?

Solution. We have

$$\begin{aligned}
&\mathbb{P}(X_1 = X_2 \text{ or } Y_1 = Y_2) \\
&\quad = \mathbb{P}(X_1 = X_2 \text{ and } Y_1 \neq Y_2) + \mathbb{P}(X_1 = X_2 \text{ and } Y_1 = Y_2) \\
&\qquad + \mathbb{P}(X_1 \neq X_2 \text{ and } Y_1 = Y_2) \\
&\quad = \mathbb{P}[(X_1, Y_1) = (X_2, Y_2) = (-1, 1)] + \mathbb{P}[(X_1, Y_1) = (X_2, Y_2) = (0, 0)] \\
&\qquad + \mathbb{P}[(X_1, Y_1) = (X_2, Y_2) = (1, 1)] \\
&\qquad + \mathbb{P}[(X_1, Y_1) = (-1, 1) \text{ and } (X_2, Y_2) = (1, 1)] \\
&\qquad + \mathbb{P}[(X_1, Y_1) = (1, 1) \text{ and } (X_2, Y_2) = (-1, 1)] \\
&\quad = 5 \cdot \tfrac{1}{9} = \tfrac{5}{9}.
\end{aligned}$$

□

(iii) The same question when dealing with a sample of size ≥ 3.

Solution. Let us first consider a sample $(X_1, Y_1), (X_2, Y_2), (X_3, Y_3)$ of size 3. There are two possibilities:

(a) there is a pair of equal (vectorial) outcomes in this sequence,

(b) the vectorial outcomes are mutually different.

In both cases there is a tie, so there is a probability 1 that a tie will occur. Of course the same applies for samples of size larger than 3. □

Exercise 10

Let $(X_1, Y_1), \ldots, (X_n, Y_n)$ be a sample from a two-dimensional population which enjoys a continuous probability distribution. Setting $D_i := r(X_i) - r(Y_i)$, prove that Spearman's rank correlation coefficient R_S can be written as

$$R_S = 1 - \frac{6 \sum_{i=1}^n D_i^2}{n(n^2 - 1)}.$$

Proof. Using Exercise 3 (iii) we see that $\sum_i (r(X_i) - \frac{1}{2}(n+1))^2 = \frac{1}{12} n(n^2 - 1)$, which shows that the denominator in the definition of R_S equals $\frac{1}{12} n(n^2 - 1)$.

According to Exercise 3 (i) we have furthermore

$$\begin{aligned}
&\sum_{i=1}^n \left(r(X_i) - \tfrac{1}{2}(n+1)\right)\left(r(Y_i) - \tfrac{1}{2}(n+1)\right) \\
&\quad = \sum_{i=1}^n r(X_i) r(Y_i) - \tfrac{1}{2}(n+1) \sum_{i=1}^n r(X_i) - \tfrac{1}{2}(n+1) \sum_{i=1}^n r(Y_i) + \tfrac{1}{4} n(n+1)^2 \\
&\quad = \sum_{i=1}^n r(X_i) r(Y_i) - \tfrac{1}{4} n(n+1)^2.
\end{aligned}$$

Hence

$$\begin{aligned} R_S &= \left(\tfrac{1}{12}n(n^2-1)\right)^{-1}\left(\sum_{i=1}^{n} r(X_i)r(Y_i) - \tfrac{1}{4}n(n+1)^2\right) \\ &= \frac{12}{n(n^2-1)}\sum_{i=1}^{n} r(X_i)r(Y_i) - 3\,\frac{n+1}{n-1}. \end{aligned}$$

On the other hand (see Exercise 3 (ii)) we also have

$$\sum_{i=1}^{n} D_i^2 = \sum_{i=1}^{n}(r(X_i)-r(Y_i))^2 = 2\,\frac{n(n+1)(2n+1)}{6} - 2\sum_{i=1}^{n} r(X_i)r(Y_i),$$

which shows that

$$\frac{6\sum_{i=1}^{n} D_i^2}{n(n^2-1)} = \frac{2(2n+1)}{n-1} - \frac{12}{n(n^2-1)}\sum_{i=1}^{n} r(X_i)r(Y_i) = 1 - R_S.$$

This completes the proof. □

Exercise 11

Let $(X_1,Y_1),\ldots,(X_n,Y_n)$ be a sample from a two-dimensional population which has a continuous distribution function. Assume that the stochastic n-vectors $(r(X_1),\ldots,r(X_n))$ and $(r(Y_1),\ldots,r(Y_n))$ are statistically independent.

(i) Prove that

$$\mathbb{P}(R_S = 1) = \frac{1}{n!}.$$

Proof. According to Exercise 10 we have $R_S = 1$ if and only if $r(X_i) = r(Y_i)$ for all $i = 1,\ldots,n$. Consequently we may write (see Appendix C, Theorem 2)

$$\begin{aligned} \mathbb{P}(R_S = 1) &= \mathbb{P}\left(r(X_i) = r(Y_i) \text{ for all } i\right) \\ &= \prod_{i=1}^{n} \mathbb{P}\left(r(X_i) = r(Y_i) \mid r(X_j) = r(Y_j) \text{ for all } j < i\right) \\ &= \prod_{i=1}^{n} \frac{1}{n-(i-1)} = \frac{1}{n!}. \end{aligned}$$

□

(ii) Prove that the largest but one possible outcome for R_S is given by

$$1 - \frac{12}{n(n^2-1)}.$$

Proof. By Exercise 10 the value of $\sum_{i=1}^{n} D_i^2$ determines completely the value of R_S. Maximizing R_S amounts to the same as minimizing $\sum_{i=1}^{n} D_i^2$. To find the smallest possible value for $\sum_{i=1}^{n} D_i^2$ we first observe that it is not possible that $D_j = 1$ for some particular j and simultaneously $D_i = 0$ for all $i \neq j$. What *is* possible is that two of the $D_1, \ldots, D_n$ equal 1 whereas the other ones are zero. For example, this occurs if $(r(X_1), \ldots, r(X_n)) = (1, 2, 3, \ldots, n)$ and $(r(Y_1), \ldots, r(Y_n)) = (2, 1, 3, \ldots, n)$. Summarizing, the largest but one possible outcome of R_S is

$$1 - \frac{6 \cdot 2}{n(n^2-1)} = 1 - \frac{12}{n(n^2-1)}.$$

□

(iii) Prove that, under the null hypothesis,

$$\mathbb{P}\left(R_S \geq 1 - \frac{12}{n(n^2-1)}\right) = \frac{1}{(n-1)!}.$$

Proof. According to (ii) we see that

$$\mathbb{P}\left(R_S \geq 1 - \frac{12}{n(n^2-1)}\right) = \mathbb{P}(R_S = 1) + \mathbb{P}\left(R_S = \frac{12}{n(n^2-1)}\right).$$

By (i) we have $\mathbb{P}(R_S = 1) = 1/n!$; hence we have to determine an explicit expression for

$$\mathbb{P}\left(R_S = \frac{12}{n(n^2-1)}\right).$$

In (ii) we learned that $R_S = 12/n(n^2-1)$ if and only if $\sum_i D_i^2 = 1^2 + 1^2 = 2$. Now the question arises in how many ways this can occur.

The expression $\sum_i D_i^2$ equals 2 if and only if the rank vectors of

$$(r(X_1), \ldots, r(X_n)) \quad \text{and} \quad (r(Y_1), \ldots, r(Y_n))$$

differ only in two components, these differences (in absolute sense) being equal to 1, whereas the remaining $n-2$ components are equal. This can occur only if $(r(Y_1), \ldots, r(Y_n))$ can be obtained from $(r(X_1), \ldots, r(X_n))$ by exchanging two components that differ (in absolute sense) only by the number 1.

How many pairs of such rank vectors are possible? Well, once the vector $(r(X_1), \ldots, r(X_n))$ is known there are $n-1$ possible arrangements for $(r(Y_1), \ldots, r(Y_n))$. Namely, in $(r(X_1), \ldots, r(X_n))$ a pair of components of type $\{1, 2\}, \ldots, \{n-1, n\}$ has to be exchanged to get $(r(Y_1), \ldots, r(Y_n))$, leaving the remaining components unaltered. Because there are $n!$ possible arrangements for the vector $(r(X_1), \ldots, r(X_n))$, we arrive at a number of $(n-1)\,n!$ possible pairs

$$\{(r(X_1), \ldots, r(X_n)),\ (r(Y_1), \ldots, r(Y_n))\}$$

such that $\sum_i D_i^2 = 2$.

Without the latter constraint there would be $(n!)^2$ different possibilities for such a pair of rank-vectors. Summarizing, it follows that

$$\mathbb{P}\left(R_S \geq 1 - \frac{12}{n(n^2-1)}\right) = \frac{(n-1)\,n!}{(n!)^2} + \frac{1}{n!} = \frac{n}{n!} = \frac{1}{(n-1)!}. \qquad \square$$

Exercise 12

Prove that Spearman's rank correlation coefficient can be written as

$$R_S = \frac{12\sum_{i=1}^{n} r(X_i)r(Y_i) - 3n(n+1)^2}{(n+1)n(n-1)}.$$

Proof. By definition of R_S we have

$$R_S = \frac{\sum_i (r(X_i) - \frac{1}{2}(n+1))(r(Y_i) - \frac{1}{2}(n+1))}{\sqrt{\sum_i (r(X_i) - \frac{1}{2}(n+1))^2}\sqrt{\sum_i (r(Y_i) - \frac{1}{2}(n+1))^2}}.$$

By Lemma VI.4.1 (or: Exercise 3 (iii)) the denominator in this expression equals $\frac{1}{12}(n+1)n(n-1)$; the numerator is given by

$$\begin{aligned}
&\sum_{i=1}^{n} \left(r(X_i)r(Y_i) - \tfrac{1}{2}(n+1)r(X_i) - \tfrac{1}{2}(n+1)r(Y_i) + \tfrac{1}{4}(n+1)^2\right) \\
&= \sum_{i=1}^{n} r(X_i)r(Y_i) - \tfrac{1}{2}(n+1)\sum_{i=1}^{n} r(X_i) - \tfrac{1}{2}(n+1)\sum_{i=1}^{n} r(Y_i) + \tfrac{1}{4}n(n+1)^2 \\
&= \sum_{i=1}^{n} r(X_i)r(Y_i) - 2\cdot\tfrac{1}{2}(n+1)\cdot\tfrac{1}{2}n(n+1) + \tfrac{1}{4}n(n+1)^2 \\
&= \sum_{i=1}^{n} r(X_i)r(Y_i) - \tfrac{1}{4}n(n+1)^2.
\end{aligned}$$

Using this, the reader can easily complete the proof. $\square$

Exercise 13

Let $(X_1, Y_1), \ldots, (X_n, Y_n)$ be a sample from a two-dimensional population having a continuous distribution function.

(i) Show that

$$\operatorname{var}(r(X_i)) = \operatorname{var}(r(Y_j)) = \frac{n^2-1}{12} \quad \text{for all } i, j.$$

Solution. Exploiting symmetry we may set

$$\operatorname{var}(r(X_i)) = \operatorname{var}(r(Y_j)) = \operatorname{var}(r(X_1))$$

for all i, j. With the help of Exercise 3 (i) we see that

$$\mathbb{E}(r(X_1)) = \sum_{k=1}^{n} k\ \mathbb{P}(r(X_1) = k) = \sum_{k=1}^{n} \frac{k}{n} = \tfrac{1}{n} \cdot \tfrac{1}{2} n(n+1) = \tfrac{1}{2}(n+1)$$

and by Exercise 3 (ii) we see that

$$\mathbb{E}(r(X_1)^2) = \sum_{k=1}^{n} \frac{1}{n} k^2 = \tfrac{1}{6}(n+1)(2n+1).$$

Therefore

$$\begin{aligned} \operatorname{var}(r(X_1)) &= \tfrac{1}{6}(n+1)(2n+1) - \tfrac{1}{4}(n+1)^2 \\ &= \tfrac{1}{12}(n+1)\left\{2(2n+1) - 3(n+1)\right\} \\ &= \tfrac{1}{12}(n+1)(n-1) = \tfrac{1}{12}(n^2-1). \end{aligned}$$

□

(ii) Under the assumption that $r(X_i)$ and $r(Y_i)$ are not correlated, prove that

$$\mathbb{E}(r(X_i)r(Y_i)) = \tfrac{1}{4}(n+1)^2.$$

Proof. We then have

$$\mathbb{E}(r(X_i)r(Y_i)) = \mathbb{E}(r(X_i))\,\mathbb{E}(r(Y_i)) = \left(\tfrac{1}{2}(n+1)\right)^2 = \tfrac{1}{4}\,(n+1)^2.$$

□

(iii) Prove that for all $i \neq j$ we have

$$\operatorname{cov}(r(X_i), r(X_j)) = \operatorname{cov}(r(Y_i), r(Y_j)) = -\tfrac{1}{2}(n+1).$$

Proof. If $i \neq j$, then

$$\operatorname{cov}(r(X_i), r(X_j)) = \mathbb{E}(r(X_i)r(X_j)) - \mathbb{E}(r(X_i))\,\mathbb{E}(r(X_j)). \qquad (*)$$

Here

$$\mathbb{E}(r(X_i))\,\mathbb{E}(r(X_j)) \overset{\text{(i)}}{=} \left(\tfrac{1}{2}(n+1)\right)^2 = \tfrac{1}{4}\,(n+1)^2$$

and furthermore

$$\begin{aligned} \mathbb{E}(r(X_i)r(X_j)) &= \sum_{k,l} kl\,\mathbb{P}(X_i = k \text{ and } X_j = l) \\ &= \sum_{k\neq l} kl\,\mathbb{P}(X_i = k)\,\mathbb{P}(X_j = l \mid X_i = k) \\ &= \sum_{k\neq l} kl\,\frac{1}{n}\,\frac{1}{n-1} \end{aligned}$$

$$
\begin{aligned}
&= \frac{1}{n(n-1)}\left(\left(\sum_{k=1}^{n}k\right)\left(\sum_{l=1}^{n}l\right)-\sum_{k=1}^{n}k^2\right)\\
&= \frac{n(n+1)}{n(n-1)}\left(\tfrac{1}{4}n(n+1)-\tfrac{1}{6}(2n+1)\right)\\
&= \frac{n+1}{12(n-1)}\left(3n(n+1)-2(2n+1)\right)\\
&= \frac{(n+1)(n-1)(3n+2)}{12(n-1)} = \tfrac{1}{12}(n+1)(3n+2).
\end{aligned}
$$

Substituting these expressions in $(*)$, we get

$$
\begin{aligned}
\operatorname{cov}(r(X_i), r(X_j)) &= \tfrac{1}{12}(n+1)(3n+2) - \tfrac{1}{4}(n+1)^2\\
&= \tfrac{1}{12}(n+1)\{3n+2-3(n+1)\} = -\tfrac{1}{12}(n+1).
\end{aligned}
$$

Of course we then also have $\operatorname{cov}(r(Y_i), r(Y_j)) = -\frac{1}{12}(n+1)$. □

(iv) If for all i there is no correlation between the variables $r(X_i)$ and $r(Y_i)$ then $\mathbb{E}(R_S) = 0$. Moreover, if the rank vectors $(r(X_1), \ldots, r(X_n))$ and $(r(Y_1), \ldots, r(Y_n))$ are statistically independent, then $\operatorname{var}(R_S) = \frac{1}{n-1}$. Prove this.

Proof. If for all i the variables $r(X_i)$ and $r(Y_i)$ are not correlated, then (using Proposition VI.4.2 together with (ii)) we may write

$$
\begin{aligned}
\mathbb{E}(R_S) &= \mathbb{E}\left(\frac{12(\sum_i r(X_i)r(Y_i)) - 3n(n+1)^2}{(n-1)n(n+1)}\right)\\
&= \frac{12}{n(n^2-1)}\,\mathbb{E}\left(\sum_{i=1}^{n} r(X_i)r(Y_i)\right) - \frac{3(n+1)}{n-1}\\
&= \frac{12}{n(n^2-1)}\cdot n \cdot \frac{(n+1)^2}{4} - \frac{3(n+1)}{n-1}\\
&= \frac{3(n+1)-3(n+1)}{n-1} = 0.
\end{aligned}
$$

Next, suppose that $(r(X_1), \ldots, r(X_n))$ and $(r(Y_1), \ldots, r(Y_n))$ are statistically independent. Then, by the foregoing, $\mathbb{E}(R_S) = 0$. Consequently

$$
\operatorname{var}(R_S) = \mathbb{E}(R_S^2) - \mathbb{E}(R_S)^2 = \mathbb{E}(R_S^2).
$$

By Proposition VI.4.2 we have

$$
R_S = \frac{12\sum_i r(X_i)r(Y_i) - 3n(n+1)^2}{(n-1)n(n+1)},
$$

so

$$
\mathbb{E}(R_S^2) = \frac{\mathbb{E}\left([12\sum_i r(X_i)r(Y_i) - 3n(n+1)^2]^2\right)}{(n-1)^2 n^2 (n+1)^2}.
$$

Here the numerator can be written as

$$144\,\mathbb{E}\left[\left(\sum_{i=1}^{n} r(X_i)r(Y_i)\right)^2\right] - 72n(n+1)^2\,\mathbb{E}\left[\sum_{i=1}^{n} r(X_i)r(Y_i)\right] + 9n^2(n+1)^4.$$

Next we write out

$$\begin{aligned}
\mathbb{E}\left[\left(\sum_{i=1}^{n} r(X_i)r(Y_i)\right)^2\right] \\
&= \mathbb{E}\left(\sum_{i,j} r(X_i)r(Y_i)r(X_j)r(Y_j)\right) \\
&= \sum_{i=1}^{n} \mathbb{E}\left(r(X_i)^2 r(Y_i)^2\right) + \sum_{i\neq j} \mathbb{E}\left(r(X_i)r(Y_i)r(X_j)r(Y_j)\right) \\
\text{(symmetry)} \quad &= n\ \mathbb{E}(r(X_1)^2)\ \mathbb{E}(r(Y_1)^2) + \\
&\quad + n(n-1)\ \mathbb{E}(r(X_1)r(X_2))\ \mathbb{E}(r(Y_1)r(Y_2)) \\
\text{((i) and (iii))} \quad &= n\ \left(\tfrac{1}{6}(n+1)(2n+1)\right)^2 + \\
&\quad + n(n-1)\ \left(\tfrac{1}{12}(n+1)(3n+2)\right)^2 \\
&= \tfrac{1}{144}n(n+1)^2\left\{4(2n+1)^2 + (n-1)(3n+2)^2\right\} \\
&= \tfrac{1}{144}n^2(n+1)^2\left\{9n^2 + 19n + 8\right\}.
\end{aligned}$$

Using (ii) we have

$$\mathbb{E}\left[\sum_{i=1}^{n} r(X_i)r(Y_i)\right] = \sum_{i=1}^{n} \mathbb{E}(r(X_i)r(Y_i)) = \tfrac{1}{4}n(n+1)^2,$$

so that

$$\begin{aligned}
\mathbb{E}\left([\ldots]^2\right) &= n^2(n+1)^2(9n^2+19n+8) - 18n^2(n+1)^4 + 9n^2(n+1)^4 \\
&= n^2(n+1)^2\left\{9n^2+19n+8-9(n+1)^2\right\} \\
&= n^2(n+1)^2\,(n-1).
\end{aligned}$$

In this way we arrive at

$$\operatorname{var}(R_S) = \mathbb{E}(R_S^2) = \frac{1}{n-1}.$$

□

Exercise 14

A two-dimensional (paired) sample shows the following outcome:

i	1	2	3	4	5	6	7	8
X_i	6.3	8.2	4.1	9.2	5.5	8.3	7.2	8.1
Y_i	7.7	6.4	2.3	8.4	6.7	3.6	7.9	9.5

(i) Determine Spearman's rank correlation coefficient.

Solution. Concerning the rank numbers, we have

i	1	2	3	4	5	6	7	8
$r(X_i)$	3	6	1	8	2	7	4	5
$r(Y_i)$	5	3	1	7	4	2	6	8

According to Proposition VI.4.2 this leads to the following outcome for R_S:

$$R_S = \frac{12\,(15+18+1+56+8+14+24+40) - 3\cdot 8\cdot 81}{7\cdot 8\cdot 9} = \frac{1}{3}.$$

(ii) At a level of significance $\alpha = 0.10$, test $H_0 : \rho(X,Y) = 0$ against $H_1 : \rho(X,Y) > 0$.

Solution. In Table IX it can be read off that under these conditions the critical region for R_S equals $[0.524, +\infty)$. Hence we maintain H_0.

As an example, we compute an approximation for the P-value, starting from the assumption that R_S is (under H_0) approximately $N(0, \frac{1}{7})$-distributed. Using Table II, this leads to

$$\mathbb{P}\,(R_S \geq 0.333) \approx \mathbb{P}\left(Z \geq \frac{0.333}{\sqrt{1/7}}\right) = 1 - 0.81 = 0.19,$$

where Z is some $N(0,1)$-distributed variable. Again it seems very reasonable to maintain H_0. □

Exercise 15

Let T_X, T_Y be the test statistics in Wilcoxon's rank-sum test.

(i) Prove that $\text{cov}(\mathfrak{r}(X_1), T_X + T_Y) = 0$.

Proof. By Proposition VI.2.2 the variable $T_X + T_Y$ is identical to the constant $T_X + T_Y = \frac{1}{2}(m+n)(m+n+1)$. Applying Lemma I.5.12 we therefore have $\text{cov}(\mathfrak{r}(X_1), T_X + T_Y) = 0$. □

(ii) Prove that $\text{cov}(\mathfrak{r}(X_1), \mathfrak{r}(X_2)) = -\frac{1}{12}(m+n+1)$.

Proof. Under H_0 the probability distribution of $\mathfrak{r}(X_1)$ is determined by

$$\mathbb{P}(\mathfrak{r}(X_1) = k) = \frac{1}{m+n} \quad \text{for } k = 1, \ldots, m+n.$$

Hence

$$\mathbb{E}(\mathfrak{r}(X_1)) = \sum_{k=1}^{m+n} k\,\frac{1}{m+n} = \tfrac{1}{2}(m+n+1)$$

and

$$\mathbb{E}(\mathfrak{r}(X_1)^2) = \sum_{k=1}^{m+n} k^2 \frac{1}{m+n} = \tfrac{1}{6}(m+n+1)(2m+2n+1).$$

This shows that

$$\begin{aligned} \text{var}(\mathfrak{r}(X_1)) &= (m+n+1)\ \left[\tfrac{1}{6}(2m+2n+1) - \tfrac{1}{4}(m+n+1)\right] \\ &= \tfrac{1}{12}(m+n+1)(m+n-1). \qquad (*) \end{aligned}$$

Exploiting symmetry we see that under H_0:

$$\text{cov}(\mathfrak{r}(X_1), \mathfrak{r}(X_2)) = \text{cov}(\mathfrak{r}(X_1), \mathfrak{r}(X_i)) = \text{cov}(\mathfrak{r}(X_1), \mathfrak{r}(Y_j))$$

for all $i \geq 2$ and $j \geq 1$. Therefore we can write

$$\begin{aligned} 0 &= \text{cov}(\mathfrak{r}(X_1), T_X + T_Y) \\ &= \text{var}(\mathfrak{r}(X_1)) + \sum_{i=2}^{m} \text{cov}(\mathfrak{r}(X_1), \mathfrak{r}(X_i)) + \sum_{j=1}^{n} \text{cov}(\mathfrak{r}(X_1), \mathfrak{r}(Y_j)) \\ &= \text{var}(\mathfrak{r}(X_1)) + (m+n-1)\,\text{cov}(\mathfrak{r}(X_1), \mathfrak{r}(X_2)). \end{aligned}$$

By $(*)$ we thus arrive at

$$\text{cov}(\mathfrak{r}(X_1), \mathfrak{r}(X_2)) = -\frac{\text{var}(\mathfrak{r}(X_1))}{m+n-1} = -\tfrac{1}{12}(m+n+1). \qquad \square$$

Exercise 16

Given are four populations. We assume that their continuous probability distributions are mutually shifted. The outcomes of four independent samples are given in the following scheme:

population				
1	54.3	59.7	49.5	
2	55.1	45.2		
3	52.1	54.3	56.9	50.6
4	44.8	47.4	54.3	

At a level of significance $\alpha = 0.05$, test whether the mutual shifts are zero or not.

Solution. To this we apply the Kruskal–Wallis test. The corresponding rank numbers (note that we have a tie so we have to apply the 'mid-rank method') are given by

population				
1	8	12	4	
2	10	2		
3	6	8	11	5
4	1	3	8	

The outcome of the Kruskal–Wallis test statistic T can now be computed as

$$\tfrac{12}{12\cdot 13}\left[3\left(\tfrac{24}{3}-\tfrac{13}{2}\right)^2+2\left(\tfrac{12}{2}-\tfrac{13}{2}\right)^2+4\left(\tfrac{30}{4}-\tfrac{13}{2}\right)^2+3\left(\tfrac{12}{3}-\tfrac{13}{2}\right)^2\right]$$

which equals $\frac{30}{13} = 2.308$. We approximate the distribution of T by the $\chi^2(3)$-distribution (see Theorem VI.5.1). We then get $g = 7.81$ as our critical value for T: therefore we accept H_0. □

Exercise 17

Prove that (under H_0) in Friedman's test the variables R_{ijk} enjoy the following four properties:

(i) $\mathbb{E}(R_{ijk}) = \frac{1}{2}(pn+1)$ for all i, j, k,

(ii) $\operatorname{var}(R_{ijk}) = \frac{1}{12}(pn+1)(pn-1)$ for all i, j, k,

(iii) $\operatorname{cov}(R_{isk}, R_{itl}) = 0$ for $s \neq t$,

(iv) $\operatorname{cov}(R_{ijk}, R_{ijl}) = -\frac{1}{12}(pn+1)$ for $k \neq l$.

Proof.

(i) Firstly,

$$\begin{aligned}\mathbb{E}(R_{ijk}) &= \sum_{l=1}^{pn} l\,\mathbb{P}(R_{ijk}=l) = \frac{1}{pn}\sum_{l=1}^{pn} l \\ &= \frac{\frac{1}{2}pn(pn+1)}{pn} = \tfrac{1}{2}(pn+1) \quad \text{for all } i,j,k.\end{aligned}$$

(ii) Furthermore

$$\mathbb{E}(R_{ijk}^2) = \frac{1}{pn}\sum_{l=1}^{pn} l^2 = \frac{\frac{1}{6}pn(pn+1)(2pn+1)}{pn} = \tfrac{1}{6}(pn+1)(2pn+1)$$

so that for all i, j, k

$$\operatorname{var}(R_{ijk}) = \tfrac{1}{6}(pn+1)(2pn+1) - \tfrac{1}{4}(pn+1)^2 = \tfrac{1}{12}(pn+1)(pn-1).$$

(iii) For arbitrary i, k, l and $s \neq t$ the variables R_{isk} and R_{itl} are statistically independent. To see this, note that we can write

$$R_{isk} = \varphi_1(X_{1s1}, \ldots, X_{1sn}, \;\ldots\;, X_{ps1}, \ldots, X_{psn})$$

and

$$R_{itl} = \varphi_2(X_{1t1}, \ldots, X_{1tn}, \;\ldots\;, X_{pt1}, \ldots, X_{ptn}),$$

where φ_1 and φ_2 are certain Borel functions. The two stochastic vectors involved are, by assumption, statistically independent. Applying Proposition I.4.2 and Proposition I.5.10 we see that $\operatorname{cov}(R_{isk}, R_{itl}) = 0$.

(iv) For fixed j we consider the sequence

$$X_{1j1}, \ldots, X_{1jn}, \ \ldots \ , X_{pj1}, \ldots, X_{pjn}.$$

Under H_0 this sequence can be regarded as a sample. Because $\sum_{c,d} R_{cjd}$ is a constant stochastic variable we have (see Exercise 15 (i))

$$\text{cov}(R_{ijk}, \textstyle\sum_{c,d} R_{cjd}) = 0.$$

Furthermore, exploiting symmetry, we can write

$$\text{cov}(R_{ijk}, R_{cjd}) = \text{cov}(R_{ijk}, R_{ejf})$$

for all $(e,f) \neq (i,k)$ and $(c,d) \neq (i,k)$. From this it follows that for an arbitrary $(e,f) \neq (i,k)$ we have

$$0 = \text{cov}(R_{ijk}, \textstyle\sum_{c,d} R_{cjd}) = \text{var}(R_{ijk}) + (pn-1)\text{cov}(R_{ijk}, R_{ejf})$$

and

$$\text{cov}(R_{ijk}, R_{ejf}) = -\frac{\text{var}(R_{ijk})}{pn-1} = -\tfrac{1}{12}(pn+1).$$

In particular, for $k \neq l$ we have:

$$\text{cov}(R_{ijk}, R_{ijl}) = -\tfrac{1}{12}(pn+1).$$

□

Exercise 18

We are dealing with a two-factor experiment in which the factors A and B can be adjusted at three and four levels respectively. At each possible simultaneous level a sample of size 3 is drawn. We get the following outcomes:

$B \downarrow \quad A \rightarrow$	1			2			3		
1	3.1	2.7	5.3	4.9	3.4	4.8	7.1	8.1	8.7
2	8.3	7.1	7.9	8.1	6.9	9.1	9.5	9.6	8.0
3	9.1	10.5	9.3	11.0	10.1	8.9	9.6	8.9	9.3
4	8.4	10.8	9.6	12.9	8.1	9.8	11.7	10.4	9.9

Using Friedman's test, test, at a level of significance $\alpha = 0.05$, whether factor A assorts effect.

Solution. The corresponding rank numbers are (again we use the 'mid-rank method') arranged in the scheme below:

	1			2			3		
1	2	1	6	5	3	4	7	8	9
2	6	2	3	5	1	7	8	9	4
3	3	8	4.5	9	7	1.5	6	1.5	4.5
4	2	7	3	9	1	4	8	6	5

It follows that

$$R_{1\bullet\bullet} = 3\tfrac{23}{24}, \quad R_{2\bullet\bullet} = 4\tfrac{17}{24} \quad \text{and} \quad R_{3\bullet\bullet} = 6\tfrac{1}{3}.$$

Thus we see that T exhibits the following outcome:

$$T = \frac{12 \cdot 4}{3(3 \cdot 3 + 1)} \sum_{i=1}^{3} \left(R_{i\bullet\bullet} - \tfrac{1}{2}(3 \cdot 3 + 1)\right)^2 = 4.717.$$

Approximating the distribution of T by a $\chi^2(2)$-distribution (see Theorem VI.6.2), we see that $g = 5.99$ presents the critical value for T if $\alpha = 0.05$. We therefore accept H_0. □

Chapter 7
Stochastic analysis and its applications in statistics

7.1 Summary of Chapter VII

7.1.1 The empirical distribution function corresponding to a sample (VII.1)

Let $X_1, \dots, X_n$ be a sample from a population that has F as its distribution function. As an estimator of F we introduce the *empirical distribution function* $\hat{F} = \hat{F}(X_1, \dots, X_n)$ via

$$\hat{F}(X_1, \dots, X_n) : x \mapsto \frac{\#\{i : X_i \leq x\}}{n} = \overline{Y},$$

where $Y_i := 1_{(-\infty,x]}(X_i)$ for $i = 1, \dots, n$. These Y_i constitute a sample from a Bernoulli distributed population with parameter $F(x)$, from which it follows that for all $x \in \mathbb{R}$

$$\mathbb{E}(\hat{F}(x)) = F(x) \quad \text{and} \quad \operatorname{var}(\hat{F}(x)) = \tfrac{1}{n} F(x)(1 - F(x)).$$

The $\hat{F}$ as just defined is both a distribution and a step function.

7.1.2 Convergence of stochastic variables (VII.2)

Suppose a probability space $(\Omega, \mathfrak{A}, \mathbb{P})$ is given on which there is defined a sequence $X, X_1, X_2, \dots$ of stochastic variables. We are looking for ways in which the sequence $X_1, X_2, \dots$ may be said to *converge* to X. In §I.8 we have already seen such a way: $X_1, X_2, \dots$ is said to converge *in distribution* to X if

$$\lim_{n \to \infty} F_{X_n}(s) = F_X(s)$$

for each continuity point s of F_X.

A second mode of convergence is the following. The sequence $X_1, X_2, \dots$ is said to converge *in probability* (or: *in measure*) to X if for all $\varepsilon > 0$

$$\mathbb{P}(\{\omega \in \Omega : |X_n(\omega) - X(\omega)| > \varepsilon\}) = \mathbb{P}(|X_n - X| > \varepsilon) \to 0.$$

Thirdly, the sequence $X_1, X_2, \ldots$ is said to converge *almost surely* (or: *strongly*) to X if $X_n(\omega) \to X(\omega)$ for 'almost every' $\omega \in \Omega$, that is if $\mathbb{P}(C) = 1$ for

$$C := \{\omega \in \Omega \, : X_n(\omega) \to X(\omega)\}.$$

Convergence almost surely is stronger than convergence in probability, which in turn is stronger than convergence in distribution. If X is a constant variable, then convergence in probability is equivalent to convergence in distribution.

Two stochastic n-vectors $\mathbf{X}$ and $\mathbf{X}'$ are by definition *twins* if they are componentwise identically distributed. Two sequences $X_1, X_2, \ldots$ and $X_1', X_2', \ldots$ are called *twin sequences* if X_i and X_i' are identically distributed for all i.

For any distribution function $F : \mathbb{R} \to [0,1]$ the *quantile function* $q_F : (0,1) \to \mathbb{R}$ is defined by

$$q_F : u \;\mapsto\; \inf\{y \, : \, u \leq F(y)\}.$$

If F is continuous and strictly increasing this q_F is its inverse. It is always so that q_F is increasing and left-continuous. We also have:

$$u \leq F(x) \;\;\Leftrightarrow\;\; q_F(u) \leq x \quad (\text{all } u \in (0,1) \text{ and } x \in \mathbb{R}).$$

With each variable $X : \Omega \to \mathbb{R}$ we associate a twin X' on $(\Omega', \mathfrak{A}', \mathbb{P}')$ where

$$\Omega' := (0,1), \quad \mathfrak{A}' := \{A \cap (0,1) : A \in \mathfrak{B}\} \quad \text{and} \quad \mathbb{P}' := \lambda|_{(0,1)}$$

by defining $X'(u) := q_F(u)$. In this way X' and X are indeed identically distributed; X' is called the *standard twin* of X.

A famous theorem due to Skorokhod states: if $X_1, X_2, \ldots$ converges to X in distribution, then the standard twin sequence $X_1', X_2', \ldots$ converges *almost surely* to the standard twin X' of X.

Using this theorem it can easily be deduced that the following two statements are equivalent:

(i) The sequence $X_1, X_2, \ldots$ converges to X in distribution,

(ii) $\mathbb{E}[\varphi(X_n)] \to \mathbb{E}[\varphi(X)]$ for each bounded and continuous $\varphi : \mathbb{R} \to \mathbb{R}$.

Of utmost importance is the *strong law of large numbers*: if $X_1, X_2, \ldots$ is a sample from a population with mean μ, then the sequence

$$\overline{X}_n := \frac{X_1 + \cdots + X_n}{n} \qquad (n = 1, 2, \ldots)$$

converges almost surely to μ. (In that case a fortiori $\overline{X}_n \to \mu$ in probability: the 'weak law of large numbers').

7.1.3 The Glivenko–Cantelli theorem (VII.3)

Let $\mathfrak{D}(\mathbb{R})$ denote the linear space of all bounded functions $\varphi : \mathbb{R} \to \mathbb{R}$ that are 'cadlag', that is, all bounded φ for which

$$\varphi(x-) \text{ exists} \quad \text{and} \quad \varphi(x+) = \varphi(x) \quad \text{for each } x \in \mathbb{R}.$$

We equip $\mathfrak{D}(\mathbb{R})$ with the supremum norm $\| \ \|_\infty$ defined by

$$\| \varphi \|_\infty := \sup_{x \in \mathbb{R}} |\varphi(x)|.$$

In this way $\mathfrak{D}(\mathbb{R})$ becomes a non-separable and complete metric space, which includes all distribution functions.

The *Glivenko–Cantelli theorem* states: if $X_1, X_2, \dots$ is a sample from a population which has F as its distribution function, then

$$\hat{F}(X_1, \dots, X_n) \to F \quad \text{almost surely in } \mathfrak{D}(\mathbb{R}).$$

That is, $\| \hat{F}(X_1, \dots, X_n) - F \|_\infty \to 0$ almost surely.

7.1.4 The Kolmogorov–Smirnov test statistic (VII.4)

Let $X_1, \dots, X_n$ be a sample from a population with unknown distribution function F. We shall use the test statistic

$$D_n = D_n(F_0) := \|\hat{F}(X_1, \dots, X_n) - F_0\|_\infty$$

to decide whether $H_0 : F = F_0$ should be accepted or $H_1 : F \neq F_0$.

It appears that the probability distribution of D_n under H_0 is the same for *all continuous* F_0. This D_n is called the *Kolmogorov–Smirnov test statistic.* Tables exist that describe its probability distribution for continuous F_0. If $\mathbb{P}(D_n > \gamma) = \alpha$ under H_0, then

$$\{F \ : \ F \text{ is continuous and } \hat{F} - \gamma \leq F \leq \hat{F} + \gamma\}$$

is called a $(1 - \alpha) \times 100\%$ *confidence region* for F.

7.1.5 Metrics on the set of distribution functions (VII.5)

We set

$$\Gamma \ := \ \{F : \mathbb{R} \to [0, 1] \mid \ F \ \text{ is a distribution function}\}$$

and it is our concern to find suitable metrics on Γ. The most obvious choice is the *norm metric* d_N, defined by $d_N(F, G) := \|F - G\|_\infty$. This metric however proves to be rather rude. An alternative is the *Lévy metric* d_L, given via

$$d_L(F, G) := \inf\{\delta \geq 0 \ : \ F(x - \delta) - \delta \leq G(x) \leq F(x + \delta) + \delta \quad (\textit{all} \ \ x \in \mathbb{R})\}.$$

In general we have

$$d_L(F,G) \leq d_N(F,G),$$

so that d_N-convergence implies d_L-convergence. If $F \in \Gamma$ is continuous, then $F_k \xrightarrow{d_N} F$ is equivalent to $F_k \xrightarrow{d_L} F$.

The metric space (Γ, d_L) has the desirable property of being both complete and separable. Besides, d_L-convergence appears to be 'identical' to convergence in distribution.

The section concludes by giving some more possibilities for metrics on Γ, like the *Prokhorov metric* d_P, the *bounded Lipschitz metric* d_{BL} and the *Skorokhod metric* d_S.

7.1.6 Smoothing techniques (VII.6)

Let F and G be arbitrary distribution functions, that is, suppose $F, G \in \Gamma$. By $F \times G$ we mean the distribution function of $X+Y$, where X and Y are independent variables having F and G for their distribution functions. This $F \times G$ is called the *distribution product* of F and G. It appears that $(\Gamma, \times)$ is a semi-group with $E := 1_{[0,+\infty)}$ as its unit element.

For each bounded Borel function $\varphi : \mathbb{R} \to \mathbb{R}$ and probability measure $\mathbb{P} : \mathfrak{B} \to [0,1]$ we define their *convolution product* $\varphi * \mathbb{P}$ by

$$\varphi * \mathbb{P} \ : \ x \ \mapsto \ \int \varphi(x-y)\, d\mathbb{P}(y).$$

If f is a density function then by $\varphi * f$ we mean $\varphi * (f\lambda)$.

If φ is continuously differentiable and φ' is bounded, then we have

$$(\varphi * \mathbb{P})' = \varphi' * \mathbb{P}.$$

Using this we find: if $G \in \Gamma$ is k times continuously differentiable with bounded derivatives and $F \in \Gamma$, then

$$(G \times F)^{(k)} = G^{(k)} * \mathbb{P}_F,$$

which for $k = 0$ reads as $G \times F = G * \mathbb{P}_F$.

For each $G \in \Gamma$ and $\zeta > 0$ we define $G_\zeta : x \mapsto G(x/\zeta)$. Then G_ζ again is a distribution function and in d_L-sense we have

$$\lim_{\zeta \downarrow 0} G_\zeta \times F = F.$$

This provides a method to approximate discontinuous (for example: empirical) distribution functions by smoother ones. We talk about 'smoothing techniques'. A nice illustration of this is the fact that for *any* $F \in \Gamma$ and $\zeta > 0$

$$\Phi_\zeta \times F = \Phi_\zeta * \mathbb{P}_F$$

is infinitely many times differentiable.

7.1.7 Robustness of statistics (VII.7)

Let $g : \mathbb{R}^n \to \mathbb{R}$ be a Borel function. The map $\tilde{g} : \Gamma \to \Gamma$ is constructed as follows. For each $F \in \Gamma$ a sequence of independent variables $X_1, \ldots, X_n$ exists having F as their common distribution function. We define $\tilde{g}(F)$ as the distribution function of $g(X_1, \ldots, X_n)$, that is

$$\tilde{g}(F) := F_{g(X_1, \ldots, X_n)}.$$

For each continuous $g : \mathbb{R}^n \to \mathbb{R}$ the map $\tilde{g} : (\Gamma, d_L) \to (\Gamma, d_L)$ is continuous, which shows that a condition like '$\tilde{g}$ is continuous' is rather weak. Instead sequences of Borel functions $g_k : \mathbb{R}^k \to \mathbb{R}$, where $k = 1, 2, \ldots$, are considered and it is investigated whether for any $F \in \Gamma$ the sequence

$$\widetilde{g_n} : (\Gamma, d_1) \to (\Gamma, d_2)$$

is equicontinuous (see Appendix E) in F. Here d_1 and d_2 denote arbitrary metrics on Γ. If the above applies, the sequence $Y_1 := g_1(X_1)$, $Y_2 := g_2(X_1, X_2)$, ... is called *(d_1, d_2)-robust in F*. Here $X_1, X_2, \ldots$ is an infinite sample from a $\mathbb{P}_F$-distributed population.

A distribution function G is called *stoutly tailed* if for all $t \neq 0$

$$\lim_{n \to \infty} \{\chi_G(t/n)\}^n = 0.$$

For corresponding variables $X_1, X_2, \ldots$ the weak law of large numbers does *not* hold. The set of smooth distribution functions which are stoutly tailed turns out to be dense in (Γ, d_L).

A distribution function G is called *tailless* if a constant $M > 0$ exists for which $G(-M) = 0$ and $G(M) = 1$. This is the case for each empirical distribution function. The set of tailless elements in Γ is also dense in (Γ, d_L).

From the fact that both the sets of stoutly tailed and tailless elements are dense in (Γ, d_L) it can be deduced (using some general theorems on equicontinuity) that the sample mean $\overline{X}$ is *not* d_L-robust for *any* $F \in \Gamma$.

7.1.8 Trimmed means, the median and robustness of these statistics (VII.8)

In order to estimate the 'middle' of a population we noticed that the mean $\overline{X}$ has the nasty property of not being robust. In this section alternatives are developed which *are* robust.

First we define the *empirical quantile function* $\hat{q}(x_1, \ldots, x_n) : (0,1) \to \mathbb{R}$, corresponding to a sample outcome $x_1, \ldots, x_n$, by being the quantile function of the empirical distribution function $\hat{F}(x_1, \ldots, x_n)$. Now for all $\alpha \in [0, \frac{1}{2})$ we set

$$g_{n,\alpha}(x_1, \ldots, x_n) := \frac{1}{1 - 2\alpha} \int_{\alpha}^{1-\alpha} \hat{q}(x_1, \ldots, x_n)(u)\, du$$

for $n = 1, 2, \dots$ and $x_1, \dots, x_n \in \mathbb{R}$. In this way $g_{n,0}(x_1, \dots, x_n) = \overline{x}$. The function $g_{n,\alpha}$ is always continuous on $\mathbb{R}^n$. For an arbitrary sample $X_1, \dots, X_n$ we define the *α-trimmed mean of the sample* by

$$(\overline{X})_\alpha = (\overline{X})_{n,\alpha} := g_{n,\alpha}(X_1, \dots, X_n).$$

This is a generalization of the mean of the sample; α is called the 'trim level'.

For a population which is $\mathbb{P}_F$-distributed we define the *α-trimmed population mean* by

$$\mu_\alpha(F) := \frac{1}{1-2\alpha} \int_\alpha^{1-\alpha} q_F(u)\, du$$

(again, for $\alpha = 0$ we get $\mu_0(F) = \mu_F$, the corresponding expectation value). In this notation the α-trimmed mean corresponding to a sample outcome $x_1, \dots, x_n$ equals $(\overline{x})_{n,\alpha} = \mu_\alpha(\hat{F}(x_1, \dots, x_n))$. The mapping μ_α is continuous on (Γ, d_L).

Here too we have a strong law of large numbers. For any $0 < \alpha < \frac{1}{2}$ and any sample $X_1, X_2, \dots$ from a $\mathbb{P}_F$-distributed population, one has

$$\lim_{n\to\infty} (\overline{X})_{n,\alpha} = \mu_\alpha(F) \quad \text{almost surely.}$$

For $\alpha = \frac{1}{2}$, naturally connected to the median, the situation is somewhat different. We define the *median* of a population with distribution function F by

$$\hat{\mu}(F) := \tfrac{1}{2} \left\{ q_F\left(\tfrac{1}{2}-\right) + q_F\left(\tfrac{1}{2}+\right) \right\}.$$

This median is called *strict* if $q_F\left(\frac{1}{2}-\right) = q_F\left(\frac{1}{2}+\right)$. The *median of the sample outcome* $x_1, \dots, x_n$ is defined as

$$\hat{x} := \hat{\mu}(\hat{F}(x_1, \dots, x_n))$$

and equals the middle value if n is odd or the mean of the two middle values if n is even. As regards the strong law of large numbers we get: if $X_1, X_2, \dots$ denotes a sample from a $\mathbb{P}_F$-distributed population, then

$$\lim_{n\to\infty} (\hat{X})_n = \hat{\mu}(F) \quad \text{almost surely}$$

if and only if F has a strict median.

We return to our search for a robust estimator of the 'middle' of a population. For each $0 < \alpha < \frac{1}{2}$ the α-trimmed mean is an estimator of $\mu_\alpha(F)$ which is d_L-robust in every $F \in \Gamma$. The median of a sample is d_L-robust in $F \in \Gamma$ if and only if F has a strict median.

7.1.9 Statistical functionals (VII.9)

A function $\Lambda : D_\Lambda \to \mathbb{R}$, where $D_\Lambda \subset \Gamma$, is called a *statistical functional* if it satisfies the following two conditions:

(i) Each empirical distribution function $\hat{F}(x_1, \dots, x_n)$ is a member of D_Λ,

(ii) The mapping $(x_1, \dots, x_n) \mapsto \Lambda(\hat{F}(x_1, \dots, x_n))$ is Borel on $\mathbb{R}^n$.

If this is the case then for all $n = 1, 2, \dots$

$$T_n := \Lambda(\hat{F}(X_1, \dots, X_n))$$

provides a sample statistic, provided that $X_1, \dots, X_n$ is a sample from a population which has F as its distribution function. Some important examples are laid down now.

Example 1. We start from the set $D_{\Lambda_1} = \{F \in \Gamma : \int |x| \, d\mathbb{P}_F(x) < +\infty\}$. Defining $\Lambda_1 : D_{\Lambda_1} \to \mathbb{R}$ by

$$\Lambda_1(F) := \int x \, d\mathbb{P}_F(x) = \mathbb{E}_F(X),$$

Λ_1 is a statistical functional for which $\Lambda_1(\hat{F}(X_1, \dots, X_n)) = \overline{X}$.

Example 2. We take $D_{\Lambda_2} = \{F \in \Gamma : \int x^2 \, d\mathbb{P}_F(x) < +\infty\}$. Setting

$$\Lambda_2(F) := \int x^2 \, d\mathbb{P}_F(x) - \Big(\int x \, d\mathbb{P}_F(x)\Big)^2$$

we get $\Lambda_2(\hat{F}(X_1, \dots, X_n)) = \frac{n-1}{n} S_n^2$.

Example 3. We take $D_{\Lambda_3} = \Gamma$ and $0 < \alpha < \frac{1}{2}$ and set

$$\Lambda_3(F) := \frac{1}{1 - 2\alpha} \int_\alpha^{1-\alpha} q_F(u) \, du = \mu_\alpha(F)$$

as in §8. This leads to $\Lambda_3(\hat{F}(X_1, \dots, X_n)) = (\overline{X})_{n,\alpha}$, the α-trimmed mean of the sample $X_1, \dots, X_n$.

Example 4 in a similar way leads to the median of the sample.

Starting from any family $\{f(\bullet, \theta) : \theta \in \Theta\}$ of density functions, we may define a characteristic corresponding to a given statistical functional Λ by setting

$$\kappa(\theta) \; := \; \Lambda(F_\theta) \quad \text{for all } \theta \in \Theta.$$

A statistical functional $\Lambda : D_\Lambda \to \mathbb{R}$ is called *continuous* in $F \in D_\Lambda$ if $\Lambda(F_n) \to \Lambda(F)$ for each sequence $F_1, F_2, \dots$ in D_Λ that converges to F in the metric d_L. This appears to be a rather strong condition on statistical functionals (the functional in Example 1 is not continuous). For this reason we impose weaker continuity conditions. We call Λ *strongly consistent* in $F \in D_\Lambda$ if

$$\Lambda(\hat{F}(X_1, \dots, X_n)) \; \to \; \Lambda(F)$$

almost surely for any sample from a $\mathbb{P}_F$-distributed population. We say that Λ is *weakly consistent* in $F \in D_\Lambda$ if the same condition holds not almost surely but merely in distribution. If Λ is continuous in F then it is strongly consistent in F. Strong consistency, in turn, of course implies weak consistency.

As to d_L-robustness of $\Lambda(\hat{F}(X_1, \dots, X_n))$ we have the following. If Λ is continuous in $F \in \Gamma$ and the map

$$(x_1, \dots, x_n) \mapsto \Lambda(\hat{F}(x_1, \dots, x_n))$$

is continuous on $\mathbb{R}^n$, the sequence

$$\Lambda(\hat{F}(X_1, \dots, X_n)) \qquad (n = 1, 2, \dots)$$

is d_L-robust in F.

Now we turn to special classes of statistical functionals. To this, we first need some general operations on distribution functions. For $a \in \mathbb{R}$ and $F \in \Gamma$ we set

$$\tau_a F \; : \; x \mapsto F(x - a).$$

Furthermore we define

$$\tilde{F}(x) := 1 - F((-x)-)$$

for all x in $\mathbb{R}$, calling F *symmetric around zero* if $F = \tilde{F}$ ('symmetric around a' if $\tau_{-a}F$ is symmetric around zero). Finally, for any $F \in \Gamma$ and $b > 0$ we set

$$\rho_b F \; : \; x \mapsto F(x/b).$$

In this way, if X is a variable that is $\mathbb{P}_F$-distributed, $\tau_a F$ is the distribution function of $X + a$, $\tilde{F}$ the one of $-X$ and $\rho_b F$ the one of bX.

We will refer to the statistical functional $\Lambda : D_\Lambda \to \mathbb{R}$ as being a *location functional* if it enjoys the following three properties:

(i) $\tau_a F, \; \tilde{F}, \; \rho_b F \in D_\Lambda$ for all $F \in D_\Lambda, \; a \in \mathbb{R}, \; b > 0$,

(ii) $\Lambda(\tau_a F) = \Lambda(F) + a, \; \Lambda(\tilde{F}) = -\Lambda(F)$ and $\Lambda(\rho_b F) = b\,\Lambda(F)$ for all $F \in D_\Lambda, \; a \in \mathbb{R}, \; b > 0$,

(iii) $F, G \in D_\Lambda$ and $F \leq G \Rightarrow \Lambda(F) \geq \Lambda(G)$.

If for such a location functional $F \in \Gamma$ is symmetric around $a \in \mathbb{R}$ we have $\Lambda(F) = a$. Examples 1, 3 and 4 present location functionals.

Example 2 presents a member of the set of so-called *dispersion functionals*, that is to say statistical functionals $\Lambda : D_\Lambda \to \mathbb{R}$ such that

(i) $\tau_a F, \; \tilde{F}, \; \rho_b F \in D_\Lambda$ for all $F \in D_\Lambda, \; a \in \mathbb{R}, \; b > 0$,

(ii) $\Lambda(\tau_a F) = \Lambda(F), \; \Lambda(\tilde{F}) = \Lambda(F)$ and $\Lambda(\rho_b F) = b\,\Lambda(F)$ for all $F \in D_\Lambda, \; a \in \mathbb{R}, \; b > 0$,

(iii) $\Lambda(F) \geq 0$ for all $F \in D_\Lambda$.

Another example of a dispersion functional is $\Lambda : \Gamma \to \mathbb{R}$ for which

$$\Lambda(F) := \tfrac{1}{4}\left\{q_F\left(\tfrac{3}{4}+\right) + q_F\left(\tfrac{3}{4}-\right) - q_F\left(\tfrac{1}{4}+\right) - q_F\left(\tfrac{1}{4}-\right)\right\},$$

the so-called *interquartile distance.*

The third and last group discussed is the class of *linear functionals.* A statistical functional $\Lambda : D_\Lambda \to \mathbb{R}$ is called 'linear' if a Borel function $g : \mathbb{R} \to \mathbb{R}$ exists for which $D_\Lambda = \{F \in \Gamma : \int |g(x)| \, d\mathbb{P}_F(x) < +\infty\}$, such that

$$\Lambda(F) = \mathbb{E}_F(g(X)) \quad \text{for each } F \in D_\Lambda \text{ and corresponding } X.$$

This condition completely determines g, $g(a)$ being identical to $\Lambda(\hat{F}(a))$ for all $a \in \mathbb{R}$. Regarding linear functionals $\Lambda : D_\Lambda \to \mathbb{R}$ we have two important theorems. Firstly, they are strongly consistent in every point of their domain. Secondly, for every $F \in D_\Lambda$ satisfying the condition $\int g^2(x) \, d\mathbb{P}_F(x) < +\infty$ the sequence

$$\sqrt{n}\left\{\Lambda(\hat{F}(X_1, \ldots, X_n)) - \Lambda(F)\right\} \qquad (n = 1, 2, \ldots)$$

converges in distribution to the $N(0, \sigma^2)$-distribution, where $\sigma^2 = \operatorname{var}(g(X_1))$. This may be useful if we want to estimate $\Lambda(F)$ for some $F \in D_\Lambda$.

7.1.10 The von Mises-derivative; influence functions (VII.10)

Let $\Lambda : D_\Lambda \to \mathbb{R}$ be a statistical functional for which D_Λ is a convex subset of Γ. We take a fixed $F \in D_\Lambda$. If for $G \in D_\Lambda$ the function

$$t \mapsto \Lambda((1-t)\,F + t\,G)$$

has a right-derivative for $t = 0$ we say that Λ is *differentiable in F in the direction of G.* We write $\Lambda'_F(G)$ for the differential quotient in question. The set of elements G for which $\Lambda'_F(G)$ exists is called the 'von Mises hull' $M_\Lambda(F)$ of F.

The statistical functional Λ is called *von Mises differentiable* in F if $\Lambda'_F : M_\Lambda(F) \to \mathbb{R}$ presents a linear statistical functional. In that case Λ'_F is called the *von Mises derivative* of Λ in F. The function $\psi_F : \mathbb{R} \to \mathbb{R}$ defined by

$$\psi_F(x) \;:=\; \Lambda'_F(E_x)$$

is called the *influence function* corresponding to Λ and F and is the unique function ψ (see §9) for which

$$\Lambda'_F(G) = \int \psi(x) \, d\mathbb{P}_G(x) \quad (\text{all } G \in M_\Lambda(F)).$$

For any *linear* statistical functional $\Lambda : D_\Lambda \to \mathbb{R}$ we have that for every $F \in D_\Lambda$ the hull $M_\Lambda(F)$ is equal to D_Λ and that

$$\Lambda'_F : G \mapsto \Lambda(G) - \Lambda(F).$$

Shorthand: $\Lambda'_F = \Lambda - \Lambda(F)$ for all $F \in D_\Lambda$. The statistical functional corresponding to the mean of a population presents an example where the above applies.

If a sequence $\Lambda_1, \dots, \Lambda_p$ of statistical functionals is given, with $\Lambda_i : D_{\Lambda_i} \to \mathbb{R}$ for all i, we may set $D_{\mathbf{\Lambda}} := D_{\Lambda_1} \cap \dots \cap D_{\Lambda_p}$ and define

$$\mathbf{\Lambda} \ : \ F \ \mapsto \ (\Lambda_1(F), \dots, \Lambda_p(F))$$

for all $F \in D_{\mathbf{\Lambda}}$. Now for each Borel function $\varphi : \mathbb{R}^p \to \mathbb{R}$ the mapping

$$F \ \mapsto \ \varphi(\mathbf{\Lambda}(F))$$

presents another statistical functional. In these terms, if $\Lambda_1, \dots, \Lambda_p$ are all linear functionals and φ is differentiable in $\mathbf{\Lambda}(F)$ for some $F \in D_{\mathbf{\Lambda}}$, the statistical functional $\varphi(\mathbf{\Lambda})$ is von Mises differentiable in F, and

$$\varphi(\mathbf{\Lambda})'_F = \sum_{i=1}^{p} \left[\frac{\partial \varphi}{\partial x_i}(\mathbf{\Lambda}(F)) \right] (\Lambda_i - \Lambda_i(F)).$$

Application of these differentiation methods can for example arise in the following way. For a lot of statistical functionals Λ it can be proved that

$$\Lambda(\hat{F}(X_1, \dots, X_n)) - \Lambda(F) \ \approx \ \Lambda'_F(\hat{F}(X_1, \dots, X_n)).$$

In such cases the sequence

$$\sqrt{n}\ \{\Lambda(\hat{F}(X_1, \dots, X_n)) - \Lambda(F)\}$$

converges in distribution to some $N(0, \sigma^2)$-distribution.

Properties as boundedness and smoothness concerning the influence function ψ_F can be interpreted as messengers of robustness.

For the Λ corresponding to the population mean, ψ_F is not bounded for any F in $D_\Lambda = \{F : \mathbb{E}_F(|X|) < +\infty\}$. The functional Λ belonging to the population median produces ψ_F's which actually *are* bounded, but discontinuous in $\Lambda(F)$. The functional Λ that generates the α-trimmed mean for $0 < \alpha < \frac{1}{2}$ produces ψ_F's which are both bounded and continuous.

7.1.11 Bootstrap methods (VII.11)

Let $g : \mathbb{R}^m \to \mathbb{R}$ be a Borel function and $X_1, \dots, X_n$ a sample from a $\mathbb{P}_F$-distributed population. The *bootstrap distribution function* corresponding to the sample outcome $x_1, \dots, x_n$ is defined by

$$\tilde{g}(\hat{F}(x_1, \dots, x_n)).$$

We are going to approximate $\tilde{g}(F)$ by $\tilde{g}(\hat{F}(X_1, \dots, X_n))$.

As most of the times we cannot precisely determine

$$\tilde{g}\left(\hat{F}(X_1(\omega), \dots, X_n(\omega))\right),$$

we are also in need of an approximation for *this* distribution function. This we obtain as follows. First we define a probability space $(\Omega^*, \mathfrak{A}^*, \mathbb{P}^*)$ by

$$\begin{cases} \Omega^* & := & \{(i_1, \ldots, i_m) \in \mathbb{N}^m \, : \, 1 \leq i_1, \ldots, i_m \leq n\}, \\ \mathfrak{A}^* & := & \{A : \, A \subset \Omega^*\}, \quad \mathbb{P}^* : A \, \mapsto \, \#(A)/n^m. \end{cases}$$

Next we construct $X_1^*, \ldots, X_m^* : \Omega^* \to \mathbb{R}$ via

$$X_k^* : (i_1, \ldots, i_m) \; \mapsto \; x_{i_k},$$

$x_1, \ldots, x_n$ denoting the fixed sample outcome. Now $X_1^*, \ldots, X_m^*$ presents a sample from a population which has $\hat{F}(x_1, \ldots, x_n)$ as its distribution function and is called the *bootstrap sample* corresponding to $x_1, \ldots, x_n$. The statistic

$$T^* \; := \; g(X_1^*, \ldots, X_m^*)$$

is called the *bootstrap variable* corresponding to $X_1, \ldots, X_n$. By construction the distribution function of T^* is $\tilde{g}(\hat{F}(x_1, \ldots, x_n))$.

For very small m and n this distribution function can be determined precisely by computation. However, the number of computations involved soon gets uncontrollably large. In such cases we use the so-called *Monte Carlo method* to estimate $F_{T^*}(s)$ for each $s \in \mathbb{R}$. This method consists in drawing a fairly large sample $Y_1^*, \ldots, Y_p^*$ from Ω^* and determining the fraction of elements $\leq s$. This fraction is then used as the estimate of $F_{T^*}(s)$ and of $[\tilde{g}(F)](s)$.

Next we turn to the asymptotic behavior of the bootstrap. If g is continuous and the sample is infinitely large, then

$$d_L\left(\tilde{g}(\hat{F}(X_1, \ldots, X_n)), \tilde{g}(F)\right) \; \to \; 0$$

almost surely. This gives a justification for using $\tilde{g}(\hat{F}(X_1, \ldots, X_n))$ as an approximation for $\tilde{g}(F)$.

Now let $\Lambda : D_\Lambda \to \mathbb{R}$ be a statistical functional and suppose $X_1, X_2, \ldots$ is a sample from a $\mathbb{P}_F$-distributed population, where $F \in D_\Lambda$. Then (as in §9)

$$T_n := \Lambda(\hat{F}(X_1, \ldots, X_n)) = g_n(X_1, \ldots, X_n) \qquad (n = 1, 2, \ldots)$$

in a natural way provides a sequence of estimators of $\theta = \Lambda(F)$ (note that meanwhile we take $m = n$). A sequence of corresponding distribution functions $G_1, G_2, \ldots$ is given by

$$G_n(x) \; := \; \mathbb{P}\left(\sqrt{n}\,(T_n - \theta) \leq x\right).$$

As before, for a fixed $\omega \in \Omega$ we observe the bootstrap variables

$$T_n^* = T_n^*(\omega) = \Lambda(\hat{F}(X_1^*, \ldots, X_n^*)) \qquad (n = 1, 2, \ldots),$$

$X_1^*, \ldots, X_n^*$ being a sample from a population which has $\hat{F}(X_1(\omega), \ldots, X_n(\omega))$ as its distribution function. For all $n = 1, 2, \ldots$ we set

$$G_n^*(\omega) \; : \; x \; \mapsto \; \mathbb{P}^* \left(\sqrt{n}\,(T_n^*(\omega) - T_n(\omega)) \leq x\right).$$

It appears to make sense to look at the stochastic variable $\|G_n^* - G_n\|_\infty$. If $\|G_n^* - G_n\|_\infty \to 0$ almost surely we say that *the bootstrap works*. There are cases known where the bootstrap does not work.

Under the assumption that the sequence $\sqrt{n}\,(T_n - \theta)$ converges in distribution to a $N(0, \sigma^2)$-distribution, the fact that a $\sigma > 0$ exists such that

$$d_L(G_n, \Phi_\sigma) \to 0 \quad \text{and} \quad d_L(G_n^*, \Phi_\sigma) \to 0 \quad \text{almost surely}$$

implies that the bootstrap works (see §6 for Φ_σ).

7.1.12 Estimation of probability densities by means of kernel densities (VII.12)

Let $X_1, \ldots, X_n$ be a sample from a population with distribution function F and probability density f. We look for ways to estimate f basing ourselves on the sample.

In case the population is discretely distributed of course we estimate f by means of the *empirical density function* $\hat{f}$ given by

$$\hat{f}(X_1, \ldots, X_n) \; : \; x \mapsto \frac{\#\{i : X_i = x\}}{n}.$$

In the case of an absolutely continuously distributed population, we take some *kernel function* k together with a *bandwidth* $\zeta > 0$ and define the *kernel density estimation* $\hat{f}_\zeta$ of f by

$$\hat{f}_\zeta(X_1, \ldots, X_n)(x) := \frac{1}{n\,\zeta} \sum_{i=1}^{n} k\left(\frac{x - X_i}{\zeta}\right).$$

We restrict ourselves to kernel functions k that are positive, continuous and symmetric around the origin, and we further require that

$$\lim_{x \to \pm\infty} k(x) = 0 \quad \text{and} \quad \int k \, d\lambda = 1.$$

Examples are the so-called *triangular kernel*, the *Epanechnikov kernel* and the *normal kernel*; each kernel function is automatically a density function.

Under these conditions, setting $k_\zeta(x) := \frac{1}{\zeta}\, k(x/\zeta)$, we have

$$\hat{f}_\zeta = (K_\zeta \times \hat{F})',$$

K_ζ denoting the distribution function corresponding to k_ζ.

This $\hat{f}_\zeta$ is a good estimator in the sense that if $\zeta_1, \zeta_2, \ldots \to 0$ and $n\,\zeta_n^2 \to +\infty$,

$$\hat{f}_{\zeta_n}(X_1, \ldots, X_n)(x) \longrightarrow f(x)$$

in probability in every continuity point x of f.

7.1.13 Estimation of probability densities by means of histograms (VII.13)

Another way in which we can estimate probability densities is by means of so-called histograms. Again we assume that a population is given which enjoys an absolutely continuous distribution, together with a sample $X_1, \ldots, X_n$. The probability density will be denoted by f.

We start from an *interval partition* $\mathcal{I}$ of $\mathbb{R}$, that is, a sequence $I_1, I_2, \ldots$ of pairwise disjoint intervals that together span $\mathbb{R}$. We write $|I_k| := \text{length}(I_k)$ for all k and suppose that

$$0 < \inf\{|I_k| \, : \, k = 1, 2, \ldots\} \leq \sup\{|I_k| \, : \, k = 1, 2, \ldots\} < +\infty.$$

Associated with this interval partition $\mathcal{I}$ and this sample, the *histogram* h is defined by

$$h(\mathcal{I}, X_1, \ldots, X_n)(x) := \sum_{k=1}^{\infty} \frac{\#\{i \, : \, X_i \in I_k\}}{n\,|I_k|} 1_{I_k}(x).$$

This h is used as an estimator for f. For any sequence $\mathcal{I}_1, \mathcal{I}_2, \ldots$ of interval partitions of $\mathbb{R}$ that satisfies the two conditions

$$\begin{cases} \sup\{(\mathcal{I}_n)_k \, : \, k = 1, 2, \ldots\} & \to & 0 \\ n \, \inf\{(\mathcal{I}_n)_k \, : \, k = 1, 2, \ldots\} & \to & +\infty \end{cases}$$

we have that

$$h(\mathcal{I}_n, X_1, \ldots, X_n)(x) \longrightarrow f(x)$$

in probability in every continuity point x of f.

7.2 Exercises to Chapter VII

Exercise 1

A sample of size 4 shows an outcome given by

$$(X_1, X_2, X_3, X_4) = (4.5, \ 1.0, \ 2.0, \ 3.5).$$

Make a sketch of the corresponding empirical distribution function.

Solution. See the figure below.

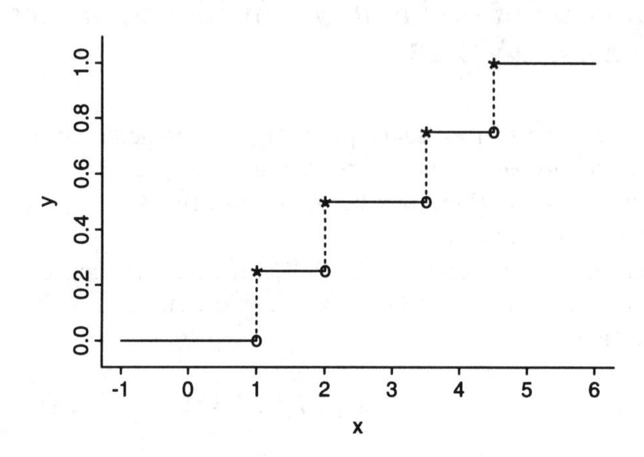

Exercise 2

Let $X_1, X_2, \ldots$ be an infinite sample from a population with probability density $f(\bullet, \theta)$, where $\theta \in \Theta$. Let $\kappa : \Theta \to \mathbb{R}$ be a characteristic of the population and let T_n be an estimator of κ, based on the finite sample $X_1, \ldots, X_n$. The sequence $T_1, T_2, \ldots$ is called 'asymptotically unbiased' if $\mathbb{E}_\theta(T_n) \to \kappa(\theta)$ for every $\theta \in \Theta$. The sequence is called 'weakly consistent' if $T_n \to \kappa(\theta)$ in probability for every $\theta \in \Theta$.

(i) Let $T_1, T_2, \ldots$ be an asymptotically unbiased sequence of stochastic variables. Suppose that $\lim_{n\to\infty} \mathrm{var}(T_n) = 0$. Prove that the sequence $T_1, T_2, \ldots$ is weakly consistent.

Proof. Choose arbitrary θ, $\delta > 0$ and $\varepsilon > 0$. Then there exist integers N, M such that

$$n \geq N \quad \Longrightarrow \quad |\mathbb{E}(T_n) - \kappa(\theta)| < \tfrac{1}{2}\varepsilon \qquad (*)$$

$$n \geq M \quad \Longrightarrow \quad \mathrm{var}(T_n) < (\tfrac{1}{2}\varepsilon)^2\,\delta \qquad (**)$$

It follows from this that for $n \geq N + M$

$$\begin{aligned}
\mathbb{P}(\,|T_n - \kappa(\theta)| \geq \varepsilon\,) &= \mathbb{P}\left(|(T_n - \mathbb{E}(T_n)) + (\mathbb{E}(T_n) - \kappa(\theta))| \geq \varepsilon\right) \\
&\leq \mathbb{P}\left(|T_n - \mathbb{E}(T_n)| \geq \tfrac{1}{2}\varepsilon \ \text{ or } \ |\mathbb{E}(T_n) - \kappa(\theta)| \geq \tfrac{1}{2}\varepsilon\right) \\
&\overset{(*)}{=} \mathbb{P}\left(|T_n - \mathbb{E}(T_n)| \geq \tfrac{1}{2}\varepsilon\right) \\
(\text{Chebychev}) \quad &\leq \frac{\mathrm{var}(T_n)}{(\frac{1}{2}\varepsilon)^2} \overset{(**)}{\leq} \delta.
\end{aligned}$$

In other words, for all $\varepsilon > 0$ we have

$$\mathbb{P}(\,|T_n - \kappa(\theta)| \geq \varepsilon\,) \to 0.$$

That is to say, $T_n \to \kappa(\theta)$ in probability. □

(ii) Let $X_1, X_2, \ldots$ be, as before, an infinite sample from a population with variance σ^2, having a fourth moment. Prove that the sequence of sample variances $\{S_n^2\}$, given by

$$S_n^2 = \frac{1}{n-1} \sum_{i=1}^{n} (X_i - \overline{X})^2$$

presents a weakly consistent sequence of estimators of σ^2.

Proof. The sequence of estimators $S_2^2, S_3^2, \ldots$ is of course asymptotically unbiased, as its elements are *all* unbiased estimators of σ^2. We are going to prove that $\text{var}(S_n^2) \to 0$. The statement to be proved then directly follows from (i).

Let μ_k, for all $k = 1, 2, \ldots$, be the kth moment belonging to the population. Note that, generally, μ_k will depend on θ. Moreover we define

$$\begin{cases} \rho_1 := \text{cov}\left({X_1}^2, {X_1}^2\right) = \mathbb{E}(X_1^4) - \mathbb{E}({X_1}^2)^2 = \mu_4 - \mu_2^2, \\ \rho_2 := \text{cov}\left({X_1}^2, X_1 X_2\right) = \mu_3\mu_1 - \mu_2\mu_1^2, \\ \rho_3 := \text{cov}\left(X_1 X_2, X_1 X_2\right) = \mu_2^2 - \mu_1^4, \\ \rho_4 := \text{cov}\left(X_1 X_2, X_1 X_3\right) = \mu_2\mu_1^2 - \mu_1^4. \end{cases}$$

Simply writing out the definition we see that

$$S_n^2 = \frac{1}{n}\sum_{i=1}^{n} X_i^2 - \frac{2}{n(n-1)} \sum_{j_1<j_2} X_{j_1} X_{j_2}.$$

Consequently (we assume that $n \geq 2$)

$$\begin{aligned} \text{var}(S_n^2) &= \text{cov}\left(S_n^2, S_n^2\right) \\ &= \frac{1}{n^2}\sum_{i_1,i_2} \text{cov}\left(X_{i_1}^2, X_{i_2}^2\right) - \frac{4}{n^2(n-1)} \sum_{i=1}^{n} \sum_{j_1<j_2} \text{cov}\left(X_i^2, X_{j_1}X_{j_2}\right) \\ &\quad + \frac{4}{n^2(n-1)^2} \sum_{j_1<j_2} \sum_{k_1<k_2} \text{cov}\left(X_{j_1}X_{j_2}, X_{k_1}X_{k_2}\right). \end{aligned}$$

Exploiting symmetry this form is easily simplified. To start with, we note that

$$\frac{1}{n^2}\sum_{i_1,i_2} \text{cov}\left(X_{i_1}^2, X_{i_2}^2\right) = \frac{1}{n^2}\sum_{i=1}^{n} \text{cov}\left(X_i^2, X_i^2\right) = \frac{1}{n}\,\rho_1.$$

Moreover we have that

$$\begin{aligned} \frac{4}{n^2(n-1)} \sum_{i=1}^{n} \sum_{j_1<j_2} \text{cov}\left(X_i^2, X_{j_1}X_{j_2}\right) &= \frac{4}{n(n-1)} \sum_{j_1<j_2} \text{cov}\left(X_1^2, X_{j_1}X_{j_2}\right) \\ &= \frac{4}{n(n-1)} \sum_{j=2}^{n} \text{cov}\left(X_1^2, X_1 X_j\right) = \frac{4}{n}\,\rho_2. \end{aligned}$$

Finally we see that

$$\begin{aligned}
&\frac{4}{n^2(n-1)^2} \sum_{j_1<j_2} \sum_{k_1<k_2} \operatorname{cov}(X_{j_1}X_{j_2}, X_{k_1}X_{k_2}) \\
&= \frac{2}{n(n-1)} \sum_{k_1<k_2} \operatorname{cov}(X_1X_2, X_{k_1}X_{k_2}) \\
&= \frac{2}{n(n-1)} \left\{ \operatorname{cov}(X_1X_2, X_1X_2) + \sum_{k=3}^{n} \operatorname{cov}(X_1X_2, X_1X_k) \right. \\
&\qquad \left. + \sum_{k=3}^{n} \operatorname{cov}(X_1X_2, X_2X_k) \right\} \\
&= \frac{2}{n(n-1)} \rho_3 + \frac{4(n-2)}{n(n-1)} \rho_4.
\end{aligned}$$

In conclusion:

$$\operatorname{var}(S_n^2) = \frac{1}{n}\rho_1 - \frac{4}{n}\rho_2 + \frac{2}{n(n-1)}\rho_3 + \frac{4(n-2)}{n(n-1)}\rho_4.$$

This proves that $\operatorname{var}(S_n^2) \to 0$, which completes the proof. □

Exercise 3

Let $(\Omega, \mathfrak{A}, \mathbb{P})$ be a probability space and let $X, X_1, X_2, \ldots : \Omega \to \mathbb{R}$ be stochastic variables.

(i) Prove that

$$\left\{ \omega \in \Omega \; : \; \lim_{n\to\infty} X_n(\omega) = X(\omega) \right\} \in \mathfrak{A}.$$

Proof. Passing from X_n to $X_n - X$ we may, without loss of generality, assume that $X = 0$.

The subset of all $\omega \in \Omega$ for which $X_n(\omega) \to 0$ can be characterized as follows:

$$\left\{ \omega \in \Omega \; : \; \lim_{n\to\infty} X_n(\omega) = 0 \right\} = \bigcap_{k=1}^{\infty} \bigcup_{n=1}^{\infty} \bigcap_{m=n}^{\infty} \left\{ \omega \in \Omega \; : \; X_m(\omega) \in [-\tfrac{1}{k}, \tfrac{1}{k}] \right\}.$$

Because $X_m^{-1}([-\frac{1}{k}, \frac{1}{k}]) \in \mathfrak{A}$, it follows from the above that the set under consideration is an element of the σ-algebra $\mathfrak{A}$. □

(ii) Prove that

$$\left\{ \omega \in \Omega \; : \; \lim_{n\to\infty} X_n(\omega) \text{ exists} \right\} \in \mathfrak{A}.$$

Proof. First we note that $\lim_{n\to\infty} X_n(\omega)$ exists if and only if the sequence $X_1(\omega), X_2(\omega), \ldots$ is Cauchy. Therefore we can characterize the set

$$\left\{\omega \in \Omega \ : \ \lim_{n\to\infty} X_n(\omega) \ \text{exists}\right\}$$

as

$$\bigcap_{k=1}^{\infty} \bigcup_{n=1}^{\infty} \bigcap_{p,q\geq n} \{\omega \in \Omega : X_p(\omega) - X_q(\omega) \in [-\tfrac{1}{k}; \tfrac{1}{k}]\}$$

(More colloquial, the sequence $X_1(\omega), X_2(\omega), \ldots$ converges if and only if for all k there is an n such that $|X_p(\omega) - X_q(\omega)| \leq \frac{1}{k}$ for $p, q \geq n$). Because $(X_p - X_q)^{-1}([-\frac{1}{k}, \frac{1}{k}]) \in \mathfrak{A}$, the above implies the statement to be proved. □

Exercise 4

(i) Give an example of a sequence of stochastic variables that converges in distribution but not in probability.

Solution. Let $X_1, X_2, \ldots$ be an independent sequence of stochastic variables, all Bernoulli-distributed with parameter $\theta = \frac{1}{2}$. Now this sequence converges (see §I.11 Exercise 29) in distribution to, say, X_{10}. However, this sequence does not converge in probability to X_{10}. Namely, if this were the case, we would have

$$\lim_{n\to\infty} \mathbb{P}\left(|X_n - X_{10}| \geq \tfrac{1}{4}\right) = 0.$$

This can impossibly be so, because

$$\lim_{n\to\infty} \mathbb{P}\left(|X_n - X_{10}| \geq \tfrac{1}{4}\right) = \mathbb{P}\left(|X_1 - X_{10}| = 1\right) = \tfrac{1}{2}.$$ □

(ii) Give an example of a sequence of stochastic variables that converges in probability but not almost surely.

Solution. An example of a sequence of stochastic variables converging in probability but not strongly is provided by the following. Let S be the unit circle in $\mathbb{R}^2$, equipped with the probability measure that assigns to each segment its length divided by 2π.

We now construct subsets $I_1, I_2, \ldots$ of S in the following way. I_1 is the segment of length 1, obtained when travelling anticlockwise, starting in the point $(1, 0)$. Going on in this way, I_2 is the segment of length $\frac{1}{2}$, obtained when starting in the end-point of I_1 and travelling in the same direction. Then I_3, of length $\frac{1}{3}$, is obtained by starting at the end-point of I_2; etcetera.

Next, set $X_1 := 1_{I_1}$, $X_2 := 1_{I_2}, \ldots$. This sequence converges in probability to 0. To see this, we notice that for all $\varepsilon > 0$ one has:

$$\lim_{n\to\infty} \mathbb{P}(|X_n - 0| \geq \varepsilon) = \lim_{n\to\infty} \mathbb{P}(1_{I_n} \geq \varepsilon) = \lim_{n\to\infty} \mathbb{P}(1_{I_n} = 1) = \lim_{n\to\infty} \frac{1}{2\pi n} = 0.$$

The sequence does not converge strongly. If it did then the only possible limit would be the constant 0. However, exploiting the fact that $1 + \frac{1}{2} + \frac{1}{3} + \cdots = +\infty$, it is easily seen there is *no* point $s \in S$ such that $X_n(s) \to 0$. □

Exercise 5

Given is a sequence $X_1, X_2, \ldots$ of stochastic variables converging in probability to X. Prove that there is a subsequence $X_{n_1}, X_{n_2}, \ldots$ that converges strongly to X.

Proof. Without loss of generality we may assume that $X = 0$. In this case, as $X_n \to 0$ in probability, there is an $n_1 \in \mathbb{N}$ such that

$$\mathbb{P}(|X_n| \geq 1) \leq \tfrac{1}{2} \quad \text{for } n \geq n_1.$$

Next, choose $n_2 > n_1$ in such a way that

$$\mathbb{P}(|X_n| \geq \tfrac{1}{2}) \leq \tfrac{1}{2^2} \quad \text{for } n \geq n_2.$$

Going on in this way we get a sequence of integers $n_1 < n_2 < \cdots$ such that

$$\mathbb{P}(|X_n| \geq \tfrac{1}{k}) \leq \tfrac{1}{2^k} \quad \text{for } n \geq n_k.$$

Now, given any $\varepsilon > 0$, there is an integer $k(\varepsilon)$ such that $\frac{1}{k} \leq \varepsilon$ for $k \geq k(\varepsilon)$. It follows that

$$\mathbb{P}(|X_{n_k}| \geq \varepsilon) \leq \mathbb{P}(|X_{n_k}| \geq \tfrac{1}{k}) \leq \tfrac{1}{2^k}$$

for $k \geq k(\varepsilon)$, so that

$$\sum_{k=1}^{\infty} \mathbb{P}(|X_{n_k}| \geq \varepsilon) < +\infty.$$

By Lemma VII.2.4 this implies that $X_{n_k} \to 0$ strongly. □

Exercise 6

Let $X_1, \ldots, X_n$ be a sample from a population with distribution function F. Suppose that $a < b$ and that $F(a) = F(b)$. Prove that $\hat{F}(X_1, \ldots, X_n)(a)$ and $\hat{F}(X_1, \ldots, X_n)(b)$ are essentially equal.

Proof. We have

$$\begin{aligned} &\mathbb{P}\left[\hat{F}(X_1, \ldots, X_n)(a) \neq \hat{F}(X_1, \ldots, X_n)(b)\right] \\ &\quad = \mathbb{P}\left[\#\{i : X_i \leq a\} < \#\{i : X_i \leq b\}\right] \\ &\quad \leq \sum_{i=1}^{n} \mathbb{P}\left[a < X_i \leq b\right] = n\left(F(b) - F(a)\right) = 0 : \end{aligned}$$

this proves the statement. □

Exercise 7

Let $X_1, \ldots, X_n$ be a sample from a population that has F as its distribution function. Suppose that $a, b \in \mathbb{R}$ and that $a < b$. Determine

$$\operatorname{cov}\left(\hat{F}(X_1, \ldots, X_n)(a)\,,\, \hat{F}(X_1, \ldots, X_n)(b)\right).$$

Solution. For every $c \in \mathbb{R}$ we may write

$$\hat{F}(X_1, \ldots, X_n)(c) = \frac{1}{n} \sum_{i=1}^{n} 1_{(-\infty, c]}(X_i).$$

Hence

$$\begin{aligned}
&\operatorname{cov}\left(\hat{F}(X_1, \ldots, X_n)(a)\,,\, \hat{F}(X_1, \ldots, X_n)(b)\right) \\
&\quad = \frac{1}{n^2} \sum_{i,j} \operatorname{cov}\left(1_{(-\infty,a]}(X_i), 1_{(-\infty,b]}(X_j)\right) \\
&\quad = \frac{1}{n^2} \sum_{i=1}^{n} \operatorname{cov}\left(1_{(-\infty,a]}(X_i), 1_{(-\infty,b]}(X_i)\right) \\
&\quad = \frac{1}{n^2} \sum_{i=1}^{n} \left\{\mathbb{E}(1_{(-\infty,a]}(X_i)) - \mathbb{E}(1_{(-\infty,a]}(X_i))\, \mathbb{E}(1_{(-\infty,b]}(X_i))\right\} \\
&\quad = \frac{1}{n^2} \sum_{i=1}^{n} (F(a) - F(a)\, F(b)) = \frac{1}{n}\, F(a)\, (1 - F(b)).
\end{aligned}$$

Exercise 8

For every distribution function $F : \mathbb{R} \to [0,1]$ the quantile function $q : (0,1) \to \mathbb{R}$ is defined as

$$q(u) := \inf\{y \,:\, u \leq F(y)\}.$$

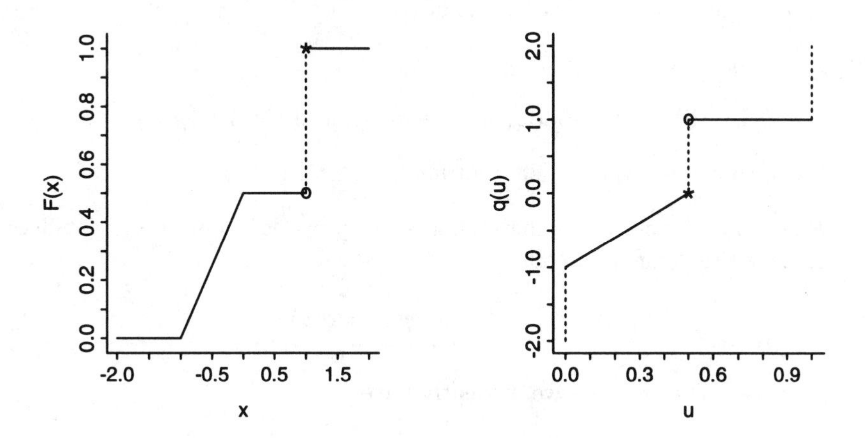

(i) Prove that q is an increasing function on $(0,1)$, taking on its values in $\mathbb{R}$.

Proof. Because of the fact that $\lim_{x\to+\infty} F(x) = 1$ we have for every $u \in (0,1)$ that $q(u) < +\infty$.

If $q(u) = -\infty$, then we would have $F(x) \geq u$ for all $x \in \mathbb{R}$. This is, however, contradictory to the property that $\lim_{x\to-\infty} F(x) = 0$.

As to the fact that q is increasing: suppose $u_1 \leq u_2$. Then of course we have the implication $u_2 \leq F(y) \;\Rightarrow\; u_1 \leq F(y)$, so that

$$\{y : u_2 \leq F(y)\} \subset \{y : u_1 \leq F(y)\}$$

and as a direct consequence $q(u_1) \leq q(u_2)$; so q is increasing. □

(ii) Check that $\{y : u \leq F(y)\} = [q(u), +\infty)$.

Proof. From the fact that F is right-continuous it follows that $q(u) \in \{y : u \leq F(y)\}$. In turn it follows from this, together with the fact that F is increasing, that

$$\{y : u \leq F(y)\} \supset [q(u), +\infty).$$

On the other hand it is trivial that

$$\{y : u \leq F(y)\} \subset [q(u), +\infty),$$

so the desired equality follows immediately. □

(iii) Prove the following equivalence:

$$u \leq F(x) \iff q(u) \leq x.$$

Proof. This is just a restatement of (ii). □

(iv) Prove that always $F(q(u)) \geq u$.

Proof. By (ii) the number $q(u)$ is an element of the set $\{y : u \leq F(y)\}$. □

(v) Prove that q is always a left-continuous function.

Proof. The function q is increasing, so to prove left-continuity it suffices to establish that for all $u \in \mathbb{R}$

$$\sup_{v<u} q(v) = q(u).$$

Because q is increasing we evidently have

$$\sup_{v<u} q(v) \;\leq\; q(u).$$

To prove the reversed equality, choose any $\lambda < q(u)$. Then $\lambda \notin \{y : u \leq F(y)\}$, which is the same as saying that $F(\lambda) < u$. Next, choose a v in such a way that $F(\lambda) < v < u$. Now $\lambda \notin \{y : v \leq F(y)\}$, that is: $\lambda < q(v)$. This shows that $\lambda < \sup_{v<u} q(v)$ for all $\lambda < q(u)$. Of course this implies that

$$\sup_{v<u} q(v) \geq q(u).$$

□

(vi) Show that for all $\varepsilon > 0$ one has $F(q(u) - \varepsilon) < u$.

Solution. By (ii) we have that $q(u) - \varepsilon \notin \{y : u \leq F(y)\}$. Hence $F(q(u) - \varepsilon) < u$. □

(vii) If F is continuous in $q(u)$ then $F(q(u)) = u$. Prove this statement.

Proof. By (iv) we always have $F(q(u)) \geq u$. In cases where F is continuous in $q(u)$, we have by (vi) that

$$F(q(u)) = \lim_{\varepsilon \downarrow 0} F(q(u) - \varepsilon) \leq u,$$

which completes the proof. □

(viii) Prove that

$$x \in [q(u), q(u+)) \Rightarrow F(x) = u.$$

Proof. First we notice that, by (ii), for $v > u$ we always have that

$$x \in [q(u), q(v)) \Longrightarrow F(x) \in [u, v) \qquad (*)$$

If $x \in [q(u), q(u+))$, then

$$x \in [q(u), q(v)) \quad \text{for all } v > u.$$

By $(*)$ this implies that $F(x) \in [u, v)$ for all $v > u$. The latter is, of course, possible if and only if $F(x) = u$. □

(ix) If X is a stochastic variable with a continuous distribution function F, then $F(X)$ is uniformly distributed on the interval $(0, 1)$. Prove this statement.

Proof. For all $u \in (0, 1)$ we have, using (iii), (vii) and the fact that F is continuous, that

$$\begin{aligned} \mathbb{P}(F(X) \geq u) &= \mathbb{P}(X \geq q(u)) = \mathbb{P}(X > q(u)) \\ &= 1 - \mathbb{P}(X \leq q(u)) = 1 - F(q(u)) = 1 - u. \end{aligned}$$

So, denoting the distribution function of the uniform distribution on $(0, 1)$ by F_0, it follows that for all $u \in \mathbb{R}$ we have

$$\mathbb{P}(F(X) < u) = F_0(u).$$

Letting $u \downarrow x$ in this equality we learn that

$$\mathbb{P}\left(F(X) \leq x\right) = F_0(x) \quad \text{for all } x \in \mathbb{R}.$$

From this it follows that $F(X)$ has F_0 as its distribution function. □

(x) If X is a stochastic variable with distribution function F and if $F(X)$ is uniformly distributed on $(0,1)$, then F is continuous. Prove this.

Proof. If F is not continuous in, say, x, then choose numbers a and b such that $F(x-) < a < b < F(x)$. It is now impossible for F to take on a value in the interval $[a,b]$. Consequently

$$\mathbb{P}\left(F(X) \in [a,b]\right) = 0.$$

This shows that $F(X)$ can impossibly be uniformly distributed on $(0,1)$, for then we would have

$$\mathbb{P}\left(F(X) \in [a,b]\right) = b - a \neq 0.$$

□

Exercise 9

From a population with a continuous distribution function F a sample $X = X_1$ of size 1 is drawn.

(i) Give a sketch of the empirical distribution function $\hat{F}$.

Solution.

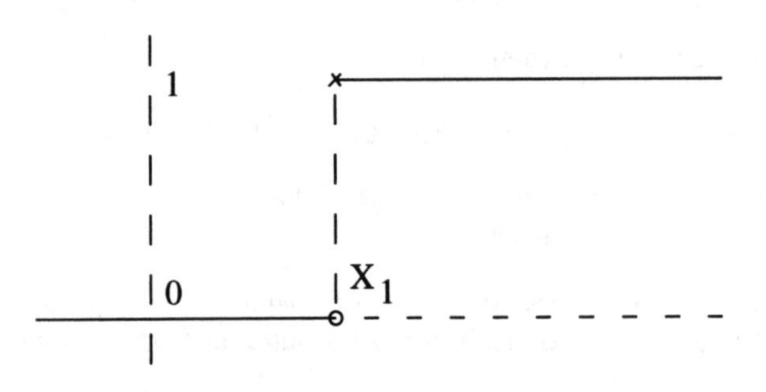

(ii) Show that

$$D_1 = \|\hat{F} - F\|_\infty = \max\left(F(X_1),\ 1 - F(X_1)\right).$$

Solution. Because F is increasing, we have $\sup_{x<X_1} |\hat{F}(x) - F(x)| = F(X_1)$ and therefore (see the figure)

$$\sup_{x \leq X_1} |\hat{F}(x) - F(x)| = \max\left(F(X_1),\ 1 - F(X_1)\right).$$

Furthermore (see figure)

$$\sup_{x \geq X_1} |\hat{F}(x) - F(x)| = 1 - F(X_1).$$

Hence

$$D_1 = \sup_x |\hat{F}(x) - F(x)| = \max(F(X_1),\ 1 - F(X_1)).$$

□

(iii) Prove that

$$\mathbb{P}(D_1 \leq a) = \begin{cases} 0 & \text{if} \quad a < \frac{1}{2}, \\ 2a - 1 & \text{if} \quad \frac{1}{2} \leq a \leq 1, \\ 1 & \text{if} \quad a > 1. \end{cases}$$

Proof. From

$$\mathbb{P}(D_1 \leq a) = \mathbb{P}[\max(F(X_1),\ 1 - F(X_1)) \leq a] = \mathbb{P}[1 - a \leq F(X_1) \leq a],$$

together with the fact that $F(X_1) \in [0, 1]$, the above follows immediately for $a < \frac{1}{2}$ and $a > 1$.

Furthermore $F(X_1)$ is (by Exercise 8 (ix)) uniformly distributed on the interval $(0, 1)$, which tells us that for $a \in [\frac{1}{2}, 1]$

$$\mathbb{P}(D_1 \leq a) = a - (1 - a) = 2a - 1.$$

□

Exercise 10

From a population with continuous distribution function F a sample X_1, X_2 of size 2 is drawn.

(i) Describe the empirical distribution function for outcomes of the form $X_1 < X_2$.

Solution. For such outcomes $\hat{F}$ assumes the shape as sketched in the figure:

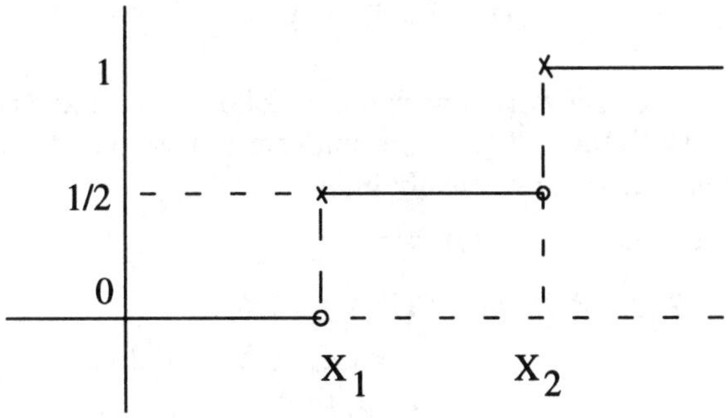

1

1/2

0

X_1 X_2

(ii) Show that for outcomes of the form $X_1 < X_2$ we have

$$D_2 = \|\hat{F} - F\|_\infty = \max\left(F(X_1),\ \tfrac{1}{2} - F(X_1),\ F(X_2) - \tfrac{1}{2},\ 1 - F(X_2)\right).$$

Solution. Quite analogous to 9 (ii): F is increasing, hence

$$\sup_{x<X_1} |\hat{F}(x) - F(x)| = F(X_1).$$

It follows from this that

$$\sup_{x\le X_1} |\hat{F}(x) - F(x)| = \max\left(F(X_1),\ |\tfrac{1}{2} - F(X_1)|\right).$$

Furthermore we have

$$\sup_{X_1\le x<X_2} |\hat{F}(x) - F(x)| = \max\left(|\tfrac{1}{2} - F(X_1)|,\ |\tfrac{1}{2} - F(X_2)|\right),$$

which leads to

$$\sup_{X_1\le x\le X_2} |\hat{F}(x) - F(x)| = \max\left(|\tfrac{1}{2} - F(X_1)|,\ |\tfrac{1}{2} - F(X_2)|,\ 1 - F(X_2)\right).$$

Finally $\sup_{X_2\le x} |\hat{F}(x) - F(x)| = 1 - F(X_2)$ and we conclude that

$$\begin{aligned} D_2 &= \|\hat{F} - F\|_\infty \\ &= \max\left(F(X_1),\ |\tfrac{1}{2} - F(X_1)|,\ |\tfrac{1}{2} - F(X_2)|,\ 1 - F(X_2)\right) \qquad (*) \end{aligned}$$

Next we note that $F(X_1) > \frac{1}{2}$ implies $F(X_2) > \frac{1}{2}$, hence if $F(X_1) > \frac{1}{2}$

$$|\tfrac{1}{2} - F(X_1)| = F(X_1) - \tfrac{1}{2} \le F(X_2) - \tfrac{1}{2} = |\tfrac{1}{2} - F(X_2)|.$$

It follows from this that in $(*)$ the modulus bars around the expression $\frac{1}{2} - F(X_1)$ can be deleted. In a similar way it is easy to see that we may replace $|\frac{1}{2} - F(X_2)|$ by $F(X_2) - \frac{1}{2}$. Summarizing we arrive at

$$D_2 = \max\left(F(X_1),\ \tfrac{1}{2} - F(X_1),\ F(X_2) - \tfrac{1}{2},\ 1 - F(X_2)\right). \qquad \square$$

(iii) Show that

$$\mathbb{P}(X_1 < X_2 \text{ and } F(X_1) \ge F(X_2)) = 0.$$

Solution. Because F is continuous it follows (see Exercise 8 (ix)) that the variables $F(X_1)$ and $F(X_2)$ are uniformly distributed on $(0,1)$. Moreover, these variables are statistically independent.

Exploiting this we may write:

$$\begin{aligned} &\mathbb{P}(X_1 < X_2 \text{ and } F(X_1) \ge F(X_2)) \\ &\qquad = \mathbb{P}(X_1 < X_2 \text{ and } F(X_1) = F(X_2)) \\ &\qquad \le \mathbb{P}(F(X_1) = F(X_2)) = 0. \end{aligned}$$

Here we used the fact that the joint probability distribution of $F(X_1)$ and $F(X_2)$ is just the Lebesgue measure on $[0,1] \times [0,1]$. Therefore the probability that the 2-vector $(F(X_1), F(X_2))$ will take on a value in the diagonal $\{(x,y) \in [0,1] \times [0,1] \, : \, x = y\}$ is zero. □

(iv) Prove that for all $a \in \mathbb{R}$

$$\begin{aligned} &\mathbb{P}\,(D_2 \leq a \text{ and } \; X_1 < X_2) \\ &\quad = \; \mathbb{P}\left(\max\left(Y_1,\; \tfrac{1}{2} - Y_1,\; Y_2 - \tfrac{1}{2},\; 1 - Y_2\right) \leq a \;\text{ and }\; Y_1 < Y_2\right) \end{aligned}$$

where Y_1 and Y_2 are independent variables, both uniformly distributed on the interval $(0,1)$.

Proof. As noticed before, the variables $Y_1 := F(X_1)$ and $Y_2 := F(X_2)$ are statistically independent and uniformly distributed on $(0,1)$.

For an arbitrary $a \in \mathbb{R}$ we therefore have

$$\begin{aligned} &\mathbb{P}\,(D_2 \leq a \text{ and } \; X_1 < X_2) \\ &\quad = \; \mathbb{P}\,(D_2 \leq a,\; X_1 < X_2 \text{ and } F(X_1) < F(X_2)) \\ &\qquad\qquad + \mathbb{P}\,(D_2 \leq a,\; X_1 < X_2 \text{ and } F(X_1) \geq F(X_2)) \\ &\quad \overset{\text{(iii)}}{=} \; \mathbb{P}\,(D_2 \leq a,\; X_1 < X_2 \text{ and } F(X_1) < F(X_2)) \\ &\quad = \; \mathbb{P}\,(D_2 \leq a \text{ and } F(X_1) < F(X_2)) \\ &\quad \overset{\text{(ii)}}{=} \; \mathbb{P}\left(\max\left(Y_1,\; \tfrac{1}{2} - Y_1,\; Y_2 - \tfrac{1}{2},\; 1 - Y_2\right) \leq a \;\text{ and }\; Y_1 < Y_2\right) \end{aligned}$$

The last but one equality follows from the fact that $F(X_1) < F(X_2) \Rightarrow X_1 < X_2$. □

(v) Show that the distribution function of the statistic D_2 is given by

$$\mathbb{P}\,(D_2 \leq a) = \begin{cases} 0 & \text{if } a < \tfrac{1}{4}, \\ 2\left(2a - \tfrac{1}{2}\right)^2 & \text{if } \tfrac{1}{4} \leq a < \tfrac{1}{2}, \\ 1 - 2(1-a)^2 & \text{if } \tfrac{1}{2} \leq a \leq 1, \\ 1 & \text{if } a > 1. \end{cases}$$

Solution. First we note that

$$\mathbb{P}\,(X_1 = X_2) \;\leq\; \mathbb{P}\,[F(X_1) = F(X_2)] = 0.$$

From this it follows that

$$\begin{aligned} \mathbb{P}\,(D_2 \leq a) &= \mathbb{P}\,(D_2 \leq a \;\text{ and }\; X_1 < X_2) + \mathbb{P}\,(D_2 \leq a \;\text{ and }\; X_2 < X_1) \\ &\overset{\text{(iv)}}{=} \; 2\,\mathbb{P}\left(\max\left(Y_1,\; \tfrac{1}{2} - Y_1,\; Y_2 - \tfrac{1}{2},\; 1 - Y_2\right) \leq a \;\text{ and }\; Y_1 < Y_2\right) \qquad (*) \end{aligned}$$

where Y_1 and Y_2 are independent and uniformly distributed on $(0,1)$.

Now

$$\begin{aligned}\mathbb{P}\left(\max\left(Y_1,\ \tfrac{1}{2}-Y_1,\ Y_2-\tfrac{1}{2},\ 1-Y_2\right)\leq a \text{ and } Y_1<Y_2\right)\\ =\mathbb{P}\left(\tfrac{1}{2}-a\leq Y_1\leq a,\ 1-a\leq Y_2\leq \tfrac{1}{2}+a,\ Y_1<Y_2\right)\end{aligned}$$

and indeed for $a<\frac{1}{4}$ this equals 0 and for $a>1$ it equals $\frac{1}{2}$.

Next we consider the case $a\in[\frac{1}{4},1]$. To this we have a look at the unit square $[0,1]\times[0,1]$ in $\mathbb{R}^2$, on which the Lebesgue measure plays the role of the probability distribution of the 2-vector (Y_1,Y_2). Define the subset A as $A:=A_1\cap A_2$, where $A_1:=\{(y_1,y_2)\in[0,1]\times[0,1]\ :\ y_1<y_2\}$ and $A_2:=[\frac{1}{2}-a,a]\times[1-a,\frac{1}{2}+a]$.

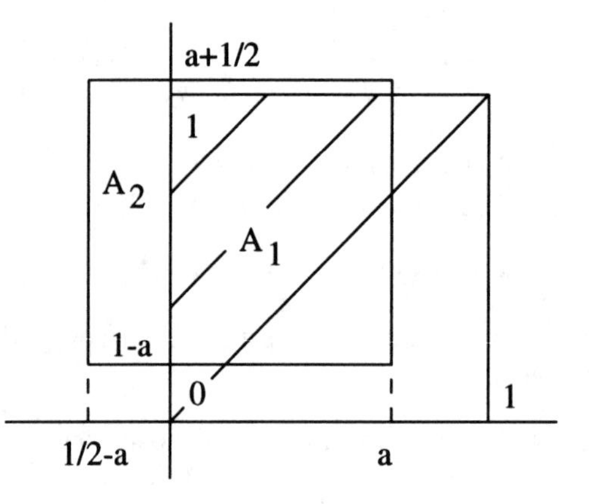

If $\frac{1}{4}\leq a<\frac{1}{2}$, then $A_2\subset A_1$ and therefore

$$\mathbb{P}\left((Y_1,Y_2)\in A_1\cap A_2\right)=\mathbb{P}\left((Y_1,Y_2)\in A_2\right)=\text{area}(A_2)=\left(2a-\tfrac{1}{2}\right)^2.$$

By $(*)$ it then follows that $\mathbb{P}\left(D_2\leq a\right)=2\left(2a-\frac{1}{2}\right)^2$.

For $\frac{1}{2}\leq a\leq 1$ we see that $\mathbb{P}((Y_1,Y_2)\in A)$ equals

$$\text{area}(A_1)-\text{area}(B)=\tfrac{1}{2}-2\left(\tfrac{1}{2}(1-a)^2\right)=\tfrac{1}{2}-(1-a)^2,$$

where B is as in the figure.

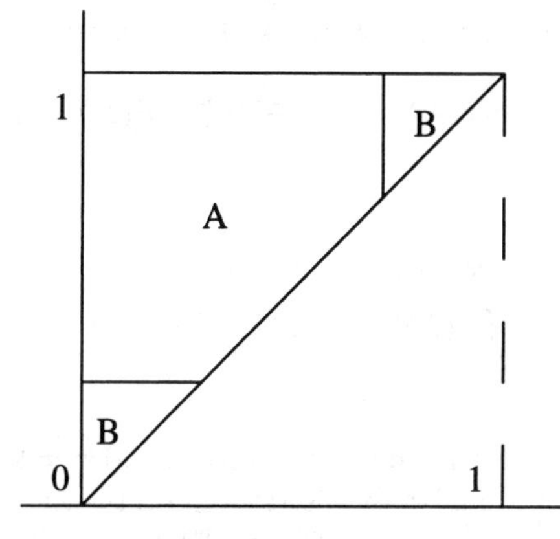

Applying $(*)$ we see that then $\mathbb{P}(D_2 \leq a) = 1 - 2(1-a)^2$.

Altogether we described the distribution function of D_2 completely. □

(vi) Check that the above is in tune with Table XI.

Solution. From Table XI it can be read off that $\mathbb{P}(D_2 > 0.7764) = 0.10$, $\mathbb{P}(D_2 > 0.8419) = 0.05$, $\mathbb{P}(D_2 > 0.9000) = 0.02$ and $\mathbb{P}(D_2 > 0.9293) = 0.01$. Using the result of (v) we can just check this. For example:

$$\mathbb{P}(D_2 > 0.9293) = 2(1 - 0.9293)^2 = 0.0100.$$ □

Exercise 11

Let $X_1, \ldots, X_n$ be a sample of size n from a population having a continuous distribution function F.

(i) Prove that

$$\mathbb{P}(D_n \leq a) = n!\, \mathbb{P}\Big(\max\big\{Y_i - \tfrac{i-1}{n},\ \tfrac{i}{n} - Y_i\ :\ i = 1, \ldots, n\big\} \leq a \text{ and } Y_1 < \cdots < Y_n\Big),$$

where the $Y_1, \ldots, Y_n$ are statistically independent and all uniformly distributed on the interval $(0,1)$.

Proof. First we note that for any outcome $x_1 < \cdots < x_n$ the supremum of $|\hat{F}(x) - F(x)|$ on the interval $[x_i, x_{i+1}]$ is given by

$$\max\left(\left|F(x_i) - \tfrac{i}{n}\right|,\ \left|F(x_{i+1}) - \tfrac{i}{n}\right|, \left|F(x_{i+1}) - \tfrac{i+1}{n}\right|\right).$$

Hence, for an outcome $x_1 < \cdots < x_n$ we have

$$D_n = \max\Big(F(x_1) - 0,\ 1 - F(x_n),\ \max_{i=1,\ldots,n-1}\{|F(x_i) - \tfrac{i}{n}|\},\ \max_{i=1,\ldots,n-1}\{|F(x_{i+1}) - \tfrac{i}{n}|\}\Big).$$

It is easy to see (just as in Exercise 10) that we may rewrite this as

$$D_n = \max\left\{F(x_i) - \tfrac{i-1}{n},\ \tfrac{i}{n} - F(x_i)\ :\ i = 1,\dots,n\right\}.$$

It follows that

$$\begin{aligned}
&\mathbb{P}\left(D_n \le a \text{ and } X_1 < \cdots < X_n\right)\\
&\quad= \mathbb{P}\left(\max\left\{F(X_i) - \tfrac{i-1}{n},\ \tfrac{i}{n} - F(X_i)\ :\ i = 1,\dots,n\right\} \le a\right.\\
&\qquad\qquad\qquad\qquad\left.\text{and } X_1 < \cdots < X_n\right).
\end{aligned}$$

Setting $Y_i := F(X_i)$ we get a statistically independent system $Y_1,\dots,Y_n$ of variables that are uniformly distributed on the interval $(0,1)$. Reasoning as in the previous exercise we arrive at the conclusion that

$$\begin{aligned}
&\mathbb{P}\left(D_n \le a \text{ and } X_1 < \cdots < X_n\right)\\
&\quad= \mathbb{P}\left(\max\left\{Y_i - \tfrac{i-1}{n},\ \tfrac{i}{n} - Y_i\ :\ i = 1,\dots,n\right\} \le a \text{ and } Y_1 < \cdots < Y_n\right)
\end{aligned}$$

for an independent system $Y_1,\dots,Y_n$ of variables that are uniformly distributed on the interval $(0,1)$.

Exploring the exchangeable role played by the $X_1,\dots,X_n$ we deduce from the above that for every permutation π of the integers $1,\dots,n$:

$$\begin{aligned}
&\mathbb{P}\left(D_n \le a \text{ and } X_{\pi(1)} < \cdots < X_{\pi(n)}\right)\\
&\quad= \mathbb{P}\left(\max\left\{Y_i - \tfrac{i-1}{n},\ \tfrac{i}{n} - Y_i\ :\ i = 1,\dots,n\right\} \le a \text{ and } Y_1 < \cdots < Y_n\right).
\end{aligned}$$

By the fact that the $X_1,\dots,X_n$ all have one and the same *continuous* distribution function F, we have

$$\mathbb{P}\left(D_n \le a\right) = \sum_{\pi} \mathbb{P}\left(D_n \le a \text{ and } X_{\pi(1)} < \cdots < X_{\pi(n)}\right),$$

where the summation is taken over the set of all permutations of $1,\dots,n$. Thus the statement follows. □

(ii) Show that

$$\mathbb{P}\left(D_n \le \frac{1}{2n}\right) = 0.$$

Solution. This is a consequence of (i), together with the fact that always

$$\max\left(F(X_1) - 0,\ \tfrac{1}{n} - F(X_1)\right) \ \ge\ \tfrac{1}{2n}.$$

□

(iii) Prove that $\mathbb{P}(D_n \le a) = n!\,(2a - \frac{1}{n})^n$ if $\frac{1}{2n} < a \le \frac{1}{n}$.

Proof. We proceed in the way of Exercise 10 (v), restricting ourselves to the case $n = 3$.

$$\begin{aligned}
&\mathbb{P}(D_3 \le a \text{ and } X_1 < X_2 < X_3) \\
&\overset{(i)}{=} \mathbb{P}\left(F(X_1) \le a,\ \tfrac{1}{3} - F(X_1) \le a,\ F(X_2) - \tfrac{1}{3} \le a,\ \tfrac{2}{3} - F(X_2) \le a,\right. \\
&\qquad \left. F(X_3) - \tfrac{2}{3} \le a,\ 1 - F(X_3) \le a,\ X_1 < X_2 < X_3\right) \\
&= \mathbb{P}\left(\tfrac{1}{3} - a \le F(X_1) \le a,\ \tfrac{2}{3} - a \le F(X_2) \le a + \tfrac{1}{3},\right. \\
&\qquad \left. 1 - a \le F(X_3) \le a + \tfrac{2}{3},\ X_1 < X_2 < X_3\right).
\end{aligned}$$

It follows from the above that (see also 10 (iv))

$$\begin{aligned}
&\mathbb{P}(D_3 \le a \text{ and } X_1 < X_2 < X_3) \\
&\quad = \mathbb{P}\left(Y_1 \in \left[\tfrac{1}{3} - a,\ a\right],\ Y_2 \in \left[\tfrac{2}{3} - a,\ a + \tfrac{1}{3}\right],\right. \\
&\qquad \left. Y_3 \in \left[1 - a,\ a + \tfrac{2}{3}\right] \text{ and } Y_1 < Y_2 < Y_3\right)
\end{aligned}$$

where Y_1, Y_2, Y_3 is a sample of size 3 from a population that is uniformly distributed on the interval $(0, 1)$.

For an arbitrary n the above reads as:

$$\begin{aligned}
&\mathbb{P}(D_n \le a \text{ and } X_1 < \cdots < X_n) \\
&= \mathbb{P}\left(Y_i \in \left[\tfrac{i}{n} - a, a + \tfrac{i-1}{n}\right] \text{ for all } i = 1, \ldots, n \text{ and } Y_1 < \cdots < Y_n\right)
\end{aligned}$$

where $Y_1, \ldots, Y_n$ is a sample of size n from a population that is uniformly distributed on the interval $(0, 1)$.

We notice that the condition $a \ge \frac{1}{2n}$ is equivalent to the condition that none of the intervals $[\frac{i}{n} - a, a + \frac{i-1}{n}]$ are empty.

Furthermore, the condition $a \le \frac{1}{n}$ implies that $a + \frac{i-1}{n} \le \frac{i+1}{n} - a$ for all i. It follows that we almost surely have $Y_1 < \cdots < Y_n$ whenever $Y_i \in [\frac{i}{n} - a, a + \frac{i-1}{n}]$ for all i. Hence

$$\begin{aligned}
\mathbb{P}(D_n \le a \text{ and } X_1 < \cdots < X_n) &= \prod_{i=1}^{n} \left(\left(a + \tfrac{i-1}{n}\right) - \left(\tfrac{i}{n} - a\right)\right) \\
&= \prod_{i=1}^{n} \left(2a - \tfrac{1}{n}\right) = \left(2a - \tfrac{1}{n}\right)^n.
\end{aligned}$$

We conclude that

$$\mathbb{P}(D_n \le a) = n!\,(2a - \tfrac{1}{n})^n.$$

□

Exercise 12

We want to test, at a 10% level of significance, whether a certain population is or is not $N(0,1)$-distributed. To this we draw a sample X_1, X_2 from the population in question. The outcome: $X_1 = 0.2$ and $X_2 = 1.1$.

(i) Sketch in one figure the distribution function Φ of the $N(0,1)$-distribution and the empirical distribution function of the sample.

Solution.

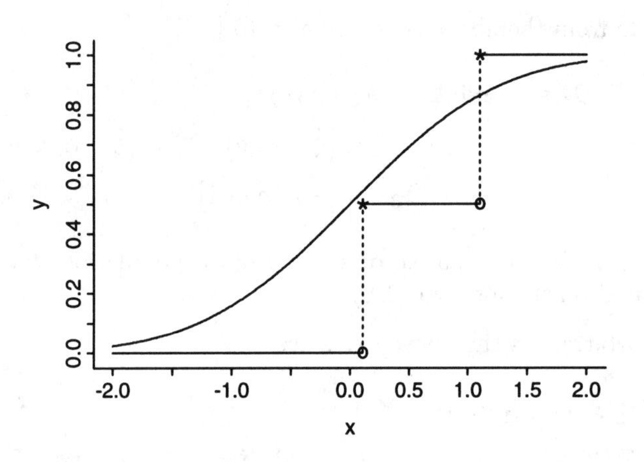

(ii) Determine the outcome of D_2.

Solution. It is clear (see also Exercise 10 (ii)) that $D_2 = \Phi(0.2) = 0.579$. □

(iii) Compute the P-value corresponding to the outcome of D_2.

Solution. Using Exercise 10 (v), this P-value appears to be

$$\mathbb{P}(D_2 > 0.579) = 1 - (1 - 2(1 - 0.579)^2) = 0.354.$$ □

(iv) Applying the Kolmogorov–Smirnov test, should we reject the null hypothesis that the population be $N(0,1)$-distributed?

Solution. The P-value exceeds the level of significance, hence we do not reject H_0. □

Exercise 13

Let $\hat{F} : \mathbb{R}^n \to \mathfrak{D}(\mathbb{R})$, where $\mathfrak{D}(\mathbb{R})$ is the linear space comprising all cadlag-functions, be defined by

$$\hat{F}(x_1, \ldots, x_n) \; : \; x \mapsto \frac{\#\{i : x_i \leq x\}}{n}.$$

(i) *Question.* Is $\hat{F}$ injective?

Solution. If $n = 1$ then $\hat{F}$ is indeed injective. However, for $n \geq 2$ we have $\hat{F}(1, 0, \ldots, 0) = \hat{F}(0, 1, 0, \ldots, 0)$ which shows that then $\hat{F}$ is not injective. □

(ii) *Question.* Is $\hat{F}$ surjective?

Solution. Note that $\hat{F}(x_1, \ldots, x_n)$ is always a step function, whereas there are elements in $\mathfrak{D}(\mathbb{R})$ that are not so. Consequently $\hat{F}$ can impossibly be surjective. □

(iii) *Question.* Is $\hat{F}$ continuous?

Solution. If $\hat{F}$ were continuous in, say, the point $\mathbf{0} \in \mathbb{R}^n$, then there would be a $\delta > 0$ such that

$$\|(y_1, \ldots, y_n)\| < \delta \Rightarrow \|\hat{F}(y_1, \ldots, y_n) - \hat{F}(0)\|_\infty < \tfrac{1}{n} \qquad (*)$$

Now the point $\left(\frac{1}{2}\delta, 0, \ldots, 0\right)$ satisfies the premise in this implication, whereas we have

$$\begin{aligned} \left\|\hat{F}\left(\tfrac{1}{2}\delta, 0, \ldots, 0\right) - \hat{F}(\mathbf{0})\right\|_\infty &\geq \left|\hat{F}\left(\tfrac{1}{2}\delta, 0, \ldots, 0\right)\left(\tfrac{1}{4}\delta\right) - \hat{F}(\mathbf{0})\left(\tfrac{1}{4}\delta\right)\right| \\ &= 1 - \left(1 - \tfrac{1}{n}\right) = \tfrac{1}{n}. \end{aligned}$$

This is contradictory to $(*)$, hence $\hat{F}$ is not continuous in $\mathbf{0}$. In a similar way it can be proved that there is no point in $\mathbb{R}^n$ in which $\hat{F} : \mathbb{R}^n \to \mathfrak{D}(\mathbb{R})$ is continuous. □

(iv) From now on we consider the case $n = 1$. Show that then we have $\hat{F}(x) = 1_{[x, +\infty)}$.

Solution. Trivial, as we have

$$\hat{F}(x)(y) = \frac{1_{x \leq y}}{1} = 1_{[x, +\infty)}(y).$$

□

(v) Prove that for all $x \neq y$ we have $\|\hat{F}(x) - \hat{F}(y)\|_\infty = 1$.

Solution. If $x \neq y$ then

$$\|\hat{F}(x) - \hat{F}(y)\|_\infty \geq \left|\hat{F}(x)\left(\frac{x+y}{2}\right) - \hat{F}(y)\left(\frac{x+y}{2}\right)\right| \stackrel{\text{(iv)}}{=} 1.$$

For arbitrary a, b in $\mathbb{R}$ we always have $\hat{F}(a)(b) \in \{0, 1\}$, from which it follows that $\|\hat{F}(x) - \hat{F}(y)\|_\infty \leq 1$, so $\|\hat{F}(x) - \hat{F}(y)\|_\infty = 1$. □

(vi) Let C be any subset of $\mathbb{R}$ and let $\mathfrak{C} := \{\hat{F}(x) : x \in C\} \subset \mathfrak{D}(\mathbb{R})$. Prove that $\mathfrak{C}$ is a closed subset of $\mathfrak{D}(\mathbb{R})$.

Proof. Suppose that $\hat{F}(x_1), \hat{F}(x_2), \dots$ is a sequence in $\mathfrak{C}$ that converges to an element $\varphi \in \mathfrak{D}(\mathbb{R})$; we have to prove that then necessarily $\varphi \in \mathfrak{C}$. As the sequence is convergent in $\mathfrak{D}(\mathbb{R})$ there is an integer N such that

$$\|\hat{F}(x_p) - \hat{F}(x_q)\|_\infty < \tfrac{1}{2} \quad \text{for all } p, q \geq N.$$

By (v) this implies that $x_p = x_q$ for all $p, q \geq N$. That is to say, the sequence $\hat{F}(x_1), \hat{F}(x_2), \dots$ is stationary. Obviously this implies that $\varphi \in \mathfrak{C}$. □

(vii) Check that $\hat{F}^{-1}(\mathfrak{C}) = C$.

Proof. This is a direct consequence of the definition of $\mathfrak{C}$.

In addition, choose a set $C \in \mathbb{R}$ that is not Borel. Then $\mathfrak{C}$ is a closed set in $\mathfrak{D}(\mathbb{R})$ and therefore surely Borel. Now $\hat{F}^{-1}(\mathfrak{C})$ is not a Borel set: this shows that $\hat{F}$ is not Borel measurable. □

Exercise 14

Give an example showing that the probability distribution of $D_n = \|\hat{F} - F\|_\infty$, under the null hypothesis that the population has $F = F_0$ as its distribution function, *does* depend on F_0 when we allow F_0 to be discontinuous.

Solution. Consider the case $n = 1$ (corresponding to a sample X_1 of size 1) and let F_0 be the distribution function belonging to the Bernoulli distribution with parameter $\theta = \frac{3}{4}$. Then F_0 is given by

$$F_0(x) = \begin{cases} 0 & \text{if} \quad x < 0, \\ \frac{1}{4} & \text{if} \quad 0 \leq x < 1, \\ 1 & \text{if} \quad x \geq 1. \end{cases}$$

With probability 1 the outcome x_1 of X_1 is 0 or 1. It follows from this that the probability distribution of D_1 is discrete. In fact we have:

$$\mathbb{P}\left(D_1 = \tfrac{3}{4}\right) = \mathbb{P}(X_1 = 0) = \tfrac{1}{4} \quad \text{and} \quad \mathbb{P}\left(D_1 = \tfrac{1}{4}\right) = \mathbb{P}(X_1 = 1) = \tfrac{3}{4}.$$

Hence D_1 can impossibly have the distribution function described in Exercise 9 for continuous F_0. □

Exercise 15

Let $X_1, X_2, \dots$ be an infinite sample from a population that is Cauchy distributed with parameters $\alpha = 0$ and $\beta = 1$. Does then the sequence

$$\overline{Y}_n := \frac{X_1 + \cdots + X_n}{n}$$

converge in probability?

Solution.

Step 1. First we notice an important permanence property of the Cauchy distribution. Namely, if for $i = 1, \dots, n$ the variable X_i is Cauchy distributed with parameters α_i and β_i and if the $X_1, \dots, X_n$ constitute a statistically independent system, then the linear combination $c_1X_1 + \cdots + c_nX_n$ enjoys a Cauchy distribution with parameters $c_1\alpha_1 + \cdots + c_n\alpha_n$ and $|c_1|\beta_1 + \cdots + |c_n|\beta_n$.

This can be deduced by examining the characteristic function of the Cauchy distribution, which we shall do now.

For each j by Appendix D the characteristic function of X_j is given by

$$\chi_{X_j}(t) = e^{i\alpha_j t} e^{-\beta_j |t|}.$$

It follows that (all $t \in \mathbb{R}$)

$$\chi_{c_j X_j}(t) = \chi_{X_j}(c_j t) = e^{i\alpha_j c_j t} e^{-\beta_j |c_j|\, |t|}$$

so that c_jX_j in turn is Cauchy distributed with $c_j\alpha_j$ and $|c_j|\,\beta_j$ as its parameters.

On the other hand, if we add up two independent Cauchy distributed variables, say Y and Z, their parameters being α_Y, β_Y and α_Z, β_Z respectively, we see that

$$\chi_{Y+Z}(t) = \chi_Y(t)\, \chi_Z(t) = e^{i(\alpha_Y + \alpha_Z)t} e^{-(\beta_Y + \beta_Z)|t|}.$$

This shows that then $Y + Z$ is Cauchy distributed with $\alpha_Y + \alpha_Z$ and $\beta_Y + \beta_Z$ as its parameters.

Using these two observations it is easily deduced that $c_1X_1 + \cdots + c_nX_n$ is again Cauchy distributed, its parameters as stated.

Step 2. If $\overline{Y}_n$ would converge in probability, then

$$\lim_{n\to\infty} \mathbb{P}\left(\overline{Y}_{2n} - \overline{Y}_n \geq \tfrac{1}{2}\right) = 0 \tag{$*$}$$

However,

$$\begin{aligned} \overline{Y}_{2n} - \overline{Y}_n &= \frac{X_1 + \cdots + X_n + X_{n+1} + \cdots + X_{2n}}{2n} - \frac{2(X_1 + \cdots + X_n)}{2n} \\ &= \frac{(X_{n+1} - X_1) + \cdots + (X_{2n} - X_n)}{2n}. \end{aligned}$$

The quantity above is (by step 1) Cauchy distributed with parameters $\alpha = 0$ and $\beta = 1$, indifferent the value of n. It follows from this that $(*)$ can impossibly be true. □

Exercise 16

If $g : \mathbb{R} \to \mathbb{R}$ is continuous and $X_k \to X$ in distribution, then also $g(X_k) \to g(X)$ in distribution. Use Theorem VII.2.11 to prove this statement.

Proof. To the sequence $X, X_1, X_2, \dots$ by Theorem VII.2.11 we choose a twin sequence $X', X_1{}', X_2{}', \dots$ such that $X_k' \to X'$ strongly. Then, g being continuous, also $g(X'_k) \to g(X')$ strongly, which of course implies that $g(X'_k) \to g(X')$ in distribution. However (see §I.11 Exercise 27) for all k the two variables $g(X_k)$ and $g(X'_k)$ are identically distributed and the same holds for $g(X)$ and $g(X')$. Hence $g(X_k) \to g(X)$ in distribution. □

Exercise 17

Let $F_1, \dots, F_n$ be an arbitrary set of distribution functions and let $q_1, \dots, q_n$ be the corresponding quantile functions. Define the probability space $(\Omega, \mathfrak{A}, \mathbb{P})$ by

$$\begin{cases} \Omega & := & (0,1)^n = (0,1) \times \cdots \times (0,1), \\ \mathfrak{A} & := & \text{the collection of all Borel sets on } (0,1)^n, \\ \mathbb{P} & := & \text{the Lebesgue measure on } (0,1)^n. \end{cases}$$

For all i we define the function $X_i : \Omega \to \mathbb{R}$ by

$$X_i(u_1, \dots, u_n) := q_i(u_i).$$

(i) Prove that X_i is a stochastic variable that has F_i as its distribution function $(i = 1, \dots, n)$.

Proof. For any Borel set A in $\mathbb{R}$ we have

$$X_i{}^{-1}(A) = (0,1)^{i-1} \times q_i{}^{-1}(A) \times (0,1)^{n-i}.$$

As q_i is increasing, by Lemma VII.2.6 $q_i{}^{-1}(A)$ is Borel in $\mathbb{R}$, so that $X_i{}^{-1}(A)$ is Borel in $\mathbb{R}^n$. This shows that X_i is a stochastic variable on $(\Omega, \mathfrak{A}, \mathbb{P})$. Furthermore, by Exercise 8 (iii),

$$\begin{aligned} \mathbb{P}(X_i \leq x) &= \mathbb{P}(\{(u_1, \dots, u_n) \in \Omega : q_i(u_i) \leq x\}) \\ &= \mathbb{P}\left((0,1)^{i-1} \times (0, F_i(x)] \times (0,1)^{n-i}\right) = F_i(x). \end{aligned}$$

□

(ii) Prove that the variables $X_1, \dots, X_n$ constitute a statistically independent system.

Proof. Let $A_1, \dots, A_n$ be Borel sets in $\mathbb{R}$. Define the Borel sets $B_1, \dots, B_n$ in $(0,1)$ by $B_i := X_i^{-1}(A_i)$. Then, as $\mathbb{P}$ denotes the Lebesgue measure on $(0,1)^n$,

$$\begin{aligned} \mathbb{P}((X_1, \dots, X_n) \in A_1 \times \cdots \times A_n) &= \mathbb{P}(B_1 \times \cdots \times B_n) \\ &= \mathbb{P}(B_1) \cdots \mathbb{P}(B_n) \\ &= \mathbb{P}(X_1 \in A_1) \cdots \mathbb{P}(X_n \in A_n), \end{aligned}$$

which shows that $X_1, \dots, X_n$ is a statistically independent system. □

(iii) Compare one thing and another with §I.11 Exercise 42.

Solution. This exercise generalizes the result of §I.11 Exercise 42. □

Exercise 18

Use the previous exercise to prove the following generalization of Skorokhod's theorem.

If $F_k \to F$ in (Γ, d_L), then there exists a stochastic n-vector $\mathbf{X} = (X_1, \dots, X_n)$ together with a sequence of n-vectors $\mathbf{X}_k = (X_{1k}, \dots, X_{nk})$ such that

- The variables $X_1, \dots, X_n$ have a common distribution function F and they form a statistically independent system,
- For fixed k the variables $X_{1k}, \dots, X_{nk}$ have a common distribution function F_k and they form a statistically independent system,
- For all $i = 1, \dots, n$ we have: $X_{ik} \to X_i$ almost surely if $k \to \infty$.

Proof. Choose $X_1, \dots, X_n$ in relation to F as we did in the previous exercise, and choose, for each fixed k, in the same way variables $X_{1k}, \dots, X_{nk}$, but then in relation to F_k. This can be done in such a way that each of these sequences forms a statistically independent system.

Re-examining the proof of Skorokhod's theorem (Theorem VII.2.11) leads to the conclusion that for each fixed i we have $X_{ik} \to X_i$ almost surely. □

Exercise 19

Set $\Gamma_c := \{F \in \Gamma \ : \ F \text{ is continuous}\}$.

(i) Prove that d_L and d_N generate the same topology on Γ_c.

Proof. This is a direct consequence of Theorem VII.5.6. □

(ii) Is the metric space (Γ_c, d_L) complete?

Solution. No. To see this, choose any discontinuous F. By Theorem VII.6.8 it is possible to choose a sequence $F_1, F_2, \dots$ in Γ_c such that $F_k \to F$ in the d_L-metric on Γ. (For example, we may take $F_k := \Phi_{1/k} \times F$ for $k = 1, 2, \dots,$ Φ as always denoting the standard normal distribution function.) Now $F_1, F_2, \dots$ is a Cauchy sequence in (Γ_c, d_L) without having a limit in it. □

(iii) Is the metric space (Γ_c, d_N) complete?

Solution. Yes. Namely, if $F_1, F_2, \dots$ is a Cauchy sequence in the d_N-metric, then for every fixed $x \in \mathbb{R}$ the sequence $F_1(x), F_2(x), \dots$ is Cauchy in $\mathbb{R}$. Hence, as $\mathbb{R}$ is complete, there is a limit $F(x)$ in $\mathbb{R}$.

The sequence $F_1, F_2, \dots$ converges uniformly to this F. To see this we note that for any $\varepsilon > 0$ there is an integer N such that

$$m, n \geq N \quad \Longrightarrow \quad |F_m(x) - F_n(x)| < \varepsilon \quad \text{for all } x \in \mathbb{R}.$$

It follows that

$$m \geq N \quad \Longrightarrow \quad |F_m(x) - F(x)| \leq 2\varepsilon \quad \text{for all } x \in \mathbb{R}.$$

This exactly proves the uniform convergence of $F_1, F_2, \ldots$ to F.

As the F_k are continuous it follows from this that F is continuous. It is left to the reader to prove that F is a distribution function. We conclude that every Cauchy sequence in (Γ_c, d_N) is convergent, that is, that (Γ_c, d_N) is complete. □

(iv) Are d_L and d_N equivalent metrics on Γ_c ?

Solution. If d_L and d_N were equivalent on Γ_c, then (see Appendix E, Proposition 6) the metric space (Γ_c, d_L) would also be complete. This is contradictory to (ii). □

Exercise 20

Prove that for all $F, G \in \Gamma$ one has $d_L(F,G) \leq d_N(F,G)$.

Proof. Set $\alpha := d_N(F,G)$. Then we have

$$G(x-\alpha) - \alpha \;\leq\; G(x) - \alpha \;\leq\; F(x) \;\leq\; G(x) + \alpha \;\leq\; G(x+\alpha) + \alpha$$

from which it immediately follows that $d_L(F,G) \leq \alpha$. □

Exercise 21

(i) Prove that if two stochastic n-vectors $(X_1, \ldots, X_n)$ and $(X'_1, \ldots, X'_n)$ are identically distributed, they are automatically twins.

Proof. Define for every $i = 1, \ldots, n$ the continuous function $g_i : \mathbb{R}^n \to \mathbb{R}$ by $g_i(x_1, \ldots, x_n) := x_i$. Now apply the result of §I.11 Exercise 27. □

(ii) Give an example of two stochastic vectors that form a twin, whereas they are not identically distributed.

Proof. Let X and Y be independent identically distributed variables. Now the pair of 2-vectors (X, X) and (X, Y) form a twin. However, they are not identically distributed (unless X is essentially a constant). □

(iii) If both the variables $X_1, \ldots, X_n$ and $X'_1, \ldots, X'_n$ form statistically independent systems, then the stochastic n-vectors $(X_1, \ldots, X_n)$ and $(X'_1, \ldots, X'_n)$ are identically distributed if and only if they are twins. Prove this.

Proof. If the two n-vectors are identically distributed then they are surely twins (see (i)). To prove the converse in the case of statistical independence of the components, we notice that in that case (in a self-explaining notation) the twin condition implies that for any $A_1, \ldots, A_n$ that are Borel in $\mathbb{R}$

$$\begin{aligned}\mathbb{P}((X_1, \ldots, X_n) \in A_1 \times \cdots \times A_n) &= \mathbb{P}(X_1 \in A_1) \cdots \mathbb{P}(X_n \in A_n) \\ &= \mathbb{P}(X_1' \in A_1) \cdots \mathbb{P}(X_n' \in A_n) \\ &= \mathbb{P}((X_1', \ldots, X_n') \in A_1 \times \cdots \times A_n).\end{aligned}$$

The above implies, however, (see Appendix B) that

$$\mathbb{P}((X_1, \ldots, X_n) \in A) = \mathbb{P}((X_1', \ldots, X_n') \in A)$$

for *all* Borel sets A in $\mathbb{R}^n$, which is the same as saying that the two vectors are identically distributed. □

Exercise 22

Question. In §VII.6 we have seen that with respect to (Γ, d_L) every $F \in \Gamma$ is the limit of a sequence $F_1, F_2, \ldots$ of continuous distribution functions. Is this also true in (Γ, d_N) ?

Solution. No. See Exercise 19: the limit of such a sequence in (Γ, d_N) will automatically be continuous. □

Exercise 23

Let $E_a := 1_{[a,+\infty)}$ for all $a \in \mathbb{R}$. Prove that

$$d_L(E_a, E_b) = \min(1, |a-b|).$$

Proof. Without loss of generality we may suppose that $a > b$ (the case that $a = b$ is trivial). We shall prove that

$$d_L(E_a, E_b) = \begin{cases} 1 & \text{if } a - b \geq 1, \\ a - b & \text{if } a - b < 1. \end{cases}$$

In each case we are looking for the infimum of the $\delta \geq 0$ for which

$$E_b(x-\delta) - \delta \overset{(1)}{\leq} E_a(x) \overset{(2)}{\leq} E_b(x+\delta) + \delta$$

for all $x \in \mathbb{R}$.

If $a - b \geq 1$, then for every $\delta < 1$ for $x := b + \delta$ inequality (1) does not apply. Therefore in this case $\delta = 1$ is the best available value for δ, so $d_L(E_a, E_b) = 1$.

If $a - b < 1$ we need to know when the equalities (1) and (2) apply. For the first one it is required that

$$E_b(x-\delta) = 1 \quad \Longrightarrow \quad E_a(x) = 1,$$

that is to say, that $x - \delta \geq b \Rightarrow x \geq a$. It is clear that for $\delta < a - b$ we get a counterexample by putting $x := b+\delta$ and that $\delta = a-b$ is the best value available.

Inequality (2) is always true for $\delta = a-b$ ($E_a(x) = 1 \Rightarrow E_b(x+\delta) = 1$ because if $x \geq a$ then $x + \delta \geq x \geq a > b$). In short: $d_L(E_a, E_b) = a - b$ if $a - b < 1$, which concludes the proof. □

Exercise 24

Let F be an arbitrary distribution function. The 'graph' $\mathcal{G}_F$ of F is understood to be the set

$$\mathcal{G}_F := \{(x, F(x)) \in \mathbb{R}^2 \ : \ x \in \mathbb{R}\}.$$

For all $a \in \mathbb{R}$ we define the 'straight line' $\mathcal{L}_a$ in $\mathbb{R}^2$ by

$$\mathcal{L}_a := \{(x, y) \in \mathbb{R}^2 \ : \ x + y = a\}.$$

(i) Can it occur that $\mathcal{L}_a \cap \mathcal{G}_F = \emptyset$?

Solution. Yes, this can very well occur if F exhibits discontinuities, as is shown in the figure.

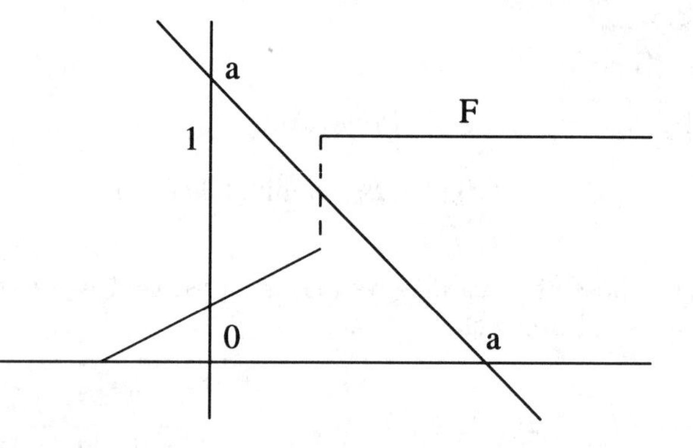

(ii) If F is continuous, then for all $a \in \mathbb{R}$ the intersection $\mathcal{L}_a \cap \mathcal{G}_F$ consists of exactly one point. Prove this.

Proof. If F is continuous, the map $x \mapsto x + F(x)$ is strictly increasing as well as continuous. The question is whether an intersection point (x, y) exists for which $y = F(x)$ as well as $x + y = a$, that is, whether an $x \in \mathbb{R}$ exists for which $x + F(x) = a$. Application of Weierstrass' theorem on intermediate values shows that there is a point (x, y) satisfying these conditions; of course it is unique. □

(iii) From here on, in case of a continuous distribution function F, we will denote the intersection point $\mathcal{L}_a \cap \mathcal{G}_F$ by $Q(F, a)$. By $\| \ \|$ we mean the Euclidian norm on $\mathbb{R}^2$. Prove that if F and G are continuous distribution functions, then

$$d_L(F,G) = \frac{1}{\sqrt{2}} \sup_{a\in\mathbb{R}} \|Q(F,a) - Q(G,a)\|.$$

Proof. We first prove that

$$d_L(F,G) \leq \frac{1}{\sqrt{2}} \sup_{a\in\mathbb{R}} \|Q(F,a) - Q(G,a)\|. \tag{1}$$

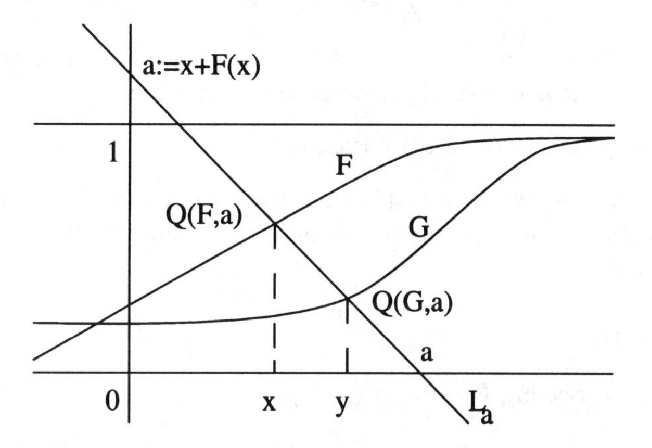

To this we denote the right side of the inequality above by λ. Now choose any $x \in \mathbb{R}$: we have to prove that

$$G(x-\lambda) - \lambda \leq F(x) \leq G(x+\lambda) + \lambda. \tag{2}$$

To this we take a look at the point $(x, F(x))$, which equals $Q(F,a)$ for $a := x + F(x)$. Let y be the point for which $(y, G(y)) = Q(G,a)$.

Now the line $\mathcal{L}_a$ contains both $(x, F(x))$ and $(y, G(y))$. Because of this we see that $x+F(x) = y+G(y)$, or: $x-y = G(y)-F(x)$. Denoting $\delta := |x-y|$, Pythagoras' rule tells us that

$$\delta^2 + \delta^2 = \|(x,F(x)) - (y,G(y))\|^2 \leq (\sqrt{2}\,\lambda)^2 = 2\lambda^2.$$

As a consequence $|F(x) - G(y)| = \delta \leq \lambda$, so

$$G(y) - \lambda \leq F(x) \leq G(y) + \lambda.$$

As $x - \lambda \leq y \leq x + \lambda$, (2) easily follows.

Next, we choose any $\alpha > d_L(F,G)$. Then for all $x \in \mathbb{R}$ we have

$$G(x-\alpha) - \alpha \leq F(x) \leq G(x+\alpha) + \alpha. \tag{3}$$

Now choose any $a \in \mathbb{R}$ and let x and y be such that

$$(x, F(x)) = Q(F, a) \quad \text{and} \quad (y, G(y)) = Q(G, a).$$

Then $x + F(x) = y + G(y)$. Together with (3) this implies that

$$(x - \alpha) + G(x - \alpha) \leq y + G(y) \leq (x + \alpha) + G(x + \alpha).$$

Because the map $x \mapsto x + G(x)$ is strictly increasing, we arrive at the conclusion that $x - \alpha \leq y \leq x + \alpha$, so that $|y - x| = |G(y) - F(x)| \leq \alpha$. This implies that

$$\frac{1}{\sqrt{2}} \|Q(F, a) - Q(G, a)\| \leq \alpha.$$

Since this holds for every $\alpha > d_L(F, G)$ and $a \in \mathbb{R}$ was taken arbitrarily, we arrive at the converse of (1), thus completing the proof. □

(iv) Give a geometrical interpretation of (iii).

Solution. The result of (iii) is that, aside from a factor $\sqrt{2}$, the Lévy metric is measuring the 'maximum' distance of $\mathcal{G}_F$ and $\mathcal{G}_G$ in the direction of the vector $(1, -1)$. □

Exercise 25

Suppose $\zeta > 0$. Setting $F_\zeta : x \mapsto F(x/\zeta)$, one has

$$\chi_{F_\zeta}(t) = \chi_F(\zeta t)$$

for all $t \in \mathbb{R}$. Prove this.

Proof. Choose a stochastic variable X that has F as its distribution function. Then ζX has F_ζ as its distribution function. Now

$$\begin{aligned} \chi_{F_\zeta}(t) = \chi_{\zeta X}(t) &= \mathbb{E}(e^{it(\zeta X)}) = \mathbb{E}(e^{i(t\zeta)X}) \\ &= \chi_X(\zeta t) = \chi_F(\zeta t), \end{aligned}$$

which proves the statement. □

Exercise 26

Prove that the Lévy metric is translation invariant, that is to say, prove that for all $a \in \mathbb{R}$

$$d_L(\tau_a F, \tau_a G) = d_L(F, G).$$

Proof. Choose $\lambda > d_L(F, G)$ and let a be a fixed real number. We then have for all $x \in \mathbb{R}$

$$G(x - \lambda) - \lambda \leq F(x) \leq G(x + \lambda) + \lambda.$$

Because this holds for all x, we may replace x by $x - a$ and thus for all $x \in \mathbb{R}$ we have

$$(\tau_a G)(x - \lambda) - \lambda \;\le\; (\tau_a F)(x) \;\le\; (\tau_a G)(x + \lambda) + \lambda.$$

It follows from this that $d_L(\tau_a F, \tau_a G) \le \lambda$, which holds for all $\lambda > d_L(F, G)$, hence

$$d_L(\tau_a F, \tau_a G) \;\le\; d_L(F, G).$$

The reversed inequality is a direct consequence of the above, since

$$d_L(F, G) = d_L(\tau_{-a}(\tau_a F), \tau_{-a}(\tau_a G)) \;\le\; d_L(\tau_a F, \tau_a G). \qquad \square$$

Exercise 27

Let $\varphi : \mathbb{R} \to [0, 1]$ be an arbitrary increasing function. Show that for the function $F : \mathbb{R} \to [0, 1]$ defined by $F(x) := \varphi(x+)$ $(x \in \mathbb{R})$ we have:

(i) F is increasing on $\mathbb{R}$,

(ii) $F(x-) = \varphi(x-)$ for all $x \in \mathbb{R}$,

(iii) $F(x+) = \varphi(x+) = F(x)$ for all $x \in \mathbb{R}$,

(iv) $\lim_{x \to \pm\infty} F(x) = \lim_{x \to \pm\infty} \varphi(x)$.

Solution. First note that F is well defined: for any $x \in \mathbb{R}$ and any sequence $x_k \downarrow x$ the sequence $\varphi(x_k)$ is decreasing and $\ge \varphi(x)$. From this it follows that surely $F(x) \ge \varphi(x)$ for all x.

(i) Suppose $x < y$. Then we have

$$F(x) \;\le\; \varphi\left(\tfrac{x+y}{2}\right) \;\le\; \varphi(y) \;\le\; F(y),$$

which shows that F is increasing.

(ii) We take a fixed $x \in \mathbb{R}$ and suppose that $\varphi(x-) < \alpha$. Now assume, to get a contradiction, that $F(x-) > \alpha$. Then $F(x - \frac{1}{n}) > \alpha$ for some $n \in \mathbb{N}$ (as F is increasing), whereas $\varphi(x - \frac{1}{2n}) < \alpha$. This cannot be so, since

$$F(x - \tfrac{1}{n}) = \varphi\left((x - \tfrac{1}{n})+\right) \;\le\; \varphi\left((x - \tfrac{1}{n}) + \tfrac{1}{2n}\right),$$

from which it follows that $F(x-) \le \alpha$. As α was taken arbitrarily it follows that $F(x-) \le \varphi(x-)$.

On the other hand,

$$F(x-) = \lim_{n\to\infty} F(x - \tfrac{1}{n}) \;\ge\; \lim_{n\to\infty} \varphi(x - \tfrac{1}{n}) = \varphi(x-).$$

Hence $F(x-) = \varphi(x-)$.

(iii) Again we fix an $x \in \mathbb{R}$. We prove that $F(x+) = \varphi(x+)$. Of course

$$F(x+) = \lim_{n\to\infty} F(x + \tfrac{1}{n}) \geq \lim_{n\to\infty} \varphi(x + \tfrac{1}{n}) = \varphi(x+).$$

Now suppose $\varphi(x+) < \alpha$ and $F(x+) > \alpha$. Then $\varphi(x + \frac{1}{n}) < \alpha$ for some integer n whereas $F(x + \frac{1}{2n}) > \alpha$. As in (ii) this leads to a contradiction.

We conclude that $F(x+) \leq \alpha$ and, in general, $F(x+) \leq \varphi(x+)$, which concludes the proof.

(iv) Finally we prove that $\lim_{x\to+\infty} F(x) = \lim_{x\to+\infty} \varphi(x)$ (for $x \to -\infty$ similar arguments can be used).

Of course both limits exist and are ≤ 1. As $F(x) \geq \varphi(x)$ for all x we surely have

$$\lim_{x\to+\infty} F(x) \geq \lim_{x\to+\infty} \varphi(x).$$

On the other hand, if $\lim_{x\to+\infty} \varphi(x) \leq \alpha$ then $\varphi \leq \alpha$ so $F \leq \alpha$ and $\lim_{x\to+\infty} F(x) \leq \alpha$. Therefore

$$\lim_{x\to+\infty} F(x) \leq \lim_{x\to+\infty} \varphi(x).$$

□

Exercise 28

For $n = 1, 2, \ldots$ we define the probability measure $\mathbb{P}_n$ by $\mathbb{P}_n := \frac{1}{n}(\delta_1 + \cdots + \delta_n)$. The distribution function belonging to $\mathbb{P}_n$ (in fact the empirical distribution function $\hat{F}(1, 2, \ldots, n)$) is denoted by F_n.

(i) Prove that for all n one has $d_L(F_n, \tau_a F_n) \leq \frac{1}{n}$ if $|a| < \frac{1}{2}$.

Proof. Drawing a picture makes it apparent that $|F_n(x) - F_n(x-a)| \leq \frac{1}{n}$ for all n and x if $|a| < \frac{1}{2}$. This shows that

$$\|F_n - \tau_a F_n\|_\infty \leq \tfrac{1}{n} \quad \text{for all } n \text{ if } |a| < \tfrac{1}{2}.$$

By Exercise 20 this of course implies that $d_L(F_n, \tau_a F_n) \leq \frac{1}{n}$. □

(ii) Prove that for all n one has $d_P(F_n, \tau_a F_n) \geq |a|$ if $|a| < \frac{1}{2}$, where d_P is the Prokhorov metric on Γ.

Proof. Define the measure $\mathbb{P}_{n,a}$ by $\mathbb{P}_{n,a} := \frac{1}{n}(\delta_{1+a} + \cdots + \delta_{n+a})$: this $\mathbb{P}_{n,a}$ is the probability measure belonging to $\tau_a F_n$.

Furthermore define the set A_0 by $A_0 := \{1, \ldots, n\}$. Then (see §5)

$$A_0^\delta = (1-\delta, 1+\delta) \cup (2-\delta, 2+\delta) \cup \cdots \cup (n-\delta, n+\delta),$$

and we therefore have

$$\mathbb{P}_{n,a}(A_0^\delta) = 0 \quad \text{if } \delta < |a| < \tfrac{1}{2}.$$

Hence

$$\begin{aligned} d_P(F_n, \tau_a F_n) &= \inf\{\delta \geq 0 \ : \ \mathbb{P}_n(A) \leq \mathbb{P}_{n,a}(A^\delta) + \delta \text{ and} \\ &\qquad \mathbb{P}_{n,a}(A) \leq \mathbb{P}_n(A^\delta) + \delta \quad \text{for all } A \in \mathfrak{B}\} \\ &\geq \inf\{\delta \geq 0 \ : \mathbb{P}_n(A_0) \leq \mathbb{P}_{n,a}(A_0^\delta) + \delta\} \\ &= \inf\{\delta \geq 0 \ : \ 1 \leq 0 + \delta\} = 1 \ \geq \ |a|. \end{aligned}$$

□

(iii) Explain that the metrics d_L and d_P are not equivalent.

Solution. Equivalence of d_L and d_P implies uniform continuity of the identity map from (Γ, d_L) onto (Γ, d_P). Letting $n \to \infty$ in (i) and (ii), we see that the latter is impossible. □

Exercise 29

Using Proposition VII.7.8 and Proposition VII.5.3, prove that the metric space (Γ, d_L) is separable.

Proof. Proposition VII.7.8 states that the set containing all empirical distribution functions is dense in (Γ, d_L). Consequently, if the set of all 'rational' empirical distribution functions $\{\hat{F}(q_1, \dots, q_n) \ : \ q_i \in \mathbb{Q} \text{ for all } i\}$ is dense inside *this* set, it is (check) a countable dense subset of (Γ, d_L). In that case (Γ, d_L) is (by definition) separable. We will prove that this is indeed the case.

To this, we take any $\varepsilon \in (0,1)$ and $(x_1, \dots, x_n) \in \mathbb{R}^n$. Using the same technique as in Exercise 23, it is not very difficult to show that

$$d_L(\hat{F}(\mathbf{a}), \hat{F}(\mathbf{b})) \leq \max\{|a_j - b_j| \ : \ j = 1, \dots, n\}$$

for all $\mathbf{a} = (a_1, \dots, a_n)$ and $\mathbf{b} = (b_1, \dots, b_n)$ in $\mathbb{R}^n$. Taking $\mathbf{a} := (x_1, \dots, x_n)$ and $\mathbf{b} := (q_1, \dots, q_n) \in \mathbb{Q}^n$ with q_j such that $|x_j - q_j| \leq \varepsilon$ $(j = 1, \dots, n)$, we see that $d_L(\hat{F}(\mathbf{a}), \hat{F}(\mathbf{b})) \leq \varepsilon$. This exactly provides us with what we need. □

Exercise 30

Let F be an arbitrary distribution function and let χ_F be its characteristic function. Prove that

(i) $\chi_F : \mathbb{R} \to \mathbb{C}$ is a continuous function,

(ii) $\chi_F(0) = 1$,

(iii) $|\chi_F(t)| \leq 1$ for all $t \in \mathbb{R}$.

Proof. First we take a stochastic variable X corresponding to the distribution function F.

(i) We show the continuity of χ_F in $t \in \mathbb{R}$. Let $t_1, t_2, \ldots$ be a sequence in $\mathbb{R}$ converging to t. For all n we have

$$\chi_F(t_n) = \mathbb{E}(e^{it_nX}) = \int e^{it_nX}\, d\mathbb{P}_F(x).$$

Furthermore of course $e^{it_n\mathbf{x}} \to e^{it\mathbf{x}}$ (by $f(\mathbf{x})$ we mean the function $x \mapsto f(x)$) and $|e^{it_nx}| \leq 1$ for all $x \in \mathbb{R}$. Now we just apply Lebesgue's theorem on dominated convergence and conclude that $\chi_F(t_n) \to \chi_F(t)$, that is, χ_F is continuous in t.

(ii) Of course we have $\chi_F(0) = \mathbb{E}(e^0) = 1$.

(iii) Finally, for any fixed $t \in \mathbb{R}$,

$$\begin{aligned} |\chi_F(t)| &= \left|\int e^{itX}\, d\mathbb{P}_F(x)\right| \\ &\leq \int |e^{itX}|\, d\mathbb{P}_F(x) = \int 1\, d\mathbb{P}_F(x) = 1. \end{aligned}$$

□

Exercise 31

Prove that for all $F, G, H \in \Gamma$ one has $d_L(F \times H, G \times H) \leq d_L(F, G)$.

Proof. Choose $\alpha > d_L(F, G)$. Then for all $x, y \in \mathbb{R}$ we have that

$$G(x - y - \alpha) - \alpha \ \leq\ F(x - y) \ \leq\ G(x - y + \alpha) + \alpha.$$

Integrating with respect to $\mathbb{P}_H(y)$ we get, applying Proposition VII.6.2:

$$(G \times H)(x - \alpha) - \alpha \ \leq\ (F \times H)(x) \ \leq\ (G \times H)(x + \alpha) + \alpha \quad \text{for all } x \in \mathbb{R}.$$

By definition of the Lévy metric this implies that $d_L(F \times H, G \times H) \leq \alpha$.

Because this holds for all $\alpha > d_L(F, G)$, we conclude that

$$d_L(F \times H, G \times H) \ \leq\ d_L(F, G).$$

□

Exercise 32

For all $n = 1, 2, \ldots$ and $F \in \Gamma$ we set $F^{\times n} := F \times \cdots \times F$ (n factors). Prove that for all $F, G \in \Gamma$ and $n = 1, 2, \ldots$ we have

$$d_L(F^{\times n}, G^{\times n}) \ \leq\ n\, d_L(F, G).$$

Proof. This can be proved by induction. For $n = 1$ the statement is trivial. If the statement is true for n, then we can pass to $n + 1$ in the following way:

$$\begin{aligned} d_L(F^{\times(n+1)}, G^{\times(n+1)}) &\leq d_L(F^{\times(n+1)}, F^{\times n} \times G) + d_L(F^{\times n} \times G, G^{\times(n+1)}) \\ &\leq d_L(F^{\times n} \times F, F^{\times n} \times G) + d_L(F^{\times n} \times G, G^{\times n} \times G) \\ (*) \quad &\leq d_L(F, G) + n\, d_L(F, G) = (n + 1)\, d_L(F, G). \end{aligned}$$

At $(*)$ we used the previous exercise together with the induction hypothesis. □

Exercise 33

For $n = 1, 2, \ldots$ we define the function $g_n : \mathbb{R}^n \to \mathbb{R}$ by

$$g_n(x_1, \ldots, x_n) := \frac{x_1 + \cdots + x_n}{n}.$$

Prove that for all n the map $\tilde{g}_n : (\Gamma, d_L) \to (\Gamma, d_L)$ is uniformly continuous.

Proof. First, for every $F \in \Gamma$ and $n = 1, 2, \ldots$, we define the distribution function $\rho_n F$ by

$$(\rho_n F)(x) := F(nx) \qquad (x \in \mathbb{R}).$$

In this notation we have

$$d_L(\rho_n F, \rho_n G) \ \le \ d_L(F, G). \tag{$*$}$$

To see this, choose any $\alpha > d_L(F, G)$. Then

$$\begin{aligned} G(n(x-\alpha)) - \alpha \ &\le \ G(nx - \alpha) - \alpha \ \le \ F(nx) \\ &\le \ G(nx + \alpha) + \alpha \ \le \ G(n(x+\alpha)) + \alpha, \end{aligned}$$

so

$$(\rho_n G)(x - \alpha) - \alpha \ \le \ (\rho_n F)(x) \ \le \ (\rho_n G)(x + \alpha) + \alpha,$$

which implies that $d_L(\rho_n F, \rho_n G) \le \alpha$. Because this holds for every $\alpha > d_L(F, G)$, we conclude that $(*)$ is true.

Next, let $F \in \Gamma$ and let $X_1, \ldots, X_n$ be an independent set of variables, all of them with distribution function F. Then the distribution function $F^{\times n}$ is, by definition, the one belonging to the variable $X_1 + \cdots + X_n$. It is now easy to see that the distribution function $\tilde{g}_n(F)$, being the one belonging to the variable $(X_1 + \cdots + X_n)/n$, is given by

$$\tilde{g}_n(F) = \rho_n(F^{\times n}).$$

Applying $(*)$ together with the previous exercise we learn that

$$d_L(\tilde{g}_n(F), \tilde{g}_n(G)) \ \le \ n \ d_L(F, G).$$

The uniform continuity of $\tilde{g}_n$ follows immediately from this. □

Exercise 34

Let $X_1, \ldots, X_n$ be a sample from an arbitrary population. We set $T_n := \max\{X_1, \ldots, X_n\}$. This statistic emanates from the continuous function $g_n : \mathbb{R}^n \to \mathbb{R}$ defined by

$$g_n(x_1, \ldots, x_n) := \max\{x_1, \ldots, x_n\}.$$

(i) Prove that for all $F \in \Gamma$ one has $\tilde{g}_n(F)(x) = F(x)^n$ (all $x \in \mathbb{R}$).

Proof. Let F denote the distribution function corresponding to the population. Then $\tilde{g}_n(F)$ is, by definition, the distribution function of the variable $T_n = \max(X_1, \dots, X_n)$. Now

$$\begin{aligned} \mathbb{P}(T_n \le x) &= \mathbb{P}(X_1 \le x, \dots, X_n \le x) \\ &= \mathbb{P}(X_1 \le x) \cdots \mathbb{P}(X_n \le x) = F(x)^n. \end{aligned}$$

□

(ii) *Question.* Do there exist populations such that the sequence $T_1, T_2, \dots$ is d_L-robust?

Solution. No. To see this, suppose that F is an element in Γ such that the sequence $\tilde{g}_1, \tilde{g}_2, \dots$ is equicontinuous in F (in the Lévy metric). We shall deduce a contradiction from this assumption.

To this, choose ε such that $0 < \varepsilon < 1$. Then there exists a $\delta > 0$ such that

$$d_L(F,G) < \delta \implies d_L(\tilde{g}_n(F), \tilde{g}_n(G)) < \varepsilon$$

simultaneously for all $n = 1, 2, \dots$. Writing this out we see that

$$G(x-\varepsilon)^n - \varepsilon \le F(x)^n \le G(x+\varepsilon)^n + \varepsilon \qquad (*)$$

for all $n = 1, 2, \dots$ and $x \in \mathbb{R}$ whenever $d_L(F,G) < \delta$.

Now choose a tailless element G in Γ such that $d_L(F,G) < \delta$. Then $(*)$ tells us that $G(x-\varepsilon)^n - \varepsilon \le F(x)^n$. Choosing an $x \in \mathbb{R}$ such that $G(x-\varepsilon) = 1$ we see that there exists an element $x \in \mathbb{R}$ such that

$$1 - \varepsilon \le \underline{\lim}_{n\to\infty} F(x)^n.$$

This implies that there is an $x \in \mathbb{R}$ such that $F(x) = 1$. This is impossible.

To see this, choose a $G \in \Gamma$ such that $G(x) < 1$ for all $x \in \mathbb{R}$ and such that $d_L(F,G) < \delta$ (for example, choose G such that $d_N(F,G) < \delta$; draw a picture). Then by $(*)$ we have $F(x)^n \le G(x+\varepsilon)^n + \varepsilon$ for all $n = 1, 2, \dots$ and $x \in \mathbb{R}$. From this it follows that

$$\overline{\lim}_{n\to\infty} F(x)^n \le \varepsilon \quad \text{for all } x \in \mathbb{R}.$$

Because $\varepsilon < 1$ this implies that $\lim_{n\to\infty} F(x)^n = 0$, which in turn implies that $F(x) < 1$ for all $x \in \mathbb{R}$. □

Exercise 35

(i) Let $F_1, F_2, \dots$ be a sequence of distribution functions converging pointwise on $\mathbb{Q}$ to a function $F_0 : \mathbb{Q} \to [0,1]$. Define $F : \mathbb{R} \to [0,1]$ by

$$F(x) := \inf\{F_0(y) : y \in \mathbb{Q} \text{ and } y > x\}.$$

Prove that F is increasing, right-continuous and that

$$F(x-) = \sup\{F_0(y) \,:\, y \in \mathbb{Q} \text{ and } y < x\}.$$

Proof. It is easily seen that both F_0 and F are increasing. To prove that F is right-continuous in, say x, choose any $\lambda > F(x)$. Then there is a rational $y > x$ such that $F(x) \leq F_0(y) < \lambda$. It follows from this that for any z strictly between x and y we have $F(z) \leq F_0(y) < \lambda$. This proves that $\inf\{F(z) \,:\, z > x\} < \lambda$. Because this holds for every $\lambda > F(x)$ we conclude that

$$F(x) \;\geq\; \inf\{F(z) \,:\, z > x\};$$

of course $F(x) \;\leq\; \inf\{F(z) \,:\, z > x\}$ so we actually have an equality here. This proves that F is right-continuous in x.

Next we notice that for $y < z < x$, where y is rational, we have $F_0(y) \;\leq\; F(z) \;\leq\; F(x-)$. It follows from this that

$$F(x-) \;\geq\; \sup\{F_0(y) \,:\, y \in \mathbb{Q} \text{ and } y < x\}. \tag{$*$}$$

To prove the converse inequality, suppose that $\lambda < F(x-)$. Then there is a real number $z < x$ such that $\lambda < F(z)$. Choosing a rational y strictly between z and x, we have $\lambda < F(z) \leq F_0(y)$, which shows that

$$\lambda \;<\; \sup\{F_0(y) \,:\, y \in \mathbb{Q} \text{ and } y < x\}.$$

Because this holds for every $\lambda < F(x-)$ we conclude that $(*)$ is also valid when replacing the inequality $\geq$ by $\leq$. □

(ii) Give an example showing that, in spite of the above, it can very well occur that F is not a distribution function.

Solution. Just consider the sequence $F_1, F_2, \ldots$ defined by

$$F_n(x) := \begin{cases} 1 & \text{if } x \geq n, \\ 0 & \text{if } x < n. \end{cases}$$

□

(iii) Prove that $F_n(x) \to F(x)$ in every point x in which F is continuous.

Proof. Let $x \in \mathbb{R}$. Then for every rational $y > x$ we have

$$\overline{\lim}_{n\to\infty} F_n(x) \;\leq\; \overline{\lim}_{n\to\infty} F_n(y) = F_0(y),$$

and it follows that

$$\overline{\lim}_{n\to\infty} F_n(x) \;\leq\; F(x). \tag{$*$}$$

Next, choose any $\lambda < F(x-)$. Then (by (i)) there is a rational $y < x$ such that $\lambda < F_0(y)$. Hence $\lambda < F_0(y) = \underline{\lim}_{n\to\infty} F_n(y) \leq \underline{\lim}_{n\to\infty} F_n(x)$. Reasoning as before we conclude that

$$F(x-) \leq \underline{\lim}_{n\to\infty} F_n(x),$$

which together with $(*)$ completes the proof. □

(iv) Show that every sequence $F_1, F_2, \ldots$ of distribution functions contains a subsequence $F_{n_1}, F_{n_2}, \ldots$ that converges pointwise on $\mathbb{Q}$.

Solution. Let $x_1, x_2, \ldots$ be a numbering of the rational numbers. Now $\{F_n(x_1) \ : \ n = 1, 2, \ldots\}$ is a sequence of real numbers in the compact interval $[0, 1]$, so there is some convergent subsequence

$$F_{n_{11}}(x_1),\ F_{n_{12}}(x_1),\ F_{n_{13}}(x_1),\ \ldots\ .$$

Next we observe the sequence $\{F_{n_{1m}}(x_2) \ : \ m = 1, 2, \ldots\}$. Again there is some convergent subsequence

$$F_{n_{21}}(x_2),\ F_{n_{22}}(x_2),\ F_{n_{23}}(x_2),\ \ldots\ .$$

Going on in this way for every $k = 1, 2, \ldots$ we get a subsequence $F_{n_{k1}}$, $F_{n_{k2}}$, $F_{n_{k3}}, \ldots$ having the following two properties:

(a) The sequence $F_{n_{k1}}(x_i), F_{n_{k2}}(x_i), F_{n_{k3}}(x_i), \ldots$ is convergent if $i \leq k$,

(b) The sequence with rank number $k+1$ is a subsequence of the one having rank number k.

From this it is clear that the subsequence $F_{n_{11}}, F_{n_{22}}, F_{n_{33}}, \ldots$ converges pointwise on $\mathbb{Q}$. □

(v) We say that a set $H \subset \Gamma$ satisfies the 'Prokhorov condition' if for every $\varepsilon > 0$ there exist two real numbers a and b such that $F(a) \leq \varepsilon$ and $1 - F(b) \leq \varepsilon$ simultaneously for *all* $F \in H$.
Prove that a closed set in (Γ, d_L) that satisfies the Prokhorov condition is compact.

Proof. Suppose H is a closed set in (Γ, d_L) satisfying the Prokhorov condition. Let $F_1, F_2, \ldots$ be an arbitrary sequence in H. Then, by (iv), there is some subsequence $F_{n_1}, F_{n_2}, \ldots$ that converges pointwise on $\mathbb{Q}$. Define F in relation to this subsequence as we did above (in (i)). Then F is increasing and right-continuous and $F_{n_k}(x) \to F(x)$ in every point x in which F is continuous. The Prokhorov condition on H turns out to be a guarantee that F is a distribution function, that is to say:

$$\lim_{x\to-\infty} F(x) = 0 \quad \text{and} \quad \lim_{x\to+\infty} F(x) = 1.$$

To see this, choose any $\varepsilon > 0$. Then there exists (by the Prokhorov condition) a number a such that $F_{n_k}(a) \leq \varepsilon$ for all $k = 1, 2, \ldots$. Next, choose $\xi < a$ in such a way that F is continuous in ξ. Then, taking the limit for $k \to \infty$ in the above inequality, we arrive at $F(\xi) \leq \varepsilon$. So to every $\varepsilon > 0$ there is a real number ξ such that $F(x) \leq \varepsilon$ for $x \leq \xi$. This means that $\lim_{x \to -\infty} F(x) = 0$. In a similar way one sees that $\lim_{x \to +\infty} F(x) = 1$.

So F is a distribution function and it follows from Theorem VII.5.7 that $F_{n_k} \to F$ in the Lévy metric. We have proved that in the closed subset H every sequence has a convergent subsequence. This implies that H is compact. □

(vi) We will proof that the converse of (v) also holds, that is to say: a compact set H in (Γ, d_L) always satisfies the Prokhorov condition. We do this by making use of Dini's theorem, which states that if a sequence of continuous functions defined on a compact space decreases pointwise to zero then this convergence is automatically uniform.

To start with, for any $\varphi : \mathbb{R} \to \mathbb{R}$ that is bounded and continuous we define an associated map $\Phi : \Gamma \to \mathbb{R}$ by

$$\Phi(F) := \mathbb{E}(\varphi(X)),$$

X denoting a stochastic variable that has F as its distribution function. Show that such a $\Phi : (\Gamma, d_L) \to \mathbb{R}$ is continuous.

Solution. This is a direct consequence of Theorem VII.2.12. □

(vii) Now for every $n = 1, 2, \ldots$ we choose a bounded continuous function $\varphi_n : \mathbb{R} \to [0, 1]$ such that $\varphi_1 \geq \varphi_2 \geq \cdots$ and

$$\varphi_n = \begin{cases} 0 & \text{on } [-n, n], \\ 1 & \text{outside } [-(n+1), (n+1)]. \end{cases}$$

Show that for all $F \in \Gamma$ and every $n = 1, 2, \ldots$ one has

$$F(-n-1) + 1 - F(n+1) \;\leq\; \Phi_n(F) \;\leq\; F(-n) + 1 - F(n).$$

Solution. For $n = 1, 2, \ldots$ we define the set A_n by $A_n := (-\infty, -n] \cup (n, +\infty)$. Now $1_{A_{n+1}} \leq \varphi_n \leq 1_{A_n}$ for all n. Therefore, if X is a stochastic variable having F as its distribution function, $1_{A_{n+1}}(X) \leq \varphi_n(X) \leq 1_{A_n}(X)$, so that

$$\mathbb{E}(1_{A_{n+1}}(X)) \;\leq\; \mathbb{E}(\varphi_n(X)) \;\leq\; \mathbb{E}(1_{A_n}(X)). \qquad (*)$$

Now $1_{A_n}(X)$ is a variable enjoying a Bernoulli distribution with parameter $\mathbb{P}(A_n)$, hence for $n = 1, 2, \ldots$ we have

$$\mathbb{E}(1_{A_n}(X)) = \mathbb{P}(A_n) = F(-n) + 1 - F(n).$$

In combination with $(*)$ this completes the proof. □

(viii) Prove that a compact set in (Γ, d_L) always satisfies the Prokhorov condition.

Proof. First we note that it follows from (vii) that the sequence $\Phi_1, \Phi_2, \ldots$ decreases pointwise to zero on Γ. By Dini's theorem this convergence is necessarily uniform on compact sets. Now apply the left-hand inequality of (vii) and conclude that a compact set in (Γ, d_L) necessarily satisfies the Prokhorov condition. □

Exercise 36

Prove that if to an element $F \in \Gamma$ there is a G such that $F \times G = E$, then $F = E_a$ for some $a \in \mathbb{R}$.

Proof. Suppose that $F \times G = E$. Then there are independent stochastic variables X and Y, having respectively F and G as their distribution functions, such that $X + Y = 0$. Now let x be any real number. Then

$$\begin{aligned} F(x) = \mathbb{P}(X \le x) &= \mathbb{P}(X \le x \text{ and } Y \ge -x) \\ &= \mathbb{P}(X \le x)\, \mathbb{P}(Y \ge -x) = F(x)\, \mathbb{P}(-X \ge -x) \\ &= F(x)\, \mathbb{P}(X \le x) = F(x)^2. \end{aligned}$$

The above implies that $F(x)$ can only be 0 or 1. This means that F is of the form E_a. □

Exercise 37

Let $(\Omega, \mathfrak{A}, \mathbb{P})$ be a probability space and let for all $n = 1, 2, \ldots$ the function $X_n : \Omega \to \mathbb{R}$ be a stochastic variable. Assume that the sequence $X_1, X_2, \ldots$ is uniformly bounded above and define $X : \Omega \to \mathbb{R}$ by

$$X(\omega) := \sup_{n=1,2,\ldots} X_n(\omega).$$

Prove that X is a stochastic variable.

Proof. Define (see also Exercise 34) for all $n = 1, 2, \ldots$ the stochastic variable Y_n as $Y_n := \max\{X_1, \ldots, X_n\}$. Now $Y_n(\omega) \to X(\omega)$ for all $\omega \in \Omega$. By Lemma VII.2.6 this implies that X is $\mathfrak{A}$-measurable, that is, X is indeed a stochastic variable. □

Exercise 38

Prove that if F and G are distribution functions, then

$$\|F - G\|_\infty = \sup_{x \in \mathbb{Q}} |F(x) - G(x)|.$$

Proof. Evidently we have

$$\|F - G\|_\infty \ge \sup_{x \in \mathbb{Q}} |F(x) - G(x)|.$$

To prove the converse inequality, choose any scalar $\lambda < \|F - G\|_\infty$. Then there is a real member $x \in \mathbb{R}$ such that $\lambda < |F(x) - G(x)|$. Now, exploiting the fact that F and G, and therefore $F - G$, are right-continuous in x, together with the fact that $\mathbb{Q}$ is dense in $\mathbb{R}$, we see that there is a rational number $y > x$ such that $\lambda < |(F - G)(y)|$. Hence $\lambda < \sup_{x \in \mathbb{Q}} |F(x) - G(x)|$. As this holds for every $\lambda < \|F - G\|_\infty$ we conclude that the converse inequality is also valid. □

Exercise 39

If $X_1, \ldots, X_n$ is a sample from a population with distribution function F, then $\|\hat{F}(X_1, \ldots, X_n) - F\|_\infty$ presents a stochastic variable. Prove this.

Proof. We know that for fixed $x \in \mathbb{R}$ the quantity

$$|\hat{F}(X_1, \ldots, X_n)(x) - F(x)|$$

presents a stochastic variable. Choosing an enumeration $x_1, x_2, \ldots$ of the rational numbers, by the previous exercise we have:

$$\|\hat{F}(X_1, \ldots, X_n) - F\|_\infty = \sup_{k=1,2,\ldots} |\hat{F}(X_1, \ldots, X_n)(x_k) - F(x_k)|.$$

Using the result of Exercise 37 this shows that $\|\hat{F}(X_1, \ldots, X_n) - F\|_\infty$ is a stochastic variable. □

Exercise 40

Let there be given a sequence $X_1, X_2, \ldots$ of variables, converging in distribution to the variable X. We denote the distribution function of X_k by F_k $(k = 1, 2, \ldots)$ and the one belonging to X by F. The corresponding quantile functions are denoted by $q_1, q_2, \ldots$ and q. Prove that for all u in $(0, 1)$ we have:

$$\underline{\lim}_{n \to \infty} q_n(u) \geq q(u-).$$

Solution. We take a fixed $u \in (0, 1)$ as well as arbitrary $\delta, \varepsilon > 0$.

First (this is possible by Lemma VII.2.9) we choose a real number x in the interval $(q(u - \delta) - \varepsilon, q(u - \delta))$ in such a way that F is continuous in x. Since $x < q(u - \delta)$, we have by Exercise 8 (iii) that $F(x) < u - \delta < u$. Moreover, as $X_n \to X$ in distribution, $F_n(x) \to F(x)$, so that an integer N exists such that $F_n(x) < u$ if $n \geq N$. Equivalently: $x < q_n(u)$ for all $n \geq N$. It follows that

$$\underline{\lim}_{n \to \infty} q_n(u) \geq x > q(u - \delta) - \varepsilon$$

for all $\delta, \varepsilon > 0$. Letting $\varepsilon \downarrow 0$ and $\delta \downarrow 0$ successively we see that

$$\underline{\lim}_{n \to \infty} q_n(u) \geq q(u-).$$

□

Exercise 41

(i) Prove that for every distribution function F and every $x \in \mathbb{R}$ one has

$$\mathbb{P}_F(\{x\}) = F(x) - F(x-).$$

Proof. Choose a fixed $x \in \mathbb{R}$ and a sequence $x_1, x_2, \ldots$ such that $x_k < x$ for all k and $x_k \uparrow x$. Then for all k we have $\mathbb{P}_F((x_k, x]) = F(x) - F(x_k)$. Furthermore

$$(x_1, x] \supset (x_2, x] \supset \ldots \quad \text{and} \quad \bigcap_{k=1}^{+\infty} (x_k, x] = \{x\}.$$

Applying §I.11 Exercise 10 we conclude that the statement is correct. □

(ii) Let $X_1, X_2, \ldots$ be an infinite sample from a population with distribution function F. Prove that for all fixed $x \in \mathbb{R}$ one has

$$\lim_{n\to\infty} \hat{F}(X_1, \ldots, X_n)(x-) = F(x-) \quad \text{strongly.}$$

Proof. Define the variables $Y_1, Y_2, \ldots$ by $Y_i := 1_{(-\infty,x)}(X_i)$. Now for each i the variable Y_i is Bernoulli distributed with (see (i)) parameter

$$\theta = \mathbb{P}(X_i < x) = F_{X_i}(x-) = F(x-).$$

Furthermore the sequence $Y_1, Y_2, \ldots$ is independent and

$$\hat{F}(X_1, \ldots, X_n)(x-) = \frac{Y_1 + \cdots + Y_n}{n}.$$

Now just apply the strong law of large numbers. □

Exercise 42

Use the Chebychev inequality to prove the weak law of large numbers in cases where the population variance exists.

Proof. Let $X_1, X_2, \ldots$ be an infinite sample from a population with expectation μ and variance σ^2. Then the variance of the variable

$$Y_n := \frac{X_1 + \cdots + X_n}{n}$$

is given by $\text{var}(Y_n) = \sigma^2/n$, and $\mathbb{E}(Y_n) = \mu$. Applying Chebychev's inequality (§I.11, Exercise 26 (ii)) we learn that for all $\varepsilon > 0$ one has

$$\mathbb{P}(|Y_n - \mu| \geq \varepsilon) \leq \frac{\sigma^2}{n\,\varepsilon^2}.$$

Hence

$$\lim_{n\to\infty} \mathbb{P}(|Y_n - \mu| \geq \varepsilon) = 0 \quad \text{for all } \varepsilon > 0,$$

which proves that $Y_n \to \mu$ in probability. □

Exercise 43

Prove that all moments of a stochastic variable with a tailless distribution function exist.

Proof. If X is a tailless stochastic variable then its probability distribution $\mathbb{P}_X$ is concentrated on some compact interval $[-a, +a]$, that is to say, $\mathbb{P}_X([-a, +a]) = 1$. Consequently

$$\begin{aligned} \mathbb{E}(|X|^n) = \int |x|^n \, d\mathbb{P}_X(x) &= \int_{[-a,+a]} |x|^n \, d\mathbb{P}_X(x) \\ &\leq a^n \, \mathbb{P}_X([-a, +a]) = a^n < +\infty. \end{aligned}$$

Therefore the nth moment exists. □

Exercise 44

Let X be a stochastic variable with distribution function F and with standard twin X'. Denote the underlying probability space of X' by $(\Omega', \mathfrak{A}', \mathbb{P}')$. On Ω' a stochastic variable X'' is defined by $X''(u) := u$ for all $u \in \Omega'$.

(i) Check that $\mathbb{P}_{X''} = \mathbb{P}'$.

Solution. For every A in $\mathfrak{A}' = \{B \cap (0,1) : B \in \mathfrak{B}\}$ we have

$$\mathbb{P}_{X''}(A) = \mathbb{P}'(X'' \in A) = \mathbb{P}'(A),$$

so $\mathbb{P}_{X''} = \mathbb{P}'$. □

(ii) Check that for all functions $g : \mathbb{R} \to \mathbb{R}$ one has

$$g(X') = (g \circ q_F)(X'') = g(q_F(X'')),$$

where q_F is the quantile function belonging to F.

Solution. This is trivial, for we have

$$g(X'(u)) = g(q_F(u)) = g(q_F(X''(u))) \quad \text{for all } u \in \Omega'.$$ □

(iii) Prove that if $g : \mathbb{R} \to \mathbb{R}$ is a Borel function then

$$\mathbb{E}(g(X)) = \mathbb{E}(g(X')) = \int_0^1 g(X'(u)) \, du.$$

Proof. The variables X and X' are identically distributed, so, if existing, $\mathbb{E}(g(X)) = \mathbb{E}(g(X'))$. Furthermore, by definition of an expectation value, we have:

$$\begin{aligned} \mathbb{E}(g(X')) &= \mathbb{E}(g(q_F(X''))) = \int g(q_F(x)) \, d\mathbb{P}_{X''}(x) \\ &\overset{\text{(i)}}{=} \int_0^1 g(q_F(u)) \, d\mathbb{P}'(u) = \int_0^1 g(X'(u)) \, du. \end{aligned}$$

The argument shows that $\mathbb{E}(g(X'))$ exists if and only if $\int_0^1 g(X'(u))\,du$ does so. If g is positive and not bounded, the value $+\infty$ might occur. □

(iv) Prove that $\mathbb{E}(X)$ exists if and only if $\int_0^1 |q_F(u)|\,du < +\infty$, in which case

$$\mathbb{E}(X) = \int_0^1 q_F(u)\,du.$$

Proof. Applying (iii) to the case where $g : \mathbb{R} \to \mathbb{R}$ is defined by $g(x) := |x|$, we see that $\mathbb{E}(|X|)$ is finite if and only if $\int_0^1 |q_F(u)|\,du < +\infty$. If so, then for $h(x) := x$ by (iii) we see that

$$\mathbb{E}(X) = \mathbb{E}(h(X)) = \int_0^1 h(q_F(u))\,du = \int_0^1 q_F(u)\,du.$$

□

Exercise 45

Suppose that for all $n = 1, 2, \ldots$ we have a sample $X_1^{(n)}, \ldots, X_n^{(n)}$ from a population with distribution function F_n. We know by Theorem VII.8.12 that if $F_n \to F$ in (Γ, d_L), then

$$\lim_{n\to\infty} d_L\left(\hat{F}(X_1^{(n)}, \ldots, X_n^{(n)})\,,\; F\right) = 0 \quad \text{strongly.} \qquad (*)$$

Give an example clarifying that (*) does not hold when we replace d_L by d_N.

Solution. Let $a_1, a_2, \ldots$ be a strictly decreasing sequence of real numbers, converging to a. Define $F_n := E_{a_n}$ and $F := E_a$. Then $F_n \to F$ in (Γ, d_L).

Now for a sample $X_1^{(n)}, \ldots, X_n^{(n)}$ from a population with distribution function F_n (so that $\mathbb{P}(X_i^{(n)} = a_n) = 1$ for $i = 1, \ldots, n$) we have

$$\hat{F}\left(X_1^{(n)}, \ldots, X_n^{(n)}\right) = F_n = E_{a_n} \quad \text{almost sure.}$$

It follows that for all $n = 1, 2, \ldots$ one has

$$d_N\left(\hat{F}(X_1^{(n)}, \ldots, X_n^{(n)})\,,\; F\right) = 1 \quad \text{almost sure,}$$

which shows that (*) is not true when we replace d_L by d_N. □

Exercise 46

Let $X_1, X_2, \ldots$ be an infinite sample from a population with distribution function F.

(i) Prove that for all $m = 1, 2, \dots$ and all $a \in \mathbb{R}$ one has

$$\mathbb{P}\left((\hat{X})_{2m+1} \le a\right) = \sum_{k=m+1}^{2m+1} \binom{2m+1}{k} F(a)^k \, (1 - F(a))^{2m+1-k}.$$

Proof. Fix m and a and set $\hat{X} := (\hat{X})_{2m+1}$. Now the event $\hat{X} \le a$ is precisely the event that in the sequence $X_1, \dots, X_{2m+1}$ there are $m + 1$ or more elements $\le a$. So we can split up this event into the mutually disjoint events $A_{m+1}, \dots, A_{2m+1}$ where A_k is defined as the event where there are exactly k elements $\le a$ in the sequence $X_1, \dots, X_{2m+1}$.

Evidently we have

$$\mathbb{P}(A_k) = \binom{2m+1}{k} F(a)^k \, (1 - F(a))^{2m+1-k}$$

and therefore

$$\mathbb{P}\left(\hat{X} \le a\right) = \sum_{k=m+1}^{2m+1} \binom{2m+1}{k} F(a)^k \, (1 - F(a))^{2m+1-k}.$$

□

(ii) Use Newton's binomium to check that

$$\sum_{k=0}^{2m+1} \binom{2m+1}{k} = 2^{2m+1}.$$

Solution. By Newton's binomium we have

$$(a + b)^{2m+1} = \sum_{k=0}^{2m+1} \binom{2m+1}{k} a^k \, b^{2m+1-k}$$

for all m, a and b. Now just set $a = b = 1$. □

(iii) Prove that for all $a \in [q_F(\frac{1}{2}), q_F(\frac{1}{2}+))$ one has (all m)

$$\mathbb{P}\left((\hat{X})_{2m+1} \le a\right) = \tfrac{1}{2}.$$

Proof. From Exercise 8 (viii) we learn that for $a \in [q_F(\frac{1}{2}), q_F(\frac{1}{2}+))$ one has $F(a) = \frac{1}{2}$. Hence, using (i), we see that for such a

$$\begin{aligned}\mathbb{P}\left((\hat{X})_{2m+1} \le a\right) &= \sum_{k=m+1}^{2m+1} \binom{2m+1}{k} \left(\frac{1}{2}\right)^{2m+1} \\ &= \frac{1}{2^{2m+1}} \sum_{k=m+1}^{2m+1} \binom{2m+1}{k}.\end{aligned}$$

As $\binom{2m+1}{k} = \binom{2m+1}{2m+1-k}$ for all k, we have

$$\sum_{k=0}^{m} \binom{2m+1}{k} = \sum_{k=m+1}^{2m+1} \binom{2m+1}{k}.$$

Since the sum of these two sums by (ii) equals 2^{2m+1}, both are equal to 2^{2m}. From this the statement follows. □

(iv) Prove that if the sequence $(\hat{X})_1, (\hat{X})_2, \ldots$ converges in distribution to a constant, then F has a strict median.

Proof. If the sequence $(\hat{X})_1, (\hat{X})_2, \ldots$ converges in distribution to a constant, say to $\alpha \in \mathbb{R}$, then for $\beta \neq \alpha$ in $\mathbb{R}$ we have

$$\mathbb{P}\left((\hat{X})_{2m+1} \leq \beta\right) \rightarrow \begin{cases} 0 & \text{if } \beta < \alpha, \\ 1 & \text{if } \beta > \alpha. \end{cases}$$

Therefore the set

$$\left\{a \,:\, \lim_{m\to\infty} \mathbb{P}\left((\hat{X})_{2m+1} \leq a\right) = \tfrac{1}{2}\right\}$$

can impossibly contain a non-empty open interval. So, by (iii), the interval $[q_F(\frac{1}{2}), q_F(\frac{1}{2}+))$ is necessarily empty, that is: $q_F(\frac{1}{2}) = q_F(\frac{1}{2}+)$. This is (by definition) the same as saying that F has a strict median. □

Exercise 47

Let F be an arbitrary distribution function.

(i) Prove that $q_F(u+) = \sup\{x \,:\, F(x-) \leq u\}$.

Proof. We prove that

$$(-\infty, q_F(u+)) \subset \{x \,:\, F(x-) \leq u\} \subset (-\infty, q_F(u+)].$$

Suppose that $F(x-) \leq u$. Choose an arbitrary $v > u$ and an arbitrary $y < x$. Then

$$F(y) \leq F(x-) \leq u < v.$$

This implies (see Exercise 8 (iii)) that $y < q_F(v)$. As this holds for every $v > u$, we have $y \leq q_F(u+)$. In turn this is true for every $y < x$, which shows that $x \leq q_F(u+)$. It thus appears that

$$\{x \,:\, F(x-) \leq u\} \subset (-\infty, q_F(u+)].$$

To prove the other inclusion, suppose that $x < q_F(u+)$. Then $x < q_F(v)$ for all $v > u$, which implies that $F(x) < v$, hence $F(x-) < v$. This being true for all $v > u$ we conclude that $F(x-) \leq u$, so

$$(-\infty, q_F(u+)) \subset \{x \,:\, F(x-) \leq u\}.$$

□

(ii) Prove that $q_{\tilde{F}}(u) = -q_F((1-u)+)$.

Proof. We may write

$$\begin{aligned} q_{\tilde{F}}(u) &= \inf\{x : \tilde{F}(x) \geq u\} = \inf\{x : 1 - F((-x)-) \geq u\} \\ &= \inf\{x : F((-x)-) \leq 1 - u\} = \inf\{-y : F(y-) \leq 1 - u\} \\ &= -\sup\{y : F(y-) \leq 1 - u\} \stackrel{\text{(i)}}{=} -q_F((1-u)+). \end{aligned}$$

□

Exercise 48

Prove that if $\Lambda : D_\Lambda \to \mathbb{R}$ is a dispersion functional and $c \in \mathbb{R}$, then $\Lambda(E_c) = 0$.

Proof. It is sufficient to prove this for $c = 0$, as for all c we have $\Lambda(E_c) = \Lambda(\tau_{-c}E_c) = \Lambda(E_0)$. Let, as usual, $E := E_0$. In this case, exploiting the fact that $\rho_a E = E$ for all $a > 0$, we arrive at the following:

$$\Lambda(E) = \Lambda(\rho_a E) = a\,\Lambda(E) \quad \text{for all } a > 0.$$

Evidently this can only be so if $\Lambda(E) = 0$. □

Exercise 49

Let $\Lambda : D_\Lambda \to \mathbb{R}$ be a linear statistical functional. Prove that the following two statements apply:

(i) If $0 \leq t_1, \ldots, t_n \leq 1$ are such that $t_1 + \cdots + t_n = 1$ and if $F_1, \ldots, F_n \in D_\Lambda$, then $t_1 F_1 + \cdots + t_n F_n \in D_\Lambda$.

(ii) With $t_1, \ldots, t_n$ and $F_1, \ldots, F_n$ as above we have:

$$\Lambda(t_1 F_1 + \cdots + t_n F_n) = t_1 \Lambda(F_1) + \cdots + t_n \Lambda(F_n).$$

Proof. Here (i) is a consequence of the fact that D_Λ is convex and can be proved rigorously by induction.

On the other hand, (ii) directly follows from the equality

$$\mathbb{P}_{t_1 F_1 + \cdots + t_n F_n} = t_1 \mathbb{P}_{F_1} + \cdots + t_n \mathbb{P}_{F_n},$$

the proof of which will conclude our argument.

First we note that under the current conditions $t_1 F_1 + \cdots + t_n F_n$ presents a distribution function. To prove that it is increasing and right-continuous is a triviality. Moreover

$$\begin{aligned} \lim_{x \to +\infty} (t_1 F_1 + \cdots + t_n F_n)(x) &= \lim_{x \to +\infty} \{t_1 F_1(x) + \cdots + t_n F_n(x)\} \\ &= \sum_{i=1}^{n} t_i \lim_{x \to +\infty} F_i(x) = \sum_{i=1}^{n} t_i = 1 \end{aligned}$$

and in the same way $\lim_{x\to-\infty}(t_1F_1+\cdots+t_nF_n)(x)=0$.

Because of this $\mathbb{P}_{t_1F_1+\cdots+t_nF_n}$ presents a probability measure $\mathbb{P}_1$ for which

$$\mathbb{P}_1((-\infty,x]) = t_1F_1(x)+\cdots+t_nF_n(x)$$

for all x in $\mathbb{R}$. On the other hand, if we write $\mathbb{P}_2 := t_1\mathbb{P}_{F_1}+\cdots+t_n\mathbb{P}_{F_n}$ we get

$$\begin{aligned}\mathbb{P}_2((-\infty,x]) &= t_1\mathbb{P}_{F_1}((-\infty,x])+\cdots+t_n\mathbb{P}_{F_n}((-\infty,x]) \\ &= t_1F_1(x)+\cdots+t_nF_n(x)\end{aligned}$$

for all x in $\mathbb{R}$. It follows that $\mathbb{P}_1=\mathbb{P}_2$. □

Exercise 50

Question. Is the interquartile distance a robust statistic?

Solution. Re-examining carefully how we arrived at the d_L-robustness properties of the median we learn that the interquartile distance, defined by the statistical functional

$$\Lambda(F) := \tfrac{1}{4}\left\{q_F(\tfrac{3}{4}-)+q_F(\tfrac{3}{4}+)-q_F(\tfrac{1}{4}-)-q_F(\tfrac{1}{4}+)\right\}$$

is d_L-robust if and only if the population has a distribution function F that satisfies

$$q_F(\tfrac{1}{4}-)=q_F(\tfrac{1}{4}+) \quad\text{and}\quad q_F(\tfrac{3}{4}-)=q_F(\tfrac{3}{4}+).$$

□

Exercise 51

We define the statistical functional $\Lambda : D_\Lambda \to \mathbb{R}$, where $D_\Lambda = \{F\in\Gamma : \int |x|\,d\mathbb{P}_F(x) < +\infty\}$ (which is, by the way, convex), as follows:

$$\Lambda(F) := \mathbb{E}(|X-\mu_\alpha(F)|),$$

where X is a variable with distribution function F and $\mu_\alpha(F)$ the α-trimmed mean of F.

(i) *Question.* Is Λ a location functional?

Solution. No, it is not. To see this we observe the impact of translations.

$$\begin{aligned}\Lambda(\tau_aF) &= \mathbb{E}(|X+a-\mu_\alpha(\tau_aF)|) = \mathbb{E}(|X+a-\mu_\alpha(F)-a|) \\ &= \mathbb{E}(|X-\mu_\alpha(F)|) = \Lambda(F).\end{aligned}$$

For location functionals the right-hand side of this equality would have been $\Lambda(F)+a$. □

(ii) *Question.* Is Λ a dispersion functional?

Solution. Yes, it is. To see this, first we have to prove that for $F \in D_\Lambda$, $a \in \mathbb{R}$ and $b > 0$ the distribution functions $\tau_a F$, $\tilde{F}$ and $\rho_b F$ all are elements of D_Λ. After that we have to verify the four characterizing properties of a dispersion functional.

As to the first point, given $F \in D_\Lambda$ with a corresponding stochastic variable X, we need to know whether it follows from $\mathbb{E}(|X|) < +\infty$ that $\mathbb{E}(|X+a|) < +\infty$, $\mathbb{E}(|-X|) < +\infty$ and $\mathbb{E}(|bX|) < +\infty$. Of course this is so.

Next we turn to the characterizing properties. As μ_α presents a location functional, we have $\mu_\alpha(\tilde{F}) = -\mu_\alpha(F)$ and for $b > 0$ we have $\mu_\alpha(\rho_b F) = b\,\mu_\alpha(F)$.

1) We already saw in (i) that $\Lambda(\tau_a F) = \Lambda(F)$.

2) Furthermore,

$$\begin{aligned}\Lambda(\tilde{F}) &= \mathbb{E}(|-X-\mu_\alpha(\tilde{F})|)\\ &= \mathbb{E}(|-X+\mu_\alpha(F)|) = \mathbb{E}(|X-\mu_\alpha(F)|) = \Lambda(F).\end{aligned}$$

3) If $b > 0$, then

$$\begin{aligned}\Lambda(\rho_b F) &= \mathbb{E}(|bX-\mu_\alpha(\rho_b F)|) = \mathbb{E}(|bX - b\mu_\alpha(F)|)\\ &= b\,\mathbb{E}(|X-\mu_\alpha(F)|) = b\,\Lambda(F).\end{aligned}$$

4) Finally, of course $\Lambda(F) \geq 0$ for all F. □

(iii) *Question.* Is, for arbitrary $a \in \mathbb{R}$, the distribution function E_a an element of the von Mises hull of E?

Solution. Yes. We define for $0 < t < 1$ the distribution function F_t as $F_t := (1-t)\,E + t\,E_a$. Then, if $a \geq 0$, q_{F_t} is given by

$$q_{F_t}(u) = \begin{cases} 0 & \text{if } 0 < u \leq 1-t \\ a & \text{if } 1-t < u < 1 \end{cases};$$

if $a < 0$ it is given by $q_{F_t}(u) = a\,1_{(0,t]}(u)$. It follows from this (all $a \in \mathbb{R}$) that $q_{F_t} = 0$ on $(\alpha, 1-\alpha)$ if $t \leq \alpha$; consequently $\mu_\alpha(F_t) = 0$ if $t \leq \alpha$. Now let X_t be a stochastic variable having F_t as its distribution function. Then for $t \leq \alpha$ we have

$$\Lambda(F_t) = \mathbb{E}(|X_t - \mu_\alpha(F_t)|) = \mathbb{E}(|X_t|) = t\,|a|.$$

Hence

$$\lim_{t\downarrow 0} \frac{\Lambda((1-t)\,E + t\,E_a) - \Lambda(E)}{t} = \lim_{t\downarrow 0} \frac{\Lambda(F_t) - 0}{t} = |a|.$$

This shows that, for all $a \in \mathbb{R}$, the distribution function E_a is an element of the von Mises hull of E. □

Exercise 52

Let $\Lambda : D_\Lambda \to \mathbb{R}$ be a statistical functional and let $X_1, X_2, \ldots$ be an infinite sample from a population with distribution function F. If there exists a $\sigma > 0$ such that the sequence

$$\sqrt{n}\ \Big\{\Lambda(\hat{F}(X_1,\ldots,X_n)) - \Lambda(F)\Big\} \qquad (n = 1, 2, \ldots)$$

converges in distribution to the $N(0, \sigma^2)$-distribution, then Λ is weakly consistent. Prove this assertion.

Proof. We first prove a more general statement, namely the following. If $T_n \to T$ in distribution and if $a_n \downarrow 0$, then $a_n T_n \to 0$ in distribution. To verify this, denote the distribution function of T_n (T) by F_n (F). Now to every $x, y > 0$ there is an integer N such that

$$\mathbb{P}(a_n T_n \le x) = \mathbb{P}(T_n \le \tfrac{x}{a_n}) = F_n(\tfrac{x}{a_n}) \ \ge\ F_n(y) \quad \text{for all } n \ge N.$$

It follows that

$$\underline{\lim}_{n\to\infty} \mathbb{P}(a_n T_n \le x) \ \ge\ \underline{\lim}_{n\to\infty} F_n(y) = F(y)$$

for all $x > 0$ and for all $y > 0$ in which F is continuous. Letting $y \to +\infty$ in the above we conclude that

$$\underline{\lim}_{n\to\infty} \mathbb{P}(a_n T_n \le x) = 1$$

for all $x > 0$. This shows that

$$\mathbb{P}(a_n T_n \le x) \ \to\ E(x) \quad \text{for all } x > 0.$$

In a similar way this can be deduced for $x < 0$. Altogether this implies that $a_n T_n \to 0$ in distribution.

Next, apply the above to the case where

$$T_n = \sqrt{n}\ \Big\{\Lambda(\hat{F}(X_1,\ldots,X_n)) - \Lambda(F)\Big\} \quad \text{and} \quad a_n = \frac{1}{\sqrt{n}}.$$

□

Exercise 53

Suppose that F is a strictly increasing distribution function. Setting $F_{t,x} := (1-t)\,F + tE_x$ for $0 \le t \le 1$, prove that

$$q_{F_{t,x}}(u) = \begin{cases} q_F(\frac{u}{1-t}) & \text{if } \ 0 < u \le (1-t)\,F(x-), \\ x & \text{if } \ (1-t)\,F(x-) < u \le (1-t)\,F(x) + t, \\ q_F(\frac{u-t}{1-t}) & \text{if } \ (1-t)\,F(x) + t < u < 1. \end{cases}$$

Proof. Let $t < 1$. By definition of the quantile function we have

$$\begin{aligned} q_{F_{t,x}}(u) &= \inf\{y : F_{t,x}(y) \geq u\} \\ &= \inf\left(\{y < x : F_{t,x}(y) \geq u\} \cup \{y \geq x : F_{t,x}(y) \geq u\}\right) \\ &= \inf\left(\{y < x : F(y) \geq \tfrac{u}{1-t}\} \cup \{y \geq x : F(y) \geq \tfrac{u-t}{1-t}\}\right). \end{aligned}$$

The set $\{y < x : F(y) \geq \frac{u}{1-t}\}$ is empty precisely then if

$$F(y) < \frac{u}{1-t} \quad \text{for all } y < x.$$

This is precisely the case if $(1-t)\,F(x-) < u$. Hence, if $0 < u < (1-t)\,F(x-)$ we have (see Exercise 8 (iii))

$$q_{F_{t,x}}(u) = \inf\{y < x : F(y) \geq \tfrac{u}{1-t}\} = q_F(\tfrac{u}{1-t}).$$

Exploiting the left-continuity of quantile functions, we conclude that the above also holds if $0 < u \leq (1-t)\,F(x-)$.

If the set $\{y < x : F(y) \geq \frac{u}{1-t}\}$ is empty then the set $\{y \geq x : F(y) \geq \frac{u-t}{1-t}\}$ cannot be empty; the latter infimum then equals

$$q_{F_{t,x}}(u) = \max\left\{x\,,\, q_F(\tfrac{u-t}{1-t})\right\}.$$

In virtue of Exercise 8 (iii) this maximum can then be specified as

$$q_{F_{t,x}}(u) = \begin{cases} x & \text{if } \; u \leq (1-t)\,F(x) + t, \\ q_F(\frac{u-t}{1-t}) & \text{if } \; u > (1-t)\,F(x) + t. \end{cases}$$

This completes the proof (for $t = 1$ the statement is trivial). □

Exercise 54

Let $X_1, X_2, \ldots$ be an infinite sample from a population with continuous distribution function. Denote the underlying probability space by $(\Omega, \mathfrak{A}, \mathbb{P})$ and define the set $A \subset \Omega$ by

$$A := \{\omega \in \Omega \;:\; \text{the sequence } X_1(\omega), X_2(\omega), \ldots \text{ contains no tie}\}.$$

Prove that $A \in \mathfrak{A}$ and that $\mathbb{P}(A) = 1$.

Proof. For finite samples this has already been proved in §VI.7 Exercise 4. So for the sets

$$A_n := \{\omega \in \Omega \;:\; \text{the sequence } X_1(\omega), \ldots, X_n(\omega) \text{ contains no tie}\}$$

we have $A_n \in \mathfrak{A}$ and $\mathbb{P}(A_n) = 1$. Because $A = \bigcap_{n=1}^{\infty} A_n$, it now follows that $A \in \mathfrak{A}$ and (see §I.11, Exercise 10) $\mathbb{P}(A) = 1$. □

Exercise 55

Prove that in the notations of §VII.11 the bootstrap distribution function can be expressed as

$$F_{T^*}(x) = \frac{1}{n^m} \sum_{i_1,\dots,i_m=1}^{n} 1_{(-\infty,x]}\left(g(X_{i_1},\dots,X_{i_m})\right).$$

Proof. Suppose that the sample $X_1,\dots,X_n$ shows the following outcome:

$$X_1 = x_1, \; X_2 = x_2, \; \dots \; , \; X_n = x_n.$$

The bootstrap variable T^* is then defined on the bootstrap probability space Ω^* by

$$T^*(i_1,\dots,i_m) := g(x_{i_1},\dots,x_{i_m}).$$

It follows that

$$\begin{aligned} \mathbb{P}(T^* \leq x) &= \frac{\#\{(i_1,\dots,i_m) \; : \; g(x_{i_1},\dots,x_{i_m}) \leq x\}}{n^m} \\ &= \frac{1}{n^m} \sum_{i_1,\dots,i_m=1}^{n} 1_{(-\infty,x]}\left(g(x_{i_1},\dots,x_{i_m})\right). \end{aligned}$$

This completes the proof. □

Exercise 56

Let $g : \mathbb{R}^2 \to \mathbb{R}$ be a Borel function. Assume that g is symmetric, that is, that $g(x,y) = g(y,x)$ for all $x, y \in \mathbb{R}$. Suppose X_1, X_2 is a sample of size 2 from an arbitrary population; set $T := g(X_1, X_2)$.

(i) Prove that for all $x \in \mathbb{R}$ one has

$$\mathbb{E}(F_{T^*}(x)) = \tfrac{1}{2}\left\{\mathbb{P}(g(X_1,X_1) \leq x) + F_T(x)\right\}.$$

Proof. By the previous exercise, elaborating the fact that g is symmetric, we can write

$$\begin{aligned} F_{T^*}(x) \;&=\; \tfrac{1}{4}\left\{1_{(-\infty,x]}(g(X_1,X_1)) + 1_{(-\infty,x]}(g(X_2,X_2)) \right. \\ &\qquad \left. + \, 2 \cdot 1_{(-\infty,x]}(g(X_1,X_2))\right\}. \qquad (*) \end{aligned}$$

Now both $1_{(-\infty,x]}(g(X_1,X_1))$ and $1_{(-\infty,x]}(g(X_2,X_2))$ are Bernoulli distributed with parameter

$$\theta = \mathbb{P}(g(X_1,X_1) \leq x) = \mathbb{P}(g(X_2,X_2) \leq x).$$

The variable $1_{(-\infty,x]}(g(X_1,X_2))$ is also Bernoulli distributed, but here the parameter is $\theta = F_T(x)$. Now just take the expectation value of the left and right side of $(*)$. □

(ii) From here on we take for g the function $(x_1, x_2) \mapsto \frac{1}{2}(x_1 + x_2)$.

Question. Is in the case of a $N(0,1)$-distributed population the statistic $F_{T^*}(x)$ an unbiased estimator of $F_T(x)$?

Solution. Denoting the $N(0,1)$-distribution function by Φ, we have

$$\mathbb{P}(g(X_1, X_1) \le x) = \mathbb{P}(X_1 \le x) = \Phi(x)$$

and

$$F_T(x) = \mathbb{P}(\tfrac{1}{2}(X_1 + X_2) \le x) = \Phi(\sqrt{2}\,x)$$

(as $\frac{1}{2}(X_1 + X_2)$ is $N(0, \frac{1}{2})$-distributed). Hence, applying (i) together with the above,

$$\mathbb{E}(F_{T^*}(x)) = \tfrac{1}{2}\,\{\Phi(x) + \Phi(\sqrt{2}\,x)\} \neq \; F_T(x)$$

for $x \neq 0$. This shows that, unless $x = 0$, $F_{T^*}(x)$ is *not* an unbiased estimator of $F_T(x)$. □

(iii) *Question.* Suppose that the population is Cauchy distributed with parameters $\alpha = 0$ and $\beta = 1$. Is then $F_{T^*}(x)$ an unbiased estimator of $F_T(x)$?

Solution. Denote the distribution function belonging to this Cauchy distribution by F_0. Then

$$\mathbb{P}(g(X_1, X_1) \le x) = \mathbb{P}(X_1 \le x) = F_0(x).$$

However, here we also have (see §II.11, Exercise 19)

$$F_T(x) = \mathbb{P}(\tfrac{1}{2}(X_1 + X_2) \le x) = \mathbb{P}(X_1 \le x) = F_0(x).$$

Using (i) together with the above we arrive at

$$\mathbb{E}(F_{T^*}(x)) = F_0(x) = F_T(x).$$

That is to say, $F_{T^*}(x)$ is an unbiased estimator of $F_T(x)$. □

(iv) Generalize (i) to samples of size 3.

Solution. We take $n = 3$. Again we start from the assumption that $g : \mathbb{R}^3 \to \mathbb{R}$ is symmetric. Reasoning as in (i) we then arrive at

$$\begin{aligned}\mathbb{E}(F_{T^*}(x)) \;&=\; \tfrac{1}{27}\,\{3\;\mathbb{P}(g(X_1, X_1, X_1) \le x) \\ &\qquad + 18\;\mathbb{P}(g(X_1, X_1, X_2) \le x) + 6\,F_T(x)\}\,.\end{aligned}$$ □

Exercise 57

A sample (X_1, X_2, X_3, X_4) from a population with unknown probability density f shows an outcome

$$X_1 = 1.0, \quad X_2 = 1.5, \quad X_3 = 2.5, \quad X_4 = 3.0.$$

Sketch the kernel density estimates of f, based on the triangular kernel with bandwidth 0.2, 0.5, 1.0 respectively.

Solution. Exploring the formula

$$\hat{f}_\zeta(x) = \frac{1}{4}\sum_{i=1}^{4} k\left(\frac{x - x_i}{\zeta}\right)$$

this amounts to the following:

(i) $\zeta = 0.2$

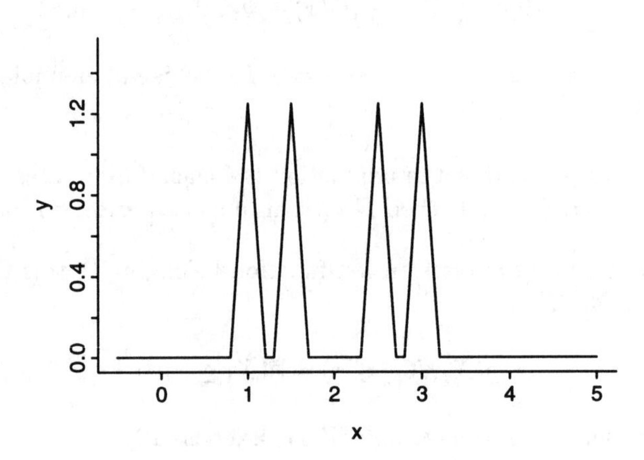

(ii) $\zeta = 0.5$

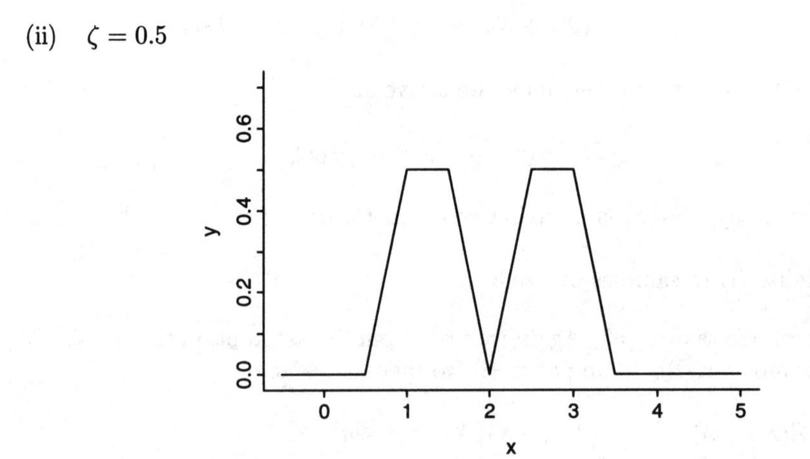

(iii) $\zeta = 1.0$

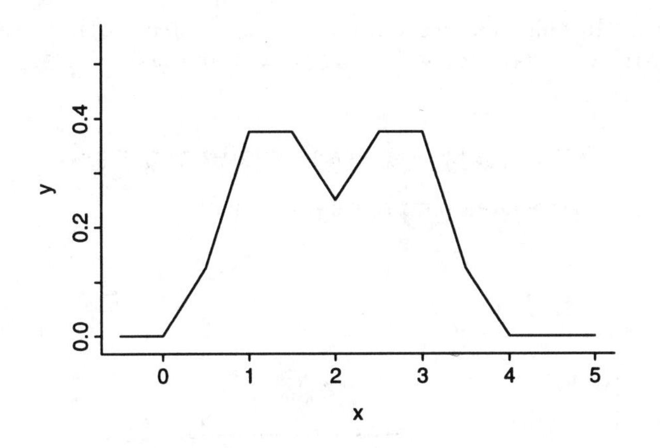

Exercise 58

A sample of size 12 exhibits the following outcome:

$$\begin{array}{cccccc} 1.92, & 3.48, & 1.12, & 1.25, & 3.02, & 1.61, \\ 3.57, & 3.81, & 1.67, & 3.37, & 1.79, & 3.21. \end{array}$$

(i) Sketch the associated histogram with respect to the interval partition $\mathcal{I}_1 = ([n, n+1))_{n \in \mathbb{Z}}$.

Solution. Here one has $\#\{i : x_i \in [1,2)\} = \#\{i : x_i \in [3,4)\} = 6$. It follows directly from this that the corresponding histogram is given by

$$h = \tfrac{6}{12 \cdot 1} 1_{[1,2)} + \tfrac{6}{12 \cdot 1} 1_{[3,4)} = \tfrac{1}{2} 1_{[1,2) \cup [3,4)}.$$

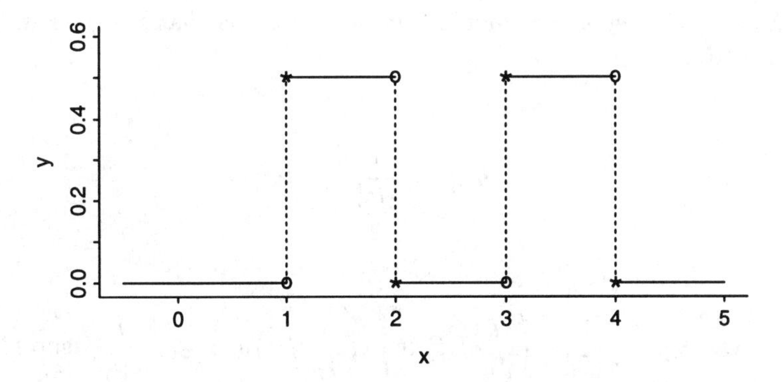

(ii) The same question as in (i), but now for the interval partition $\mathcal{I}_2 = \big([n-\frac{1}{2}, n+\frac{1}{2})\big)_{n\in\mathbb{Z}}$

Solution. In this case we have $\#\{i : x_i \in [0.5, 1.5)\} = 2$, $\#\{i : x_i \in [1.5, 2.5)\} = 4$, $\#\{i : x_i \in [2.5, 3.5)\} = 4$ and $\#\{i : x_i \in [3.5, 4.5)\} = 2$. Hence

$$h = \tfrac{1}{6}\,1_{[0.5,1.5)} + \tfrac{1}{3}\,1_{[1.5,2.5)} + \tfrac{1}{3}\,1_{[2.5,3.5)} + \tfrac{1}{6}\,1_{[3.5,4.5)}.$$

Drawing the associated picture we arrive at:

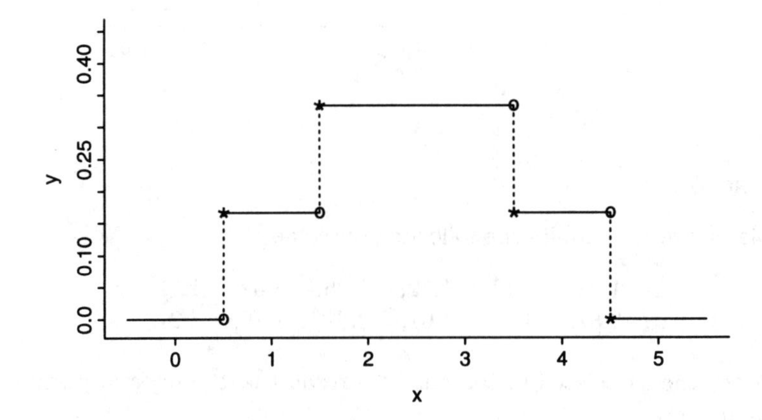

Exercise 59

Let $\mathcal{I} = (I_p)_{p=1,2,\ldots}$ be an interval partition of $\mathbb{R}$ and, for $p = 1, 2, \ldots$, write

$$c_p = c_p(X_1, \ldots, X_n) := \frac{1}{n\,|I_p|}\sum_{i=1}^{n} 1_{I_p}(X_i),$$

where $X_1, \ldots, X_n$ denotes a sample from a population that has f as its density function. Prove that

(i)

$$\mathbb{E}(c_p) = \frac{1}{|I_p|}\int_{I_p} f(t)\,dt,$$

(ii)

$$\operatorname{var}(c_p) = \frac{1}{n\,|I_p|^2}\left(\int_{I_p} f(t)\,dt\right)\left(\int_{I_p^c} f(t)\,dt\right) \le \frac{1}{n\,|I_p|}\sup_{t\in I_p} f(t).$$

Proof. As noted in §VII.13, $1_{I_p}(X_i)$ for all i enjoys a Bernoulli distribution with parameter $\theta = \int_{I_p} f(t)\,dt$.

(i) Because this holds for every p we have

$$\mathbb{E}(c_p) = \frac{1}{n\,|I_p|} \sum_{i=1}^{n} \int_{I_p} f(t)\,dt = \frac{1}{|I_p|} \int_{I_p} f(t)\,dt.$$

(ii) We also obtain:

$$\operatorname{var}(c_p) = \frac{1}{n^2|I_p|^2} \sum_{i=1}^{n} \operatorname{var}(1_{I_p}(X_i)) = \frac{n}{n^2|I_p|^2}\,\theta\,(1-\theta).$$

Here $1 - \theta = \int_{\mathbb{R}\setminus I_p} f(t)\,dt = \int_{I_p^c} f(t)\,dt$, so indeed we have

$$\operatorname{var}(c_p) = \frac{1}{n\,|I_p|^2} \left(\int_{I_p} f(t)\,dt\right) \left(\int_{I_p^c} f(t)\,dt\right). \qquad (*)$$

For any fixed p we have $\sup_{t\in I_p} f(t)$ (finite or infinite), and

$$\int_{I_p} f(t)\,dt \le \left(\sup_{t\in I_p} f(t)\right) \int_{I_p} 1\,dt = |I_p| \sup_{t\in I_p} f(t).$$

Estimating $(*)$ from above by this inequality, together with the trivial $\int_{I_p^c} f(t)\,dt \le 1$, the required inequality emerges to be correct. □

Chapter 8
Vectorial statistics

8.1 Summary of Chapter VIII

8.1.1 Linear algebra (VIII.1)

The opening section of Chapter VIII gives a summary of some well-known facts from linear algebra. The following basic concepts are needed.

The $q \times p$-matrix of a linear operator $\mathbf{T} : \mathbb{R}^p \to \mathbb{R}^q$ is systematically denoted by $[\mathbf{T}]$. The matrix entry in row no. i and column no. j is denoted by $[\mathbf{T}]_{ij}$. We denote the determinant of $\mathbf{T}$ by $\det(\mathbf{T})$, its trace by $\mathrm{tr}(\mathbf{T})$. On $\mathbb{R}^p$ the standard inner product is given by

$$\langle \mathbf{x}, \mathbf{y} \rangle := x_1 y_1 + \cdots + x_n y_n.$$

The *adjoint* of a linear operator $\mathbf{T} : \mathbb{R}^p \to \mathbb{R}^q$ is the unique linear operator $\mathbf{T}^* : \mathbb{R}^q \to \mathbb{R}^p$ that satisfies

$$\langle \mathbf{T}\mathbf{x}, \mathbf{y} \rangle = \langle \mathbf{x}, \mathbf{T}^*\mathbf{y} \rangle \qquad (\text{all } \mathbf{x} \in \mathbb{R}^p, \mathbf{y} \in \mathbb{R}^q).$$

If $\mathbf{T} = \mathbf{T}^*$ then $\mathbf{T}$ is called *self-adjoint.* By an extremely important theorem in linear algebra, such operators can be diagonalized. An operator $\mathbf{T}$ on $\mathbb{R}^p$ is said to be *of positive type* if $\langle \mathbf{T}\mathbf{x}, \mathbf{x} \rangle \geq 0$ for all $\mathbf{x} \in \mathbb{R}^p$. If $\mathbf{T}$ is self-adjoint and of positive type then we can talk about $\sqrt{\mathbf{T}}$. If $\mathbf{T}^* = \mathbf{T}^{-1}$ we speak about an *orthogonal linear operator* (or, about a 'rotation around the origin').

To every linear form $l : \mathbb{R}^p \to \mathbb{R}$ there is a unique vector $\mathbf{y} \in \mathbb{R}^p$ such that

$$l(\mathbf{x}) = \langle \mathbf{x}, \mathbf{y} \rangle \quad \text{for all } \mathbf{x} \in \mathbb{R}^p.$$

To every bilinear form $B : \mathbb{R}^p \times \mathbb{R}^p \to \mathbb{R}$ there is a unique linear operator $\mathbf{T} : \mathbb{R}^p \to \mathbb{R}^p$ such that

$$B(\mathbf{x}, \mathbf{y}) = \langle \mathbf{T}\mathbf{x}, \mathbf{y} \rangle \quad \text{for all } \mathbf{x}, \mathbf{y} \in \mathbb{R}^p.$$

The *orthogonal projection* on a linear subspace $\mathfrak{M}$ of $\mathbb{R}^p$ will systematically be denoted by $\mathbf{P}_{\mathfrak{M}}$. For any two vectors $\mathbf{x}, \mathbf{y} \in \mathbb{R}^p$ the linear operator $\mathbf{x} \otimes \mathbf{y}$ is defined by

$$(\mathbf{x} \otimes \mathbf{y})(\mathbf{z}) := \langle \mathbf{z}, \mathbf{y} \rangle \, \mathbf{x}.$$

The space $\mathbb{R}^p \times \cdots \times \mathbb{R}^p$ (n Cartesian factors) is equipped with an inner product in the following way:

$$\langle (x_1, \dots, x_n), (y_1, \dots, y_n) \rangle := \langle x_1, y_1 \rangle + \cdots + \langle x_n, y_n \rangle .$$

When identifying $\mathbb{R}^p \times \cdots \times \mathbb{R}^p$ with $\mathbb{R}^{np}$, this is of course the standard inner product. Given the linear operators $\mathbf{T}_1, \dots, \mathbf{T}_n$ on $\mathbb{R}^p$, a linear operator $\tilde{\mathbf{T}}$ is defined by

$$\tilde{\mathbf{T}}(\mathbf{x}_1, \dots, \mathbf{x}_n) := (\mathbf{T}_1\mathbf{x}_1, \dots, \mathbf{T}_n\mathbf{x}_n).$$

We denote $\tilde{\mathbf{T}} = \mathbf{T}_1 \oplus \cdots \oplus \mathbf{T}_n$.

On the linear space $\mathcal{L}(\mathbb{R}^p)$ consisting of all linear operators on $\mathbb{R}^p$, the so-called *Hilbert–Schmidt inner product* is defined by

$$\langle \mathbf{S}, \mathbf{T} \rangle := \operatorname{tr}(\mathbf{T}^*\mathbf{S}) = \sum_{i,j=1}^{p} [\mathbf{S}]_{ij} [\mathbf{T}]_{ij}.$$

When identifying $\mathcal{L}(\mathbb{R}^p)$ with $\mathbb{R}^{p^2}$ this is just the standard inner product.

8.1.2 The expectation vector and the covariance operator of a stochastic vector (VIII.2)

Let $\mathbf{X} = (X_1, \dots, X_p)$ be a stochastic p-vector for which $\mathbb{E}(X_i)$ exists for $i = 1, \dots, p$. We define the *expectation vector* $\mathbb{E}(\mathbf{X})$ of $\mathbf{X}$ by

$$\mathbb{E}(\mathbf{X}) := (\mathbb{E}(X_1), \dots, \mathbb{E}(X_p)).$$

For all $\mathbf{a} \in \mathbb{R}^p$ we write $\langle \mathbf{X}, \mathbf{a} \rangle := a_1X_1 + \cdots + a_pX_p$. In these notations $\mathbb{E}(\mathbf{X})$ exists precisely if $\mathbb{E}(\langle \mathbf{X}, \mathbf{a} \rangle)$ does so for every $\mathbf{a} \in \mathbb{R}^p$. In that case it equals the unique vector $\boldsymbol{\mu} \in \mathbb{R}^p$ that satisfies

$$\mathbb{E}(\langle \mathbf{X}, \mathbf{a} \rangle) = \langle \boldsymbol{\mu}, \mathbf{a} \rangle \quad \text{for all } \mathbf{a} \in \mathbb{R}^p.$$

If $\mathbf{T} : \mathbb{R}^p \to \mathbb{R}^q$ denotes a linear operator and if $\mathbb{E}(\mathbf{X})$ exists, then for the stochastic q-vector $\mathbf{TX}$ we have:

$$\mathbb{E}(\mathbf{TX}) = \mathbf{T}\,\mathbb{E}(\mathbf{X}).$$

Moreover: if $\mathbf{X}$ and $\mathbf{Y}$ denote stochastic p-vectors of which the expectation values exist, then

$$\begin{cases} \mathbb{E}(\alpha\mathbf{X}) = \alpha\,\mathbb{E}(\mathbf{X}),\ \mathbb{E}(\mathbf{X} + \mathbf{a}) = \mathbb{E}(\mathbf{X}) + \mathbf{a} \quad \text{and} \\ \mathbb{E}(\mathbf{X} + \mathbf{Y}) = \mathbb{E}(\mathbf{X}) + \mathbb{E}(\mathbf{Y}). \end{cases}$$

For every stochastic p-vector $\mathbf{X} = (X_1, \dots, X_p)$ satisfying $\mathrm{var}(X_i) < +\infty$ for all i, the so-called *covariance operator* $\mathcal{C}(\mathbf{X}) : \mathbb{R}^p \to \mathbb{R}^p$ exists. By definition this is the linear operator having

$$\Sigma(\mathbf{X}) := \begin{pmatrix} \mathrm{cov}\,(X_1, X_1) & \cdots & \mathrm{cov}\,(X_1, X_p) \\ \mathrm{cov}\,(X_2, X_1) & \cdots & \mathrm{cov}\,(X_2, X_p) \\ \vdots & & \vdots \\ \mathrm{cov}\,(X_p, X_1) & \cdots & \mathrm{cov}\,(X_p, X_p) \end{pmatrix}$$

for its matrix. It is also denoted by $\mathbf{C}$ if no confusion can arise. This covariance operator *exists* if and only if for all $\mathbf{a} \in \mathbb{R}^p$ the variance of $\langle \mathbf{X}, \mathbf{a} \rangle$ is finite. In that case it is characterized by being the unique linear map $\mathbf{C} : \mathbb{R}^p \to \mathbb{R}^p$ for which

$$\mathrm{cov}(\langle \mathbf{X}, \mathbf{a} \rangle, \langle \mathbf{X}, \mathbf{b} \rangle) = \langle \mathbf{Ca}, \mathbf{b} \rangle$$

for all $\mathbf{a}, \mathbf{b} \in \mathbb{R}^p$. Covariance operators are always selfadjoint and of positive type. For each vector $\mathbf{a} \in \mathbb{R}^p$ and each linear operator $\mathbf{T} : \mathbb{R}^p \to \mathbb{R}^q$ we have

$$\mathcal{C}(\mathbf{X} + \mathbf{a}) = \mathcal{C}(\mathbf{X}) \quad \text{and} \quad \mathcal{C}(\mathbf{TX}) = \mathbf{T}\mathcal{C}(\mathbf{X})\mathbf{T}^*.$$

The stochastic p-vector $\mathbf{X} = (X_1, \dots, X_p)$ is said to have *uncorrelated components* if $\mathrm{cov}\,(X_i, X_j) = 0$ for all $i \neq j$. This happens precisely if there is a sequence $\lambda_1, \dots, \lambda_p \geq 0$ for which

$$\mathcal{C}(\mathbf{X}) = \mathrm{diag}(\lambda_1, \dots, \lambda_p).$$

For each p-vector $\mathbf{X}$ an orthogonal linear operator $\mathbf{Q} : \mathbb{R}^p \to \mathbb{R}^p$ exists for which $\mathbf{QX}$ has uncorrelated components.

If the covariance operator $\mathcal{C}(\mathbf{X})$ of $\mathbf{X}$ is not invertible, $\mathbf{X}$ almost surely attains a value in the affine subspace $\boldsymbol{\mu} + \mathrm{Im}\mathbf{C}$ of $\mathbb{R}^p$. Suppose this is not so and the eigenvalues of $\mathcal{C}(\mathbf{X})$ are mutually different. Then we may reorder them in such a way that $\lambda_1 > \lambda_2 > \cdots > \lambda_p > 0$; this determines the corresponding orthonormal basis for $\mathbb{R}^p$ of eigenvectors, $\{\mathbf{v}_1, \dots, \mathbf{v}_p\}$. Denoting $Y_i := \langle \mathbf{X}, \mathbf{v}_i \rangle$ for $i = 1, \dots, p$ we obtain the system $\mathbf{Y} = (Y_1, \dots, Y_p)$ of *principal components*. By construction $\mathbf{Y}$ has uncorrelated components.

The *total variance* of a stochastic p-vector $\mathbf{X}$ is defined by

$$\mathrm{var}(\mathbf{X}) := \mathrm{tr}(\mathcal{C}(\mathbf{X})).$$

For each orthogonal $\mathbf{Q} : \mathbb{R}^p \to \mathbb{R}^p$ and $\mathbf{a} \in \mathbb{R}^p$ we have:

$$\begin{cases} \mathrm{var}(\mathbf{X} + \mathbf{a}) &= \mathrm{var}(\mathbf{X}) \quad \text{and} \\ \mathrm{var}(\mathbf{QX}) &= \mathrm{var}(\mathbf{X}). \end{cases}$$

In other words, total variance is a characteristic which is invariant under 'Euclidian movements'. As far as it makes sense, one has:

$$\mathrm{var}(\mathbf{X}) = \mathbb{E}\left(\|\mathbf{X} - \boldsymbol{\mu}\|^2\right) = \mathbb{E}\left(\|\mathbf{X}\|^2\right) - \|\boldsymbol{\mu}\|^2.$$

Therefore, if $\operatorname{var}(\mathbf{X}) = 0$ then $\mathbf{X}$ will almost surely show an outcome equal to $\boldsymbol{\mu}$.

In a canonic way the above can be generalized to variables assuming values in any finite-dimensional space equipped with an arbitrary inner product, or even to infinite-dimensional Hilbert spaces.

8.1.3 Vectorial samples (VIII.3)

A *vectorial sample* of size n by definition is a statistically independent set $\mathbf{X}_1, \dots, \mathbf{X}_n$ of stochastic vectors sharing a common probability distribution, which is called the *distribution of the (vectorial) population.*

The sample mean $\overline{\mathbf{X}}$ is defined by

$$\overline{\mathbf{X}} := \frac{\mathbf{X}_1 + \cdots + \mathbf{X}_n}{n}.$$

If $\boldsymbol{\mu}$ and $\mathbf{C}$ denote the expectation vector and covariance operator associated with the population, we have:

$$\mathbb{E}\left(\overline{\mathbf{X}}\right) = \boldsymbol{\mu} \quad \text{and} \quad \mathcal{C}(\overline{\mathbf{X}}) = \frac{1}{n}\mathbf{C}.$$

The so-called *sample covariance operator* $\mathbf{S}^2 : \Omega \to \mathcal{L}(\mathbb{R}^p)$ is defined by

$$\mathbf{S}^2 := \frac{1}{n-1}\sum_{i=1}^{n}(\mathbf{X}_i - \overline{\mathbf{X}}) \otimes (\mathbf{X}_i - \overline{\mathbf{X}}),$$

where $(\Omega, \mathfrak{A}, \mathbb{P})$ is the underlying probability space. The variable $\mathbf{S}^2$ turns out to be an unbiased estimator for $\mathbf{C}$.

8.1.4 The vectorial normal distribution (VIII.4)

A real-valued stochastic variable X is said to be *Gaussian distributed* if it is either normally distributed or constant. A stochastic p-vector $\mathbf{E}$ in turn is called *elementarily Gaussian distributed* if its components $E_1, \dots, E_p$ are statistically independent and Gaussian distributed. A stochastic p-vector $\mathbf{X}$ finally is called *Gaussian distributed* if an orthogonal $\mathbf{Q} : \mathbb{R}^p \to \mathbb{R}^p$ exists such that $\mathbf{QX}$ is elementarily Gaussian distributed.

We call two stochastic vectors $\mathbf{X}$ and $\mathbf{Y}$ *uncorrelated* if $\operatorname{cov}(X_i, Y_j) = 0$ for all i, j. Now let $\mathbf{X}$ be a Gaussian distributed p-vector and let $\mathbf{T}_1, \dots, \mathbf{T}_r$ be arbitrary linear operators. The system $\{\mathbf{T}_1\mathbf{X}, \dots, \mathbf{T}_r\mathbf{X}\}$ is statistically independent if and only if for all $i \neq j$ the vectors $\mathbf{T}_i\mathbf{X}$ and $\mathbf{T}_j\mathbf{X}$ are uncorrelated.

The probability distribution of a p-vector $\mathbf{X}$ that is Gaussian distributed is completely determined by its expectation vector $\boldsymbol{\mu}$ and its covariance operator $\mathbf{C}$;

it can, and will, therefore be denoted by $N(\boldsymbol{\mu}, \mathbf{C})$. The probability distribution of such an $N(\boldsymbol{\mu}, \mathbf{C})$-distributed $\mathbf{X}$, provided $\mathbf{C}^{-1}$ exists, is given by

$$f_{\mathbf{X}}(\mathbf{x}) = \frac{1}{(2\pi)^{\frac{1}{2}p}\sqrt{\det \mathbf{C}}} \exp\left[-\tfrac{1}{2}\left\langle \mathbf{C}^{-1}(\mathbf{x}-\boldsymbol{\mu}), (\mathbf{x}-\boldsymbol{\mu})\right\rangle\right].$$

If $\mathbf{X}$ is $N(\boldsymbol{\mu}, \mathbf{C})$-distributed and $\mathbf{a} \in \mathbb{R}^p$, then $\mathbf{X}+\mathbf{a}$ is $N(\boldsymbol{\mu}+\mathbf{a}, \mathbf{C})$-distributed. If $\mathbf{T} : \mathbb{R}^p \to \mathbb{R}^q$ is a linear operator then $\mathbf{TX}$ is $N(\mathbf{T}\boldsymbol{\mu}, \mathbf{TCT}^*)$-distributed. As a consequence, each linear combination $\sum_i a_i X_i$ of the components of $\mathbf{X}$ enjoys a Gaussian distribution.

An $N(\boldsymbol{\mu}, \mathbf{C})$-distributed $\mathbf{X}$ is called *normally* distributed if the operator $\mathbf{C}$ is invertible. In that case $\mathbf{C}^{-\frac{1}{2}} := \sqrt{\mathbf{C}^{-1}}$ exists and the *standardized form*

$$\tilde{\mathbf{X}} := \mathbf{C}^{-\frac{1}{2}}(\mathbf{X}-\boldsymbol{\mu})$$

of $\mathbf{X}$ enjoys a $N(\mathbf{0}, \mathbf{I}_p)$-distribution.

Suppose $\mathfrak{M}_1, \ldots, \mathfrak{M}_r$ are mutually orthogonal linear subspaces of $\mathbb{R}^p$ with corresponding projections $\mathbf{P}_1, \ldots, \mathbf{P}_r$. Cochran's theorem states that for any $N(\mathbf{0}, \mathbf{I}_p)$-distributed $\mathbf{X}$, the variables

$$\mathbf{P}_1\mathbf{X}, \ldots, \mathbf{P}_r\mathbf{X}$$

form an independent system and that each variable $\|\mathbf{P}_i\mathbf{X}\|^2$ is χ^2-distributed with $\dim(\mathfrak{M}_i)$ degrees of freedom. A consequence of this is that for each normally distributed p-vector $\mathbf{X}$ the variable

$$\left\langle \mathbf{C}^{-1}(\mathbf{X}-\boldsymbol{\mu}), (\mathbf{X}-\boldsymbol{\mu})\right\rangle$$

enjoys a χ^2-distribution with p degrees of freedom.

Finally the method of *dimension reduction*. Suppose $\mathbf{X}$ is an $N(\boldsymbol{\mu}, \mathbf{C})$-distributed p-vector. Let $\lambda_1 \geq \cdots \geq \lambda_p \geq 0$ and let $\mathbf{v}_1, \ldots, \mathbf{v}_p$ be corresponding eigenvalues and vectors of $\mathbf{C}$. Then for $Y_i := \langle \mathbf{X}, \mathbf{v}_i \rangle$, the $Y_1, \ldots, Y_p$ form an independent system of variables for which

$$\mathbf{X} = Y_1\mathbf{v}_1 + \cdots + Y_p\mathbf{v}_p.$$

Now for all i we have $\mathrm{var}(Y_i) = \lambda_i$. Next, among the variables $Y_1, \ldots, Y_p$ the ones of very small variance are identified with constants. When applying this, a certain percentage of the total variance of $\mathbf{X}$ is lost. This percentage can be expressed in $\lambda_1, \ldots, \lambda_p$.

8.1.5 Conditional probability distributions that emanate from Gaussian ones (VIII.5)

We start with a general remark. Suppose $\mathbf{X}$ and $\mathbf{Y}$ are stochastic vectors of size p and q respectively. Then there is always an (essentially) unique privileged family

$\{\mathbb{P}_{\mathbf{X}}(\bullet \mid \mathbf{Y} = \mathbf{y}) \,:\, \mathbf{y} \in \mathbb{R}^q\}$ of Borel probability measures, such that for all Borel sets A and B in $\mathbb{R}^p$ and $\mathbb{R}^q$ we have:

$$\mathbb{P}_{(\mathbf{X},\mathbf{Y})}(A \times B) \;=\; \int_B \mathbb{P}_{\mathbf{X}}(A \mid \mathbf{Y} = \mathbf{y})\, d\mathbb{P}_{\mathbf{Y}}(\mathbf{y}).$$

For all $\mathbf{y} \in \mathbb{R}^q$ this $\mathbb{P}_{\mathbf{X}}(\bullet \mid \mathbf{Y} = \mathbf{y})$ is called the *conditional probability distribution of* $\mathbf{X}$*, given by* $\mathbf{Y} = \mathbf{Y}$. Now let $\mathbf{Z}$ be a $N(\boldsymbol{\mu}, \mathbf{C})$-distributed stochastic p-vector. Given two linear subspaces $\mathfrak{M}$ and $\mathfrak{N}$ of $\mathbb{R}^p$ we look at $\mathbf{X} := \mathbf{P}_{\mathfrak{M}}\mathbf{Z}$ and $\mathbf{Y} := \mathbf{P}_{\mathfrak{N}}\mathbf{Z}$, so-called 'marginals' of $\mathbf{Z}$. We are interested in the conditional probability distribution of $\mathbf{X}$, given $\mathbf{Y} = \mathbf{y}$.

To this we need the following. Let $\mathbf{V}$ be a stochastic p-vector and $\mathbf{W}$ a stochastic q-vector: suppose all Cartesian coordinates have finite variance. We define the *covariance operator* $\mathcal{C}(\mathbf{V}, \mathbf{W}) : \mathbb{R}^p \to \mathbb{R}^q$ of $\mathbf{V}$ and $\mathbf{W}$ as the unique linear operator for which

$$\operatorname{cov}\left(\langle \mathbf{V}, \mathbf{a}\rangle, \langle \mathbf{W}, \mathbf{b}\rangle\right) \;=\; \langle \mathcal{C}(\mathbf{V}, \mathbf{W})\, \mathbf{a}, \mathbf{b}\rangle$$

for all $\mathbf{a} \in \mathbb{R}^p$, $\mathbf{b} \in \mathbb{R}^q$. As to its matrix we have that $[\mathcal{C}(\mathbf{V}, \mathbf{W})]_{i,j} = \operatorname{cov}(V_j, W_i)$ for all i and j. Now we return to the marginals $\mathbf{X}$ and $\mathbf{Y}$. Assuming finite variance for all coordinates it appears that there always exists a linear operator $\mathbf{T} : \mathbb{R}^p \to \mathbb{R}^p$ such that $\mathcal{C}(\mathbf{Y}, \mathbf{X} - \mathbf{T}\mathbf{Y}) = \mathbf{0}$. For such $\mathbf{T}$ we have the next result: the conditional probability distribution of $\mathbf{X}$, given $\mathbf{Y} = \mathbf{y}$, is the $N(\tilde{\boldsymbol{\mu}}, \tilde{\mathbf{C}})$-distribution with

$$\tilde{\boldsymbol{\mu}} \equiv \tilde{\boldsymbol{\mu}}(\mathbf{y}) = \boldsymbol{\mu}_{\mathbf{X}} + \mathbf{T}(\mathbf{y} - \boldsymbol{\mu}_{\mathbf{Y}}) \quad \text{and} \quad \tilde{\mathbf{C}} = \mathcal{C}(\mathbf{X}) - \mathbf{T}\mathcal{C}(\mathbf{X}, \mathbf{Y}).$$

Here $\boldsymbol{\mu}_{\mathbf{X}}$ and $\boldsymbol{\mu}_{\mathbf{Y}}$ are the expectation vectors of $\mathbf{X}$ and $\mathbf{Y}$ respectively.

8.1.6 Vectorial samples from Gaussian distributed populations (VIII.6)

Let $\mathbf{X}_1, \ldots, \mathbf{X}_n$ be a vectorial sample from a $N(\boldsymbol{\mu}, \mathbf{C})$-distributed population. Then $\overline{\mathbf{X}}$ is $N(\boldsymbol{\mu}, \frac{1}{n}\mathbf{C})$-distributed and

$$n\left\langle \mathbf{C}^{-1}(\overline{\mathbf{X}} - \boldsymbol{\mu}), (\overline{\mathbf{X}} - \boldsymbol{\mu})\right\rangle$$

enjoys a χ^2-distribution with p degrees of freedom. Basing ourselves on this we may construct confidence regions for $\boldsymbol{\mu}$ in cases where $\mathbf{C}$ is known.

For a p-dimensional sample $\mathbf{X}_1, \ldots, \mathbf{X}_n$ from a $N(\mathbf{0}, \mathbf{C})$-distributed population, the probability distribution of

$$\sum_{i=1}^{n} \mathbf{X}_i \otimes \mathbf{X}_i$$

is defined to be the *Wishart distribution* with parameters n and $\mathbf{C}$, or, briefly: the $W(n, \mathbf{C})$*-distribution.* For any p-dimensional vectorial sample $\mathbf{X}_1, \ldots, \mathbf{X}_n$ from a

$N(\boldsymbol{\mu}, \mathbf{C})$-distributed population, $\mathbf{S}^2$ enjoys a $W(n-1, \mathbf{C})$-distribution. Under the same assumptions, with $\mathbf{C}$ invertible and $n \geq p+1$, the variable

$$\frac{n(n-p)}{p(n-1)} \langle \mathbf{S}^{-2}(\overline{\mathbf{X}} - \boldsymbol{\mu}), (\overline{\mathbf{X}} - \boldsymbol{\mu}) \rangle$$

enjoys an F-distribution with p and $n-p$ degrees of freedom in the numerator and denominator respectively.

The compound statistic $(\overline{\mathbf{X}}, \mathbf{S}^2)$ is sufficient for the compound parameter $(\boldsymbol{\mu}, \mathbf{C})$; the maximum likelihood estimator for $(\boldsymbol{\mu}, \mathbf{C})$ is

$$\left(\overline{\mathbf{X}}\, ,\ \frac{1}{n} \sum_{i=1}^{n} (\mathbf{X}_i - \overline{\mathbf{X}}) \otimes (\mathbf{X}_i - \overline{\mathbf{X}}) \right).$$

8.1.7 Vectorial versions of the fundamental limit theorems (VIII.7)

Let $\mathbf{X}, \mathbf{X}_1, \mathbf{X}_2, \ldots$ be a sequence of stochastic p-vectors on a measure space $(\Omega, \mathfrak{A}, \mathbb{P})$. The sequence $\mathbf{X}_1, \mathbf{X}_2, \ldots$ is said to converge *almost surely* (*strongly*) to $\mathbf{X}$ if there exists an $A \in \mathfrak{A}$ such that $\mathbb{P}(A) = 1$ and such that

$$\lim_{n\to\infty} \mathbf{X}_n(\omega) = \mathbf{X}(\omega) \quad \text{for all } \omega \in A.$$

This turns out to be equivalent to each of the following two conditions:

(i) $\langle \mathbf{X}_n, \mathbf{v} \rangle \to \langle \mathbf{X}, \mathbf{v} \rangle$ almost surely for all $\mathbf{v} \in \mathbb{R}^p$,

(ii) $\langle \mathbf{X}_n, \mathbf{e}_i \rangle \to \langle \mathbf{X}, \mathbf{e}_i \rangle$ almost surely for all $i = 1, \ldots, p$.

The vectorial version of the strong law of large numbers states that for any infinite sample $\mathbf{X}_1, \mathbf{X}_2, \ldots$ from a population with mean $\boldsymbol{\mu}$ the sequence of sample means

$$(\overline{\mathbf{X}})_n = \frac{\mathbf{X}_1 + \cdots + \mathbf{X}_n}{n}$$

converges strongly to $\boldsymbol{\mu}$ if $n \to \infty$.

The sequence $\mathbf{X}_1, \mathbf{X}_2, \ldots$ is said to converge *in probability* to $\mathbf{X}$ if for each $\varepsilon > 0$ one has

$$\lim_{n\to\infty} \mathbb{P}(\|\mathbf{X}_n - \mathbf{X}\| \geq \varepsilon) = 0.$$

Here we have the two equivalent statements:

(i) $\langle \mathbf{X}_n, \mathbf{v} \rangle \to \langle \mathbf{X}, \mathbf{v} \rangle$ in probability for all $\mathbf{v} \in \mathbb{R}^p$,

(ii) $\langle \mathbf{X}_n, \mathbf{e}_i \rangle \to \langle \mathbf{X}, \mathbf{e}_i \rangle$ in probability for all $i = 1, \ldots, p$.

The sequence $\mathbf{X}_1, \mathbf{X}_2, \ldots$ is said to converge *in distribution* to $\mathbf{X}$ if for each continuous and bounded $\varphi : \mathbb{R}^p \to \mathbb{R}$

$$\mathbb{E}[\varphi(\mathbf{X}_n)] \;\to\; \mathbb{E}[\varphi(\mathbf{X})].$$

If this is the case, then the sequence $g(\mathbf{X}_1), g(\mathbf{X}_2), \ldots$ converges to $g(\mathbf{X})$ for every continuous $g : \mathbb{R}^p \to \mathbb{R}^q$. Convergence in distribution is *not* equivalent to componentwise convergence in distribution. However, it *is* equivalent to the following:

$$\langle \mathbf{X}_n, \mathbf{v}\rangle \;\to\; \langle \mathbf{X}, \mathbf{v}\rangle \quad \text{in distribution for } \textit{all}\ \mathbf{v} \in \mathbb{R}^p.$$

The *characteristic function* of the stochastic p-vector $\mathbf{X}$ is the map $\chi_{\mathbf{X}} : \mathbb{R}^p \to \mathbb{C}$ defined by

$$\chi_{\mathbf{X}}(\mathbf{v}) \;:=\; \mathbb{E}\left(e^{i\langle \mathbf{X}, \mathbf{v}\rangle}\right).$$

The sequence $\mathbf{X}_1, \mathbf{X}_2, \ldots$ converges in distribution to $\mathbf{X}$ if and only if we have pointwise convergence of $\chi_{\mathbf{X}_n}$ to $\chi_{\mathbf{X}}$.

As in the scalar case, convergence almost surely implies convergence in probability, which in turn implies convergence in distribution.

The vectorial form of the central limit theorem states: if $\mathbf{X}_1, \mathbf{X}_2, \ldots$ is an infinite sample from a population that enjoys a probability distribution with mean $\boldsymbol{\mu}$ and covariance operator $\mathbf{C}$, then the sequence $\sqrt{n}\,(\overline{\mathbf{X}} - \boldsymbol{\mu})$ converges in distribution to the $N(\mathbf{0}, \mathbf{C})$-distribution.

Concerning convergence in distribution of sequences of stochastic variables we have to count with the following phenomenon. If both $X_n \to X$ and $Y_n \to Y$ in distribution, then there is no guarantee at all that

$$X_n + Y_n \;\to\; X + Y, \quad \text{nor that} \quad X_n Y_n \;\to\; XY$$

in distribution. However, 'Slutsky's theorem' states that the above *is* true if the limit variable Y is a constant.

8.1.8 Normal correlation analysis (VIII.8)

Suppose a bivariate population is given which is normally distributed. Denoting the two components by X and Y, we construct a parametric method to test whether X and Y are statistically independent.

In this case five parameters are relevant: μ_X, μ_Y, σ_X, σ_Y and ρ, where

$$\rho = \frac{\operatorname{cov}(X, Y)}{\sigma_X \sigma_Y}.$$

Given a sample $(X_1, Y_1), \ldots, (X_n, Y_n)$, the maximum likelihood estimator of ρ appears to be

$$\hat{\rho} = \frac{\sum_{i=1}^n (X_i - \overline{X})(Y_i - \overline{Y})}{\sqrt{\sum_{i=1}^n (X_i - \overline{X})^2}\sqrt{\sum_{i=1}^n (Y_i - \overline{Y})^2}}.$$

As this equals Pearson's product moment correlation coefficient of the cloud of points $(X_1, Y_1), \ldots, (X_n, Y_n)$ we denote it by R_P.

Under the constraint $\rho = 0$ the statistic

$$T = \frac{\sqrt{n-2}\, R_P}{\sqrt{1 - R_P^2}}$$

enjoys a t-distribution with $n - 2$ degrees of freedom. Moreover, for large n, no matter the value of ρ, the statistic

$$Z = \tfrac{1}{2} \log \left(\frac{1 + R_P}{1 - R_P} \right)$$

is approximately $N(\mu, \sigma^2)$-distributed, with

$$\mu = \tfrac{1}{2} \log \left(\frac{1 + \rho}{1 - \rho} \right) \quad \text{and} \quad \sigma^2 = \tfrac{1}{n-3}.$$

This statistic is called 'Fisher's Z'.

8.1.9 Multiple regression analysis (VIII.9)

The method of *multiple analysis of regression* is expounded. We start from the assumption that for every $\mathfrak{x} \in \mathbb{R}^m$ there is a stochastic variable $Y_{\mathfrak{x}}$ and that this assignment is such that:

(i) For every fixed sequence $\mathfrak{x}_1, \ldots, \mathfrak{x}_n$ the $Y_{\mathfrak{x}_1}, \ldots, Y_{\mathfrak{x}_n}$ form an independent system of variables.

(ii) There exist constants $\alpha \in \mathbb{R}$ and $\boldsymbol{\beta} \in \mathbb{R}^m$ such that

$$\mathbb{E}(Y_{\mathfrak{x}}) = \alpha + \langle \boldsymbol{\beta}, \mathfrak{x} \rangle \quad \text{for all } \mathfrak{x} \in \mathbb{R}^m.$$

(iii) For every fixed sequence $\mathfrak{x}_1, \ldots, \mathfrak{x}_n$ we have

$$\operatorname{var}(Y_{\mathfrak{x}_1}) = \ldots = \operatorname{var}(Y_{\mathfrak{x}_n}) = \sigma^2.$$

Under the additional assumption that for each $\mathfrak{x}$ the variable $Y_{\mathfrak{x}}$ is normally distributed we speak about *normal* regression analysis. In order to simplify typography we write $Y_1, \ldots, Y_n$ instead of $Y_{\mathfrak{x}_1}, \ldots, Y_{\mathfrak{x}_n}$.

We define $\mathbf{e} := (1, \ldots, 1) \in \mathbb{R}^n$ and $\mathbf{x}_j := ((\mathfrak{x}_1)_j, \ldots, (\mathfrak{x}_n)_j)$ $(j = 1, \ldots, m)$, and suppose that $\mathbf{e}, \mathbf{x}_1, \ldots, \mathbf{x}_m$ is an independent set of vectors in $\mathbb{R}^n$, of which $\mathfrak{M}$ denotes the linear span. By $\mathbf{M}$ we mean the $n \times m$-matrix that has $\mathbf{e}, \mathbf{x}_1, \ldots, \mathbf{x}_m$ as its columns. Now the least squares estimate of $(\alpha, \boldsymbol{\beta})$, which we denote by $(\hat{\alpha}, \hat{\boldsymbol{\beta}})$, equals $\mathbf{TY} = \mathbf{T}(Y_1, \ldots, Y_n)$ where $\mathbf{T} : \mathbb{R}^n \to \mathbb{R}^{m+1}$ is the linear operator corresponding to the matrix $(\mathbf{M}^t\mathbf{M})^{-1}\mathbf{M}^t$. As a consequence

$$\mathbb{E}\left[(\hat{\alpha}, \hat{\boldsymbol{\beta}})\right] = (\alpha, \boldsymbol{\beta}) \quad \text{and} \quad [\mathcal{C}(\hat{\alpha}, \hat{\boldsymbol{\beta}})] = \sigma^2(\mathbf{M}^t\mathbf{M})^{-1}.$$

Defining SSE by

$$\text{SSE} := \|\mathbf{Y} - \mathbf{P}_{\mathfrak{M}}\mathbf{Y}\|^2,$$

the statistic $\text{SSE}/(n-(m+1))$ provides an unbiased estimator for σ^2.

We may try to predict the Y_i via less than m controlled variables, say by $k < m$. Then (after renumbering) we are concerned with a set $\mathbf{e}, \mathbf{x}_1, \ldots, \mathbf{x}_k$: we write $\mathfrak{N}$ for its linear span. In all cases we have $\text{SSE}_{\mathfrak{N}} > \text{SSE}$, where SSE is the sum of squares of errors of the full model. For $k = 0$ we have $\text{SSE}_{\mathfrak{N}} = S_{\mathbf{YY}}$.

Starting from the assumptions of *normal* analysis of regression, SSE/σ^2 appears to be χ^2-distributed with $n-(m+1)$ degrees of freedom. If the reduced model applies, the statistic

$$T = \frac{(\text{SSE}_{\mathfrak{N}} - \text{SSE})/(m-k)}{\text{SSE}/(n-m-1)}$$

enjoys an F-distribution with $m-k$ degrees of freedom in the numerator and $n-m-1$ in the denominator. If the reduced model fits well then T can be expected to show small outcomes.

8.1.10 The multiple correlation coefficient (VIII.10)

Let $(\mathfrak{x}_1, y_1), \ldots, (\mathfrak{x}_n, y_n)$ be a cloud of points in $\mathbb{R}^{m+1}$. Assuming the model of multiple regression

$$y = \alpha + \beta_1 x_1 + \cdots + \beta_m x_m$$

holds, we define the *multiple correlation coefficient* $R \in [0,1]$ belonging to this cloud of points via

$$R^2 := 1 - \frac{\text{SSE}}{S_{\mathbf{YY}}}.$$

This offers a generalization of Pearson's product moment correlation coefficient.

A geometrical interpretation for R is given. To this, denote the linear span of the vector $\mathbf{e} = (1, \ldots, 1)$ by $\mathfrak{E}$. When equipping the quotient space $\mathbb{R}^n/\mathfrak{E}$ with a suitable inner product, R equals the cosine of the angle between $[\mathbf{y}]$ and $[\mathfrak{M}]$.

Moreover we have that R is invariant under scale transformations. That is, using $\tilde{\mathfrak{x}}_i := \mathbf{A}\mathfrak{x}_i + \mathbf{b}$ instead of $\mathfrak{x}_i$ and $\tilde{y}_i := py_i + q$ instead of y_i for all i, the multiple correlation coefficient stays the same (here $\mathbf{A}$ denotes an invertible linear operator, $\mathbf{b} \in \mathbb{R}^m$ and $p \neq 0$).

If we take $y = \alpha$ as our model in the case of normal analysis of regression, the statistic

$$T = \frac{R^2/m}{(1-R^2)/(n-m-1)}$$

enjoys an F-distribution with m and $n-m-1$ degrees of freedom in numerator and denominator respectively.

Analogously, one may define (with $\mathfrak{R}$ as in §8) the *partial correlation coefficient* $R_{\mathfrak{R}}$ by

$$R_{\mathfrak{R}}^2 := 1 - \frac{\mathrm{SSE}}{\mathrm{SSE}_{\mathfrak{R}}}.$$

Then the statistic

$$T = \frac{R_{\mathfrak{R}}^2/(m-k)}{(1-R_{\mathfrak{R}}^2)/(n-m-1)},$$

under the reduced model, enjoys an F-distribution with $m-k$ and $n-m-1$ degrees of freedom in numerator and denominator respectively.

8.2 Exercises to Chapter VIII

Exercise 1

Suppose $\mathbf{v}_1, \dots, \mathbf{v}_m$ is an independent set of vectors in $\mathbb{R}^p$, of which $\mathfrak{M}$ denotes the linear span. Let the linear operator $\mathbf{T} : \mathbb{R}^m \to \mathbb{R}^p$ be given via

$$\mathbf{T}\,\mathbf{e}_i := \mathbf{v}_i \qquad (i = 1, \dots, m)$$

(that is, the matrix $\mathbf{M}$ of $\mathbf{T}$, with respect to the natural bases, has as its ith column $\mathbf{v}_i$).

(i) Prove that the linear operator $\mathbf{T}^*\mathbf{T}$ is invertible.

Proof. First we prove that $\mathbf{T}$ is injective. To this it suffices to prove that the equality $\mathbf{Tx} = \mathbf{0}$ is possible for $\mathbf{x} = \mathbf{0}$ only. So suppose that $\mathbf{Tx} = \mathbf{0}$ and set $\mathbf{x} := x_1\mathbf{e}_1 + \cdots + x_m\mathbf{e}_m$. Then

$$\mathbf{Tx} = \sum_{i=1}^{m} x_i\,\mathbf{Te}_i = \sum_{i=1}^{m} x_i\,\mathbf{v}_i = \mathbf{0},$$

hence $x_1 = \cdots = x_m = 0$ (as the $\mathbf{v}_1, \dots, \mathbf{v}_m$ are linearly independent). So we necessarily have $\mathbf{x} = \mathbf{0}$. This proves the injectivity of $\mathbf{T}$.

Next we prove that $\mathbf{T}^*\mathbf{T} : \mathbb{R}^m \to \mathbb{R}^m$ is injective. Suppose that $\mathbf{T}^*\mathbf{Tx} = \mathbf{0}$: then

$$0 = \langle \mathbf{T}^*\mathbf{Tx}, \mathbf{y} \rangle = \langle \mathbf{Tx}, \mathbf{Ty} \rangle$$

for all $\mathbf{y} \in \mathbb{R}^m$. For $\mathbf{y} := \mathbf{x}$ we get $\|\mathbf{Tx}\|^2 = 0$ or $\mathbf{Tx} = \mathbf{0}$, which in turn implies that $\mathbf{x} = \mathbf{0}$. This proves the injectivity of $\mathbf{T}^*\mathbf{T}$. As $\mathbf{T}^*\mathbf{T}$ is a map from $\mathbb{R}^m$ to $\mathbb{R}^m$ it immediately follows that it is surjective. Therefore it is bijective, that is, invertible. □

(ii) Show that $\mathfrak{M} = \{\mathbf{Tz} : \mathbf{z} \in \mathbb{R}^m\}$.

Solution.

$$\begin{aligned}
\mathfrak{M} &= \operatorname{span}\{\mathbf{v}_1, \dots, \mathbf{v}_m\} \\
&= \{a_1\mathbf{v}_1 + \dots + a_m\mathbf{v}_m : a_1, \dots, a_m \in \mathbb{R}\} \\
&= \{a_1\,\mathbf{Te}_1 + \dots + a_m\,\mathbf{Te}_m : a_1, \dots, a_m \in \mathbb{R}\} \\
&= \left\{\mathbf{T}\left(\sum_{i=1}^{m} a_i\mathbf{e}_i\right) : a_1, \dots, a_m \in \mathbb{R}\right\} \\
&= \{\mathbf{Tz} : \mathbf{z} \in \mathbb{R}^m\}.
\end{aligned}$$

□

(iii) Show that the orthogonal projection $\mathbf{P}_{\mathfrak{M}}$ is given by

$$\mathbf{P}_{\mathfrak{M}} = \mathbf{T}\,(\mathbf{T}^*\mathbf{T})^{-1}\,\mathbf{T}^*.$$

Proof. First we notice that any $\mathbf{x} \in \mathbb{R}^p$ can be decomposed as

$$\mathbf{x} = \mathbf{P}_{\mathfrak{M}}\mathbf{x} + (\mathbf{x} - \mathbf{P}_{\mathfrak{M}}\mathbf{x}).$$

Here $\mathbf{x} - \mathbf{P}_{\mathfrak{M}}\mathbf{x}$ is always an element of $\mathfrak{M}^\perp$. By (ii) we have $\mathbf{Ty} \in \mathfrak{M}$ for all $\mathbf{y} \in \mathbb{R}^m$. Consequently

$$\langle \mathbf{T}^*(\mathbf{x} - \mathbf{P}_{\mathfrak{M}}\mathbf{x}), \mathbf{y}\rangle = \langle \mathbf{x} - \mathbf{P}_{\mathfrak{M}}\mathbf{x}, \mathbf{Ty}\rangle = 0 \quad \text{for all } \mathbf{y} \in \mathbb{R}^m.$$

From this (for example by taking $\mathbf{y} := \mathbf{T}^*(\mathbf{x} - \mathbf{P}_{\mathfrak{M}}\mathbf{x})$) it follows that for all $\mathbf{x} \in \mathbb{R}^p$

$$\mathbf{T}^*(\mathbf{x} - \mathbf{P}_{\mathfrak{M}}\mathbf{x}) = \mathbf{0}, \quad \text{or:} \quad \mathbf{T}^*\mathbf{x} = \mathbf{T}^*\mathbf{P}_{\mathfrak{M}}\mathbf{x}.$$

Using (ii), for every $\mathbf{x}$ we have a $\mathbf{y}$ such that $\mathbf{P}_{\mathfrak{M}}\mathbf{x} = \mathbf{Ty}$. This leads to

$$\mathbf{T}^*\mathbf{x} = \mathbf{T}^*\mathbf{Ty}, \quad \text{or:} \quad (\mathbf{T}^*\mathbf{T})^{-1}\mathbf{T}^*\mathbf{x} = \mathbf{y}.$$

In turn it follows that for all $\mathbf{x} \in \mathbb{R}^p$

$$\mathbf{T}(\mathbf{T}^*\mathbf{T})^{-1}\mathbf{T}^*\mathbf{x} = \mathbf{Ty} = \mathbf{P}_{\mathfrak{M}}\mathbf{x},$$

so that $\mathbf{P}_{\mathfrak{M}} = \mathbf{T}(\mathbf{T}^*\mathbf{T})^{-1}\mathbf{T}^*$. □

(iv) Prove that if $\mathbf{v}_1, \dots, \mathbf{v}_m$ forms an *orthonormal* system,

$$\mathbf{P}_{\mathfrak{M}}\mathbf{x} = \sum_{i=1}^{m} \langle \mathbf{x}, \mathbf{v}_i\rangle\, \mathbf{v}_i.$$

Proof. First we determine $\mathbf{T}^* : \mathbb{R}^p \to \mathbb{R}^m$. We note that there exists a sequence $\mathbf{v}_{m+1}, \dots, \mathbf{v}_p$ which makes $\mathbf{v}_1, \dots, \mathbf{v}_p$ an orthonormal basis for $\mathbb{R}^p$. It is easily verified that

$$\mathbf{T}^* : \begin{cases} \mathbf{v}_i \mapsto \mathbf{e}_i & \text{for } i = 1, \dots, m, \\ \mathbf{v}_i \mapsto \mathbf{0} & \text{for } i = m+1, \dots, p. \end{cases}$$

Now $\mathbf{T}^*\mathbf{T} : \mathbf{e}_i \mapsto \mathbf{e}_i$ for all $i = 1, \ldots, m$ and thus $\mathbf{T}^*\mathbf{T} = \mathbf{I}_m$. So $\mathbf{P}_{\mathfrak{M}} = \mathbf{T}\mathbf{T}^*$;

$$\mathbf{T}\mathbf{T}^* : \begin{cases} \mathbf{v}_i \mapsto \mathbf{v}_i & \text{for } i = 1, \ldots, m, \\ \mathbf{v}_i \mapsto \mathbf{0} & \text{for } i = m+1, \ldots, p. \end{cases}$$

Therefore

$$\begin{aligned} \mathbf{P}_{\mathfrak{M}}\mathbf{x} = \mathbf{P}_{\mathfrak{M}}\left(\sum_{i=1}^{p} \langle \mathbf{x}, \mathbf{v}_i \rangle \mathbf{v}_i\right) &= \sum_{i=1}^{p} \langle \mathbf{x}, \mathbf{v}_i \rangle (\mathbf{T}\mathbf{T}^*)(\mathbf{v}_i) \\ &= \sum_{i=1}^{m} \langle \mathbf{x}, \mathbf{v}_i \rangle \mathbf{v}_i. \end{aligned}$$

□

Exercise 2

Prove that for all $\mathbf{x}, \mathbf{y}, \mathbf{z} \in \mathbb{R}^p$ we have

$$(\alpha\mathbf{x} + \beta\mathbf{y}) \otimes \mathbf{z} = \alpha(\mathbf{x} \otimes \mathbf{z}) + \beta(\mathbf{y} \otimes \mathbf{z})$$

and

$$\mathbf{z} \otimes (\alpha\mathbf{x} + \beta\mathbf{y}) = \alpha(\mathbf{z} \otimes \mathbf{x}) + \beta(\mathbf{z} \otimes \mathbf{y}).$$

Proof. For all $\mathbf{v} \in \mathbb{R}^p$ we have

$$\begin{aligned} [(\alpha\mathbf{x} + \beta\mathbf{y}) \otimes \mathbf{z}](\mathbf{v}) &= \langle \mathbf{z}, \mathbf{v} \rangle (\alpha\mathbf{x} + \beta\mathbf{y}) \\ &= \alpha \langle \mathbf{z}, \mathbf{v} \rangle \mathbf{x} + \beta \langle \mathbf{z}, \mathbf{v} \rangle \mathbf{y} \\ &= [\alpha(\mathbf{x} \otimes \mathbf{z})](\mathbf{v}) + [\beta(\mathbf{y} \otimes \mathbf{z})](\mathbf{v}) \\ &= [\alpha(\mathbf{x} \otimes \mathbf{z}) + \beta(\mathbf{y} \otimes \mathbf{z})](\mathbf{v}). \end{aligned}$$

This shows that $(\alpha\mathbf{x} + \beta\mathbf{y}) \otimes \mathbf{z} = \alpha(\mathbf{x} \otimes \mathbf{z}) + \beta(\mathbf{y} \otimes \mathbf{z})$. In exactly the same way it can be shown that

$$\mathbf{z} \otimes (\alpha\mathbf{x} + \beta\mathbf{y}) = \alpha(\mathbf{z} \otimes \mathbf{x}) + \beta(\mathbf{z} \otimes \mathbf{y}).$$

□

Exercise 3

Prove that for each pair $\mathbf{S}, \mathbf{T} \in \mathcal{L}(\mathbb{R}^p)$

$$\operatorname{tr}(\mathbf{T}^*\mathbf{S}) = \sum_{j=1}^{p} \langle \mathbf{S}\mathbf{e}_j, \mathbf{T}\mathbf{e}_j \rangle = \sum_{i,j=1}^{p} [\mathbf{S}]_{ij}[\mathbf{T}]_{ij}.$$

Moreover, show that

$$\langle \mathbf{S}, \mathbf{T} \rangle := \operatorname{tr}(\mathbf{T}^*\mathbf{S})$$

defines an inner product on $\mathcal{L}(\mathbb{R}^p)$.

Proof. By definition

$$\operatorname{tr}(\mathbf{T}^*\mathbf{S}) = \sum_{i=1}^{p} \langle \mathbf{T}^*\mathbf{S}\mathbf{e}_i, \mathbf{e}_i \rangle = \sum_{i=1}^{p} \langle \mathbf{S}\mathbf{e}_i, \mathbf{T}\mathbf{e}_i \rangle .$$

Furthermore

$$\langle \mathbf{S}\mathbf{e}_i, \mathbf{T}\mathbf{e}_i \rangle = \sum_{j=1}^{p} [\mathbf{S}]_{ji} [\mathbf{T}]_{ji}$$

and hence $\operatorname{tr}(\mathbf{T}^*\mathbf{S}) = \sum_{i,j} [\mathbf{S}]_{ji}[\mathbf{T}]_{ji}$. This shows the first point.

Now we show that $\langle \ , \ \rangle$ defines an inner product on $\mathcal{L}(\mathbb{R}^p)$. Clearly for any $\mathbf{S} \in \mathcal{L}(\mathbb{R}^p)$

$$\langle \mathbf{S}, \mathbf{S} \rangle = \sum_{i,j} [\mathbf{S}]_{ij}^2 \ \geq \ 0 \quad \text{whereas} \quad \langle \mathbf{S}, \mathbf{S} \rangle = 0 \Leftrightarrow \mathbf{S} = \mathbf{0}.$$

Furthermore, for any $\mathbf{S}, \mathbf{T} \in \mathcal{L}(\mathbb{R}^p)$

$$\langle \mathbf{S}, \mathbf{T} \rangle = \sum_{i=1}^{p} \langle \mathbf{S}\mathbf{e}_i, \mathbf{T}\mathbf{e}_i \rangle = \sum_{i=1}^{p} \langle \mathbf{T}\mathbf{e}_i, \mathbf{S}\mathbf{e}_i \rangle = \langle \mathbf{T}, \mathbf{S} \rangle .$$

Finally, for all $\alpha, \beta \in \mathbb{R}$ and $\mathbf{S}_1, \mathbf{S}_2, \mathbf{T} \in \mathcal{L}(\mathbb{R}^p)$,

$$\langle \alpha \mathbf{S}_1 + \beta \mathbf{S}_2, \mathbf{T} \rangle = \cdots = \alpha \langle \mathbf{S}_1, \mathbf{T} \rangle + \beta \langle \mathbf{S}_2, \mathbf{T} \rangle .$$

Summarizing, $\langle \ , \ \rangle$ is an inner product. □

Exercise 4

Show that for arbitrary $\mathbf{x}, \mathbf{y}, \mathbf{a}, \mathbf{b} \in \mathbb{R}^p$ and $\mathbf{T} \in \mathcal{L}(\mathbb{R}^p)$ we have:

(i) $\langle \mathbf{T}, \mathbf{x} \otimes \mathbf{y} \rangle = \langle \mathbf{T}\mathbf{y}, \mathbf{x} \rangle$,

(ii) $\langle \mathbf{T}, \mathbf{e}_i \otimes \mathbf{e}_j \rangle = [\mathbf{T}]_{ij}$,

(iii) $\langle \mathbf{x} \otimes \mathbf{y}, \mathbf{a} \otimes \mathbf{b} \rangle = \langle \mathbf{y}, \mathbf{b} \rangle \langle \mathbf{x}, \mathbf{a} \rangle$,

(iv) $\|\mathbf{x} \otimes \mathbf{y}\| = \|\mathbf{y}\| \, \|\mathbf{x}\|$,

(v) $(\mathbf{x} \otimes \mathbf{y})^* = (\mathbf{y} \otimes \mathbf{x})$.

Proof. For all $\mathbf{x}, \mathbf{y} \in \mathbb{R}^p$ we have

$$\begin{aligned} \langle \mathbf{T}, \mathbf{x} \otimes \mathbf{y} \rangle &= \sum_{i=1}^{p} \langle \mathbf{T}\mathbf{e}_i, (\mathbf{x} \otimes \mathbf{y})\mathbf{e}_i \rangle = \sum_{i=1}^{p} \langle \mathbf{y}, \mathbf{e}_i \rangle \langle \mathbf{T}\mathbf{e}_i, \mathbf{x} \rangle \\ &= \left\langle \mathbf{T} \left(\sum_{i=1}^{p} \langle \mathbf{y}, \mathbf{e}_i \rangle \mathbf{e}_i \right), \mathbf{x} \right\rangle = \langle \mathbf{T}\mathbf{y}, \mathbf{x} \rangle . \end{aligned}$$

This proves (i).

As an application, $\langle \mathbf{T}, \mathbf{e}_i \otimes \mathbf{e}_j \rangle = \langle \mathbf{T}\mathbf{e}_j, \mathbf{e}_i \rangle = [\mathbf{T}]_{ij}$ (by definition), which proves (ii).

Also, for all $\mathbf{x}, \mathbf{y}, \mathbf{a}, \mathbf{b} \in \mathbb{R}^p$ we get

$$\langle \mathbf{x} \otimes \mathbf{y}, \mathbf{a} \otimes \mathbf{b} \rangle = \langle (\mathbf{x} \otimes \mathbf{y})\mathbf{b}, \mathbf{a} \rangle = \langle \mathbf{y}, \mathbf{b} \rangle \langle \mathbf{x}, \mathbf{a} \rangle .$$

This proves (iii).

As a consequence, for all $\mathbf{x}, \mathbf{y} \in \mathbb{R}^p$

$$\|\mathbf{x} \otimes \mathbf{y}\| = \sqrt{\langle \mathbf{x} \otimes \mathbf{y}, \mathbf{x} \otimes \mathbf{y} \rangle} = \sqrt{\langle \mathbf{y}, \mathbf{y} \rangle \langle \mathbf{x}, \mathbf{x} \rangle} = \|\mathbf{y}\| \, \|\mathbf{x}\|.$$

This proves (iv).

Finally, for all $\mathbf{z}_1, \mathbf{z}_2 \in \mathbb{R}^p$

$$\begin{aligned} \langle (\mathbf{x} \otimes \mathbf{y})^* \mathbf{z}_1, \mathbf{z}_2 \rangle &= \langle \mathbf{z}_1, (\mathbf{x} \otimes \mathbf{y})\mathbf{z}_2 \rangle = \langle \mathbf{y}, \mathbf{z}_2 \rangle \langle \mathbf{z}_1, \mathbf{x} \rangle \\ &= \langle \langle \mathbf{z}_1, \mathbf{x} \rangle \mathbf{y}, \mathbf{z}_2 \rangle = \langle (\mathbf{y} \otimes \mathbf{x})\mathbf{z}_1, \mathbf{z}_2 \rangle \end{aligned}$$

so $(\mathbf{x} \otimes \mathbf{y})^* = (\mathbf{y} \otimes \mathbf{x})$, which proves (v). □

Exercise 5

Show that the p^2 operators $\mathbf{e}_i \otimes \mathbf{e}_j$ form an orthonormal basis of $\mathcal{L}(\mathbb{R}^p)$, equipped with the Hilbert–Schmidt inner product.

Solution. The orthonormality of $\{\mathbf{e}_i \otimes \mathbf{e}_j \,:\, i, j = 1, \dots, p\}$ is easily derived from Exercise 4, part (iii). This immediately shows that the system is also independent. As each $\mathbf{T} \in \mathcal{L}(\mathbb{R}^p)$ may be decomposed as

$$\mathbf{T} = \sum_{i,j=1}^{p} [\mathbf{T}]_{ij} (\mathbf{e}_i \otimes \mathbf{e}_j)$$

(check this), the system indeed forms an orthonormal basis. □

Exercise 6

Let $\Phi : \mathcal{L}(\mathbb{R}^p) \to \mathcal{L}(\mathbb{R}^p)$ be the map defined by

$$\Phi(\mathbf{T}) := \mathbf{T}^*.$$

Prove that Φ denotes an orthogonal linear operator on $\mathcal{L}(\mathbb{R}^p)$ and determine its trace.

Solution. On $\mathcal{L}(\mathbb{R}^p)$ we use the Hilbert–Schmidt inner product. Exercise 5 shows that $\{\mathbf{e}_i \otimes \mathbf{e}_j \,:\, i, j = 1, \dots, p\}$ provides an orthonormal basis for $\mathcal{L}(\mathbb{R}^p)$. Clearly Φ is a linear operator (as $\Phi(\mathbf{T} + \mathbf{S}) = (\mathbf{T} + \mathbf{S})^* = \mathbf{T}^* + \mathbf{S}^* = \Phi(\mathbf{T}) + \Phi(\mathbf{S})$ etcetera). By definition, Φ is an orthogonal operator if and only if

$$\{\Phi(\mathbf{e}_i \otimes \mathbf{e}_j) \,:\, i, j = 1, \dots, p\}$$

provides an orthonormal basis for $\mathcal{L}(\mathbb{R}^p)$. This indeed is the case, for by Exercise 4 (v) we have:

$$\Phi(\mathbf{e}_i \otimes \mathbf{e}_j) = (\mathbf{e}_i \otimes \mathbf{e}_j)^* = \mathbf{e}_j \otimes \mathbf{e}_i \quad \text{for all } i, j.$$

Using the fact that $\{\mathbf{e}_i \otimes \mathbf{e}_j : i, j = 1, \ldots, p\}$ is an orthonormal system, we get:

$$\begin{aligned} \operatorname{tr}(\Phi) &:= \sum_{i,j=1}^{p} \langle \Phi(\mathbf{e}_i \otimes \mathbf{e}_j), \mathbf{e}_i \otimes \mathbf{e}_j \rangle = \sum_{i,j=1}^{p} \langle \mathbf{e}_j \otimes \mathbf{e}_i, \mathbf{e}_i \otimes \mathbf{e}_j \rangle \\ &= \sum_{i=1}^{p} \langle \mathbf{e}_i \otimes \mathbf{e}_i, \mathbf{e}_i \otimes \mathbf{e}_i \rangle = \sum_{i=1}^{p} 1^2 = p. \end{aligned}$$

□

Exercise 7

We start with a stochastic 2-vector $\mathbf{X}$, of which the covariance operator $\mathbf{C}$ is given by the matrix

$$[\mathbf{C}] = \begin{pmatrix} 66 & -12 \\ -12 & 59 \end{pmatrix}.$$

(i) Determine $\operatorname{var}(\mathbf{X})$.

Solution. Of course $\operatorname{var}(\mathbf{X}) = \operatorname{tr}(\mathbf{C}) = 66 + 59 = 125$. □

(ii) Express the principal components of $\mathbf{X}$ in terms of X_1 and X_2.

Solution. First we find the eigenvalues belonging to $\mathbf{C}$. The equalities $\lambda_1 + \lambda_2 = \operatorname{tr}(\mathbf{C}) = 125$ and $\lambda_1\lambda_2 = \det(\mathbf{C}) = 66 \cdot 59 - (-12)^2 = 3750$ are obvious and lead to the second degree equation $\lambda_1^2 - 125\lambda_1 + 3750 = 0$. This results in

$$\lambda_1 = 75 \quad \text{and} \quad \lambda_2 = 50 \quad \text{(or vice versa)}.$$

Next we determine the corresponding eigenvectors. We start with $\mathbf{u}_1$; this vector must satisfy the equation

$$\left[\begin{pmatrix} 66 & -12 \\ -12 & 59 \end{pmatrix} - \begin{pmatrix} 75 & 0 \\ 0 & 75 \end{pmatrix}\right] \begin{pmatrix} u_{11} \\ u_{12} \end{pmatrix} = \begin{pmatrix} 0 \\ 0 \end{pmatrix},$$

which turns out to be equivalent to the condition that $3u_{11} + 4u_{12} = 0$. We could take for example $\mathbf{w}_1 = (4, -3)$. As we prefer a normalized vector, we take instead the vector

$$\mathbf{u}_1 = \frac{\mathbf{w}_1}{\|\mathbf{w}_1\|} = \left(\tfrac{4}{5}, -\tfrac{3}{5}\right).$$

In the same way we get $\mathbf{u}_2 = (\frac{3}{5}, \frac{4}{5})$.

Thus the principal components, up to sign, are given by

$$Y_1 = \langle \mathbf{X}, \mathbf{u}_1 \rangle = \tfrac{4}{5}X_1 - \tfrac{3}{5}X_2 \quad \text{and} \quad Y_2 = \langle \mathbf{X}, \mathbf{u}_2 \rangle = \tfrac{3}{5}X_1 + \tfrac{4}{5}X_2$$

respectively. □

Exercise 8

Let $\lambda_1, \ldots, \lambda_p$ be the eigenvalues belonging to $\mathcal{C}(\mathbf{X})$, $\mathbf{X}$ denoting a stochastic p-vector. Prove that for all $\mathbf{a} \in \mathbb{R}^p$ we have

$$\|\mathbf{a}\|^2 \min_{j=1,\ldots,p}(\lambda_j) \leq \operatorname{var}(\langle \mathbf{X}, \mathbf{a}\rangle) \leq \|\mathbf{a}\|^2 \max_{j=1,\ldots,p}(\lambda_j).$$

Proof. We take an orthonormal basis $\mathbf{u}_1, \ldots, \mathbf{u}_p$ for $\mathbb{R}^p$ consisting of eigenvectors of $\mathcal{C}(\mathbf{X})$, the vector $\mathbf{u}_i$ corresponding to λ_i. Any $\mathbf{a} \in \mathbb{R}^p$ can then be decomposed as $a_1\mathbf{u}_1 + \cdots + a_p\mathbf{u}_p$, and so

$$\|\mathbf{a}\|^2 = \langle \mathbf{a}, \mathbf{a}\rangle = \sum_{i,j=1}^{p} a_i a_j \langle \mathbf{u}_i, \mathbf{u}_j\rangle = \sum_{i=1}^{p} a_i^2.$$

Furthermore

$$\begin{aligned}
\operatorname{var}(\langle \mathbf{X}, \mathbf{a}\rangle) &= \operatorname{cov}(\langle \mathbf{X}, \mathbf{a}\rangle, \langle \mathbf{X}, \mathbf{a}\rangle) = \langle \mathcal{C}(\mathbf{X})\mathbf{a}, \mathbf{a}\rangle \\
&= \sum_{i,j=1}^{p} a_i a_j \langle \mathcal{C}(\mathbf{X})\mathbf{u}_i, \mathbf{u}_j\rangle \\
&= \sum_{i,j=1}^{p} \lambda_i a_i a_j \langle \mathbf{u}_i, \mathbf{u}_j\rangle = \sum_{i=1}^{p} \lambda_i a_i^2.
\end{aligned}$$

Evidently we can now write

$$\sum_{i=1}^{p} \lambda_i a_i^2 \leq \sum_{i=1}^{p} \max_{j=1,\ldots,p}(\lambda_j)\, a_i^2 = \max_{j=1,\ldots,p}(\lambda_j)\, \|\mathbf{a}\|^2$$

and

$$\sum_{i=1}^{p} \lambda_i a_i^2 \geq \sum_{i=1}^{p} \min_{j=1,\ldots,p}(\lambda_j)\, a_i^2 = \min_{j=1,\ldots,p}(\lambda_j)\, \|\mathbf{a}\|^2.$$

□

Exercise 9

Let $\mathbf{X}$ be a stochastic p-vector having $\mathbf{C}$ as its covariance operator. Suppose that the eigenvalues of $\mathbf{C}$ are given by

$$\lambda_1 > \lambda_2 > \cdots > \lambda_p$$

and that $\{\mathbf{u}_1, \ldots, \mathbf{u}_p\}$ is a corresponding orthonormal basis of eigenvectors.

(i) Under the condition $\|\mathbf{a}\|^2 \leq 1$, prove that the variable $\langle \mathbf{X}, \mathbf{a}\rangle$ attains maximum variance for $\mathbf{a} = \pm\mathbf{u}_1$.

Proof. In Exercise 8 it turned out that

$$\begin{aligned}\operatorname{var}(\langle \mathbf{X}, \mathbf{a}\rangle) = \sum_{i=1}^{p} \lambda_i a_i^2 \ &\leq \ \|\mathbf{a}\|^2 \max_{j=1,\dots,p} (\lambda_j) \\ &= \ \left(\sum_{i=1}^{p} a_i^2\right) \left[\max_{j=1,\dots,p} (\lambda_j)\right],\end{aligned}$$

where $\mathbf{a} = a_1\mathbf{u}_1 + \cdots + a_p\mathbf{u}_p$. In this situation this means that

$$\operatorname{var}(\langle \mathbf{X}, \mathbf{a}\rangle) \ \leq \ \lambda_1\|\mathbf{a}\|^2 \ \leq \ \lambda_1.$$

The first inequality gives an equality only if $a_2 = \cdots = a_p = 0$, since $\lambda_1 > \cdots > \lambda_p$; the second one if and only if $\sum_{i=1}^{p} a_i^2 = 1$. So $\mathbf{a} = \pm\mathbf{u}_1$ are the only two vectors in $\mathbb{R}^p$ for which $\operatorname{var}(\langle \mathbf{X}, \mathbf{a}\rangle) = \lambda_1$. □

(ii) When imposing the condition

$$\operatorname{cov}(\langle \mathbf{X}, \mathbf{a}\rangle, \langle \mathbf{X}, \mathbf{u}_1\rangle) = 0, \quad \text{together with} \quad \|\mathbf{a}\|^2 \ \leq \ 1,$$

the variable $\langle \mathbf{X}, \mathbf{a}\rangle$ is of maximal variance for $\mathbf{a} = \pm\mathbf{u}_2$. Prove this.

Proof. Taking an arbitrary $\mathbf{a} \in \mathbb{R}^p$ it can be decomposed as $\mathbf{a} = a_1\mathbf{u}_1 + \cdots + a_p\mathbf{u}_p$. Thus we can write $\|\mathbf{a}\|^2 = a_1^2 + \cdots + a_p^2$ and

$$\begin{aligned}\operatorname{cov}(\langle \mathbf{X}, \mathbf{a}\rangle, \langle \mathbf{X}, \mathbf{u}_1\rangle) \ &= \ \langle \mathbf{C}\mathbf{a}, \mathbf{u}_1\rangle = \langle \mathbf{C}\mathbf{u}_1, \mathbf{a}\rangle \\ &= \ \lambda_1 \langle \mathbf{u}_1, \mathbf{a}\rangle = \lambda_1 a_1.\end{aligned}$$

So the constraints $\operatorname{cov}(\langle \mathbf{X}, \mathbf{a}\rangle, \langle \mathbf{X}, \mathbf{u}_1\rangle) = 0$ and $\|\mathbf{a}\|^2 \leq 1$ are equivalent to $a_1 = 0$ and $\sum_{i=2}^{p} a_i^2 \leq 1$. From the proof of Exercise 8 we learn that

$$\operatorname{var}(\langle \mathbf{X}, \mathbf{a}\rangle) = \sum_{i=1}^{p} \lambda_i a_i^2 = \sum_{i=2}^{p} \lambda_i a_i^2,$$

so

$$\operatorname{var}(\langle \mathbf{X}, \mathbf{a}\rangle) \ \leq \ \sum_{i=2}^{p} \lambda_2 a_i^2 = \lambda_2\|\mathbf{a}\|^2 \ \leq \ \lambda_2.$$

The value λ_2 is of course taken on if and only if $|a_2| = 1$. □

(iii) Characterize the principal components of $\mathbf{X}$, using (i) and (ii).

Solution. Along the same lines as (ii), it can be shown that, given the constraint $\|\mathbf{a}\| \leq 1$ together with

$$\operatorname{cov}(\langle \mathbf{X}, \mathbf{a}\rangle, \langle \mathbf{X}, \mathbf{u}_i\rangle) = 0 \quad \text{for } i = 1, \dots, k,$$

$\operatorname{var}(\langle \mathbf{X}, \mathbf{a}\rangle)$ is maximal precisely for $\mathbf{a} = \pm\mathbf{u}_{k+1}$ (all $k < p$). Evidently this presents an alternative way to characterize the $(k+1)^{\text{th}}$ principal component of $\mathbf{X}$. □

Exercise 10

Prove that $\operatorname{var}(\mathbf{X}) = \mathbb{E}\left(\|\mathbf{X}-\boldsymbol{\mu}\|^2\right) = \mathbb{E}\left(\|\mathbf{X}\|^2\right) - \|\boldsymbol{\mu}\|^2$.

Proof. Let $\mathbf{X}$ be a stochastic p-vector and let $\boldsymbol{\mu} := \mathbb{E}(\mathbf{X})$. Then, by definition,

$$\begin{aligned}
\operatorname{var}(\mathbf{X}) &= \sum_{i=1}^{p} \operatorname{var}(X_i) = \sum_{i=1}^{p} \mathbb{E}\left[(X_i - \mathbb{E}(X_i))^2\right] \\
&= \mathbb{E}\left[\sum_{i=1}^{p} (X_i - \mu_i)^2\right] = \mathbb{E}\left(\langle \mathbf{X}-\boldsymbol{\mu}, \mathbf{X}-\boldsymbol{\mu}\rangle\right) = \mathbb{E}\left(\|\mathbf{X}-\boldsymbol{\mu}\|^2\right).
\end{aligned}$$

This proves the first equality. Furthermore (see Proposition VIII.2.1)

$$\begin{aligned}
\mathbb{E}\left(\langle \mathbf{X}-\boldsymbol{\mu}, \mathbf{X}-\boldsymbol{\mu}\rangle\right) &= \mathbb{E}\left(\langle \mathbf{X}, \mathbf{X}\rangle - \langle \mathbf{X}, \boldsymbol{\mu}\rangle - \langle \boldsymbol{\mu}, \mathbf{X}\rangle + \langle \boldsymbol{\mu}, \boldsymbol{\mu}\rangle\right) \\
&= \mathbb{E}\left(\|\mathbf{X}\|^2\right) - 2\,\mathbb{E}\left(\langle \mathbf{X}, \boldsymbol{\mu}\rangle\right) + \|\boldsymbol{\mu}\|^2 \\
&= \mathbb{E}\left(\|\mathbf{X}\|^2\right) - 2\,\langle \mathbb{E}(\mathbf{X}), \boldsymbol{\mu}\rangle + \|\boldsymbol{\mu}\|^2 \\
&= \mathbb{E}\left(\|\mathbf{X}\|^2\right) - \|\boldsymbol{\mu}\|^2.
\end{aligned}$$

□

Exercise 11

A vectorial sample $\mathbf{X}_1, \ldots, \mathbf{X}_4$ shows the following outcome:

$$\begin{cases} \mathbf{X}_1 = (2,4,1), & \mathbf{X}_2 = (3,3,2), \\ \mathbf{X}_3 = (3,4,1), & \mathbf{X}_4 = (4,1,2). \end{cases}$$

Determine the matrix of $\mathbf{S}^2$.

Solution. Clearly $\overline{\mathbf{X}} = (3, 3, \frac{3}{2})$. Therefore

$$\begin{cases} \mathbf{X}_1 - \overline{\mathbf{X}} = (-1, 1, -\frac{1}{2}), & \mathbf{X}_2 - \overline{\mathbf{X}} = (0,\ 0,\ \frac{1}{2}), \\ \mathbf{X}_3 - \overline{\mathbf{X}} = (\ 0,\ 1, -\frac{1}{2}), & \mathbf{X}_4 - \overline{\mathbf{X}} = (1, -2, \frac{1}{2}). \end{cases}$$

It follows that the matrix of the outcome of $\mathbf{S}^2$ is given by

$$\begin{aligned}
[\mathbf{S}^2] &= \left[\frac{1}{4-1}\sum_{i=1}^{4} (\mathbf{X}_i - \overline{\mathbf{X}}) \otimes (\mathbf{X}_i - \overline{\mathbf{X}})\right] \\
&= \frac{1}{3}\left[\begin{pmatrix} 1 & -1 & \frac{1}{2} \\ -1 & 1 & -\frac{1}{2} \\ \frac{1}{2} & -\frac{1}{2} & \frac{1}{4} \end{pmatrix} + \begin{pmatrix} 0 & 0 & 0 \\ 0 & 0 & 0 \\ 0 & 0 & \frac{1}{4} \end{pmatrix} + \right. \\
&\qquad \left. \begin{pmatrix} 0 & 0 & 0 \\ 0 & 1 & -\frac{1}{2} \\ 0 & -\frac{1}{2} & \frac{1}{4} \end{pmatrix} + \begin{pmatrix} 1 & -2 & \frac{1}{2} \\ -2 & 4 & -1 \\ \frac{1}{2} & -1 & \frac{1}{4} \end{pmatrix}\right] \\
&= \frac{1}{3}\begin{pmatrix} 2 & -3 & 1 \\ -3 & 6 & -2 \\ 1 & -2 & 1 \end{pmatrix}.
\end{aligned}$$

□

Exercise 12

A vectorial sample of size n consists of the p-vectors $\mathbf{X}_1, \ldots, \mathbf{X}_n$.

(i) Prove that $\dim(\operatorname{Ker}(\mathbf{S}^2)) \geq \max(p - n, 0)$.

Proof. As $\mathbf{S}^2 = \frac{1}{n-1}\sum_{i=1}^n (\mathbf{X}_i - \overline{\mathbf{X}}) \otimes (\mathbf{X}_i - \overline{\mathbf{X}})$, it follows that

$$\begin{aligned} \mathbf{S}^2\mathbf{a} &= \sum_{i=1}^n \frac{1}{n-1} \langle \mathbf{X}_i - \overline{\mathbf{X}}, \mathbf{a} \rangle \ (\mathbf{X}_i - \overline{\mathbf{X}}) \\ &\in \operatorname{span}\{\mathbf{X}_1 - \overline{\mathbf{X}}, \ldots, \mathbf{X}_n - \overline{\mathbf{X}}\} \end{aligned}$$

for all $\mathbf{a} \in \mathbb{R}^p$. Therefore $\dim(\operatorname{Im}(\mathbf{S}^2)) \leq n$. Consequently

$$\dim(\operatorname{Ker}(\mathbf{S}^2)) = p - \dim(\operatorname{Im}(\mathbf{S}^2)) \ \geq \ p - n,$$

which leads directly to the inequality asked for. □

(ii) *Question.* Is it possible for $\mathbf{S}^2$ to be invertible while $p > n$?

Solution. No. $\mathbf{S}^2$ is invertible if and only if its kernel equals $\{\mathbf{0}\}$. By (i) this is impossible for $p > n$. □

Exercise 13

Let $\{\mathbf{X}_1, \ldots, \mathbf{X}_n\}$ be a statistically independent set of stochastic p-vectors. Suppose that $\mathbf{X}_i$ is $N(\boldsymbol{\mu}_i, \mathbf{C}_i)$-distributed (all i). Then the stochastic vector $\tilde{\mathbf{X}} := (\mathbf{X}_1, \ldots, \mathbf{X}_n)$, taking on values in $\mathbb{R}^p \times \cdots \times \mathbb{R}^p$, is $N(\tilde{\boldsymbol{\mu}}, \tilde{\mathbf{C}})$-distributed where

$$\tilde{\boldsymbol{\mu}} := (\boldsymbol{\mu}_1, \ldots, \boldsymbol{\mu}_n) \quad \text{and} \quad \tilde{\mathbf{C}} := \mathbf{C}_1 \oplus \cdots \oplus \mathbf{C}_n.$$

Prove this statement.

Proof. For all i there exists an orthogonal linear operator $\mathbf{Q}_i : \mathbb{R}^p \to \mathbb{R}^p$ such that $\mathbf{Q}_i\mathbf{X}_i = \mathbf{E}_i$, where $\mathbf{E}_i$ is some elementarily Gaussian distributed p-vector. It is easily verified that $\tilde{\mathbf{Q}} := \mathbf{Q}_1 \oplus \cdots \oplus \mathbf{Q}_n$ is an orthogonal operator on $\mathbb{R}^p \times \cdots \times \mathbb{R}^p$. Furthermore we have:

$$\tilde{\mathbf{Q}}(\mathbf{X}_1, \ldots, \mathbf{X}_n) = (\mathbf{E}_1, \ldots, \mathbf{E}_n).$$

The vector $(\mathbf{E}_1, \ldots, \mathbf{E}_n)$ is of course elementarily Gaussian distributed: therefore the above implies that $(\mathbf{X}_1, \ldots, \mathbf{X}_n)$ is Gaussian distributed. Next, let $\tilde{\mathbf{a}} = (\mathbf{a}_1, \ldots, \mathbf{a}_n)$ and $\tilde{\mathbf{b}} = (\mathbf{b}_1, \ldots, \mathbf{b}_n)$ be arbitrary elements in $\mathbb{R}^p \times \cdots \times \mathbb{R}^p$. Then

$$\begin{aligned} \operatorname{cov}\left(\langle \tilde{\mathbf{X}}, \tilde{\mathbf{a}} \rangle, \langle \tilde{\mathbf{X}}, \tilde{\mathbf{b}} \rangle\right) &= \sum_{i,j} \operatorname{cov}\left(\langle \mathbf{X}_i, \mathbf{a}_i \rangle, \langle \mathbf{X}_j, \mathbf{b}_j \rangle\right) \\ &= \sum_{i=1}^n \operatorname{cov}\left(\langle \mathbf{X}_i, \mathbf{a}_i \rangle, \langle \mathbf{X}_i, \mathbf{b}_i \rangle\right) \\ &= \sum_{i=1}^n \langle \mathbf{C}_i\mathbf{a}_i, \mathbf{b}_i \rangle = \langle \tilde{\mathbf{C}}\tilde{\mathbf{a}}, \tilde{\mathbf{b}} \rangle. \end{aligned}$$

This completes the proof. □

Exercise 14

Let $\mathbf{X}_1, \ldots, \mathbf{X}_n$ be a vectorial sample from a p-dimensional population which enjoys a Gaussian distribution. Prove that

$$\overline{\mathbf{X}} \quad \text{and} \quad (\mathbf{X}_1 - \overline{\mathbf{X}}, \ldots, \mathbf{X}_n - \overline{\mathbf{X}})$$

are statistically independent.

Solution. First we define the stochastic pn-vector $\mathbf{X}$ by

$$\mathbf{X} := (\mathbf{X}_1, \ldots, \mathbf{X}_n) = (X_{11}, \ldots, X_{1p}, \ \ldots \ , X_{n1}, \ldots, X_{np}).$$

Exercise 13 tells us that $\mathbf{X}$ is Gaussian distributed. Now define linear operators $\mathbf{T}_1 : \mathbb{R}^{pn} \to \mathbb{R}^p$ and $\mathbf{T}_2 : \mathbb{R}^{pn} \to \mathbb{R}^{pn}$ by

$$\mathbf{T}_1(\mathbf{x}_1, \ldots, \mathbf{x}_n) := \overline{\mathbf{x}} \quad \text{and} \quad \mathbf{T}_2(\mathbf{x}_1, \ldots, \mathbf{x}_n) := (\mathbf{x}_1 - \overline{\mathbf{x}}, \ldots, \mathbf{x}_n - \overline{\mathbf{x}}).$$

We see that in this way, in a self-explaining notation,

$$\begin{cases} \mathbf{T}_1\mathbf{X} = \left(\overline{X}_{\bullet 1}, \ldots, \overline{X}_{\bullet p}\right) \quad \text{and} \\ \mathbf{T}_2\mathbf{X} = \left(X_{11} - \overline{X}_{\bullet 1}, \ldots, X_{1p} - \overline{X}_{\bullet p}, \ \ldots \ , X_{n1} - \overline{X}_{\bullet 1}, \ldots, X_{np} - \overline{X}_{\bullet p}\right). \end{cases}$$

As X_{mi} and X_{kl} are statistically independent for all $m \neq k$ and all $\mathbf{X}_i$ are identically distributed, for any i, k, l we have:

$$\operatorname{cov}\left(\overline{X}_{\bullet i}, X_{kl}\right) = \tfrac{1}{n} \operatorname{cov}\left(X_{ki}, X_{kl}\right) = \tfrac{1}{n} \operatorname{cov}\left(X_{1i}, X_{1l}\right).$$

As a consequence

$$\begin{aligned} \operatorname{cov}\left(\overline{X}_{\bullet i}, \overline{X}_{\bullet l}\right) &= \frac{1}{n} \sum_{k=1}^{n} \operatorname{cov}\left(\overline{X}_{\bullet i}, X_{kl}\right) \\ &= \frac{1}{n^2} \sum_{k=1}^{n} \operatorname{cov}\left(X_{1i}, X_{1l}\right) = \tfrac{1}{n} \operatorname{cov}\left(X_{1i}, X_{1l}\right), \end{aligned}$$

so $\operatorname{cov}\left(\overline{X}_{\bullet i}, X_{kl} - \overline{X}_{\bullet l}\right) = 0$ for all i, k, l, which proves that $\mathbf{T}_1\mathbf{X}$ and $\mathbf{T}_2\mathbf{X}$ are uncorrelated.

Application of Theorem VIII.4.2 now shows that $\mathbf{T}_1\mathbf{X}$ and $\mathbf{T}_2\mathbf{X}$ are statistically independent, that is to say, the stochastic p-vector $\overline{\mathbf{X}}$ and the system of p-vectors $(\mathbf{X}_1 - \overline{\mathbf{X}}, \ldots, \mathbf{X}_n - \overline{\mathbf{X}})$ are statistically independent. □

Exercise 15

Let $\mathbf{X}_1$ and $\mathbf{X}_2$ denote statistically independent stochastic p-vectors, of which $\mathbf{X}_1$ is $N(\boldsymbol{\mu}_1, \mathbf{C}_1)$-distributed and $\mathbf{X}_2$ is $N(\boldsymbol{\mu}_2, \mathbf{C}_2)$-distributed. Prove that $\mathbf{X}_1 + \mathbf{X}_2$ is $N(\boldsymbol{\mu}_1 + \boldsymbol{\mu}_2, \mathbf{C}_1 + \mathbf{C}_2)$-distributed.

Proof. Application of Exercise 13 shows that the stochastic $2p$-vector $\mathbf{X} = (\mathbf{X}_1, \mathbf{X}_2)$ is Gaussian distributed. This exercise also shows that its expectation vector equals $\boldsymbol{\mu} = (\boldsymbol{\mu}_1, \boldsymbol{\mu}_2)$ and that its covariance operator is $\mathbf{C} = \mathbf{C}_1 \oplus \mathbf{C}_2$.

Define $\mathbf{T} : \mathbb{R}^p \times \mathbb{R}^p \to \mathbb{R}^p$ by $\mathbf{T}(\mathbf{x}_1, \mathbf{x}_2) := \mathbf{x}_1 + \mathbf{x}_2$. Then by Theorem VIII.4.5 $\mathbf{T}(\mathbf{X}_1, \mathbf{X}_2) = \mathbf{X}_1 + \mathbf{X}_2$ enjoys a $N(\mathbf{T}(\boldsymbol{\mu}_1, \boldsymbol{\mu}_2), \mathbf{T}(\mathbf{C}_1 \oplus \mathbf{C}_2)\mathbf{T}^*)$-distribution. It is easily verified that $\mathbf{T}^*\mathbf{x} = (\mathbf{x}, \mathbf{x})$. From the equality

$$\begin{aligned} \mathbf{T}(\mathbf{C}_1 \oplus \mathbf{C}_2)\mathbf{T}^*\mathbf{x} &= \mathbf{T}(\mathbf{C}_1\mathbf{x}, \mathbf{C}_2\mathbf{x}) \\ &= \mathbf{C}_1\mathbf{x} + \mathbf{C}_2\mathbf{x} = (\mathbf{C}_1 + \mathbf{C}_2)\ \mathbf{x} \end{aligned}$$

it follows that the covariance operator equals $\mathbf{C}_1 + \mathbf{C}_2$. As $\mathbf{T}(\boldsymbol{\mu}_1, \boldsymbol{\mu}_2) = \boldsymbol{\mu}_1 + \boldsymbol{\mu}_2$ the variable $\mathbf{X}_1 + \mathbf{X}_2$ is indeed $N(\boldsymbol{\mu}_1 + \boldsymbol{\mu}_2, \mathbf{C}_1 + \mathbf{C}_2)$-distributed. □

Exercise 16

Suppose $\mathbf{X}$ is a normally distributed stochastic p-vector. Let $\mathbf{T} : \mathbb{R}^p \to \mathbb{R}^q$ be a linear operator. Is $\mathbf{TX}$ necessarily normally distributed?

Solution. No. $\mathbf{TX}$ is normally distributed if and only if $\mathbf{TCT}^*$ is invertible. For example for $\mathbf{T} = \mathbf{0}$ this is evidently not the case. □

Exercise 17

Suppose the 2-vector (X, Y) is normally distributed with expectation vector (μ_X, μ_Y) and covariance operator $\mathbf{C}$ with

$$[\mathbf{C}] = \begin{pmatrix} \sigma_X^2 & \rho\sigma_X\sigma_Y \\ \rho\sigma_X\sigma_Y & \sigma_Y^2 \end{pmatrix}.$$

Prove that, given $Y = y$, the variable X is $N(\mu_X + \rho\,\frac{\sigma_X}{\sigma_Y}\,(y - \mu_Y),\ (1 - \rho^2)\sigma_X^2)$-distributed.

Proof. The inverse $\mathbf{C}^{-1}$ of $\mathbf{C}$ is characterized by

$$[\mathbf{C}^{-1}] = \frac{1}{1 - \rho^2} \begin{pmatrix} 1/\sigma_X^2 & -\rho/(\sigma_X\sigma_Y) \\ -\rho/(\sigma_X\sigma_Y) & 1/\sigma_Y^2 \end{pmatrix}.$$

Furthermore, $\det(\mathbf{C}) = (1 - \rho^2)\sigma_X^2\sigma_Y^2$. Of course Y is $N(\mu_Y, \sigma_Y^2)$-distributed. Starting from the density function of (X, Y), given in Theorem VIII.4.12, we get:

$$\begin{aligned} &\frac{f_{(X,Y)}(x, y)}{f_Y(y)} \\ &= \frac{[(2\pi)^2 \det(\mathbf{C})]^{-\frac{1}{2}} \exp\left[-\frac{1}{2}\left\langle \mathbf{C}^{-1}(x - \mu_X, y - \mu_Y), (x - \mu_X, y - \mu_Y)\right\rangle\right]}{[2\pi\sigma_Y^2]^{-\frac{1}{2}} \exp\left[-\frac{1}{2}\{(y - \mu_Y)/\sigma_Y\}^2\right]}. \end{aligned}$$

Here

$$\frac{[(2\pi)^2 \det(\mathbf{C})]^{-\frac{1}{2}}}{[2\pi\sigma_Y^2]^{-\frac{1}{2}}} = [2\pi(1 - \rho^2)\sigma_X^2]^{-\frac{1}{2}}.$$

Moreover,

$$(1-\rho^2)\left\langle \mathbf{C}^{-1}(x-\mu_X, y-\mu_Y), (x-\mu_X, y-\mu_Y)\right\rangle$$
$$= \frac{(x-\mu_X)^2}{\sigma_X^2} - \rho\,\frac{(y-\mu_Y)(x-\mu_X)}{\sigma_X\sigma_Y} - \rho\,\frac{(x-\mu_X)(y-\mu_Y)}{\sigma_X\sigma_Y} + \frac{(y-\mu_Y)^2}{\sigma_Y^2}.$$

Hence the argument of the exponential in $f_{(X,Y)}(x,y)/f_Y(y)$ equals

$$-\frac{1}{2(1-\rho^2)}\left\{\left(\frac{x-\mu_X}{\sigma_X}\right)^2 - 2\rho\left(\frac{x-\mu_X}{\sigma_X}\right)\left(\frac{y-\mu_Y}{\sigma_Y}\right) + \rho^2\left(\frac{y-\mu_Y}{\sigma_Y}\right)^2\right\}$$

$$= -\frac{1}{2(1-\rho^2)}\left\{\left(\frac{x-\mu_X}{\sigma_X}\right) - \rho\left(\frac{y-\mu_Y}{\sigma_Y}\right)\right\}^2$$
$$= -\frac{1}{2(1-\rho^2)}\left\{\frac{x-(\mu_X+\rho\frac{\sigma_X}{\sigma_Y}(y-\mu_Y))}{\sigma_X}\right\}^2.$$

From here on it is easily verified that the conditional density

$$f_{X|Y=y}(x) = \frac{f_{(X,Y)}(x,y)}{f_Y(y)}$$

equals the density function of the $N(\mu_X+\rho\frac{\sigma_X}{\sigma_Y}(y-\mu_Y),\ (1-\rho^2)\sigma_X^2)$-distribution.

□

Exercise 18

In Example 2 of §6, a sample $\mathbf{X}_1, \ldots, \mathbf{X}_5$ is drawn from a 3-dimensional population. Using $\mathbf{S}^2$ as an estimator of $\mathbf{C}$, it turns out that a 95% confidence region for $\boldsymbol{\mu}$ is of type

$$G = \left\{\boldsymbol{\mu}\in\mathbb{R}^3 : \frac{5\cdot 2}{3\cdot 4}\left\langle \mathbf{S}^{-2}(\boldsymbol{\mu}-\overline{\mathbf{x}}), (\boldsymbol{\mu}-\overline{\mathbf{x}})\right\rangle < g\right\}$$

for some constant g.

The statistic $\frac{5\cdot 2}{3\cdot 4}\left\langle \mathbf{S}^{-2}(\boldsymbol{\mu}-\overline{\mathbf{x}}), (\boldsymbol{\mu}-\overline{\mathbf{x}})\right\rangle$ a priori enjoys an F-distribution with 3 versus 2 degrees of freedom. The eigenvalues $\lambda_1, \lambda_2, \lambda_3$ and eigenvectors $\mathbf{v}_1, \mathbf{v}_2, \mathbf{v}_3$ for $\mathbf{S}^2$ are given by: $\lambda_1 = 0.331$, $\lambda_2 = 0.607$, $\lambda_3 = 0.031$,

$$\begin{cases} \mathbf{v}_1 = (-0.035, 0.927, -0.373), \\ \mathbf{v}_2 = (0.939, 0.158, 0.304), \\ \mathbf{v}_3 = (-0.341, 0.340, 0.876). \end{cases}$$

Moreover, $\overline{\mathbf{x}} = (4.91, 3.15, 1.31)$. Describe G in 'full detail'.

Solution. The table for the F-distribution teaches us that $g = 19.16$. We note that $\mathbf{S}^2 = \mathbf{Q}^*\mathbf{\Lambda}\mathbf{Q}$ where $\mathbf{Q} : \mathbf{v}_i \mapsto \mathbf{e}_i$ for $i = 1,2,3$ and $\mathbf{\Lambda} = \mathrm{diag}(\lambda_1, \lambda_2, \lambda_3)$: hence $\mathbf{S}^{-2} = \mathbf{Q}^{-1}\mathbf{\Lambda}^{-1}(\mathbf{Q}^*)^{-1} = \mathbf{Q}^*\mathbf{\Lambda}^{-1}\mathbf{Q}$. So

$$\begin{aligned} G &= \{\boldsymbol{\nu} + \overline{\mathbf{x}} \in \mathbb{R}^3 : \langle \mathbf{S}^{-2}\boldsymbol{\nu}, \boldsymbol{\nu}\rangle < \tfrac{6}{5}g\} \\ &= \{\boldsymbol{\nu} \in \mathbb{R}^3 : \langle \mathbf{S}^{-2}\boldsymbol{\nu}, \boldsymbol{\nu}\rangle < \tfrac{6}{5}g\} \; + \; \overline{\mathbf{x}}, \end{aligned}$$

while, as $\boldsymbol{\nu} = \sum_{i=1}^{3} \langle \boldsymbol{\nu}, \mathbf{v}_i\rangle \, \mathbf{v}_i$ for all $\boldsymbol{\nu}$,

$$\begin{aligned} \langle \mathbf{S}^{-2}\boldsymbol{\nu}, \boldsymbol{\nu}\rangle &= \langle \mathbf{\Lambda}^{-1}\mathbf{Q}\boldsymbol{\nu}, \mathbf{Q}\boldsymbol{\nu}\rangle \\ &= \left\langle \mathbf{\Lambda}^{-1}\sum_{i=1}^{3} \langle \boldsymbol{\nu}, \mathbf{v}_i\rangle \, \mathbf{e}_i, \sum_{i=1}^{3} \langle \boldsymbol{\nu}, \mathbf{v}_i\rangle \, \mathbf{e}_i \right\rangle = \sum_{i=1}^{3} \frac{1}{\lambda_i} \langle \boldsymbol{\nu}, \mathbf{v}_i\rangle^2 . \end{aligned}$$

We get:

$$G = \left\{ \sum_{i=1}^{3} \xi_i \mathbf{v}_i \; : \; \sum_{i=1}^{3} \frac{1}{\lambda_i}\xi_i^2 < 22.99 \right\} \; + \; \overline{\mathbf{x}}.$$

This is an ellipsoid with principal axes in the directions $\mathbf{v}_1$, $\mathbf{v}_2$ and $\mathbf{v}_3$. The $\mathbf{v}_i$-axis is of length $\sqrt{22.99\,\lambda_i}$. □

Exercise 19

A sample $\mathbf{X}_1, \ldots, \mathbf{X}_{16}$ is drawn from a $N(\boldsymbol{\mu}, \mathbf{C})$-distributed population of which $\boldsymbol{\mu}$ is unknown, whereas $\mathbf{C}$ is known:

$$[\mathbf{C}] = \begin{pmatrix} 14 & -4 & -4 \\ -4 & 11 & -1 \\ -4 & -1 & 11 \end{pmatrix}.$$

The outcome of $\overline{\mathbf{X}}$ is $\overline{\mathbf{x}} = (2.1, \; 3.8, \; 1.7)$. Construct a 95% confidence region for $\boldsymbol{\mu}$.

Solution. The statistic $16 \langle \mathbf{C}^{-1}(\overline{\mathbf{X}} - \boldsymbol{\mu}), (\overline{\mathbf{X}} - \boldsymbol{\mu})\rangle$ enjoys a χ^2-distribution with 3 degrees of freedom, so the confidence interval in question is (see Exercise 18) of type

$$G = \left\{ \sum_{i=1}^{3} \xi_i \mathbf{v}_i \; : \; \sum_{i=1}^{3} \frac{1}{\lambda_i}\xi_i^2 < \frac{7.81}{16} \right\} + \overline{\mathbf{x}}.$$

Calculation of the eigenvalues $\lambda_1, \lambda_2, \lambda_3$ and the eigenvectors $\mathbf{v}_1, \mathbf{v}_2, \mathbf{v}_3$ is elementary linear algebra:

$$\begin{cases} \lambda_1 = 18, & \mathbf{v}_1 = \frac{1}{6}\sqrt{6}\,(-2, 1, \; 1), \\ \lambda_2 = 12, & \mathbf{v}_2 = \frac{1}{2}\sqrt{2}\,(\;0, \; 1, -1), \\ \lambda_3 = 6, & \mathbf{v}_3 = \frac{1}{3}\sqrt{3}\,(\;1, \; 1, \; 1). \end{cases}$$

□

Exercise 20

Let $\mathbf{X}$ be a $N(\boldsymbol{\mu}, \mathbf{C})$-distributed stochastic 3-vector, for which $\boldsymbol{\mu} = (3, 1, 2)$ and

$$[\mathbf{C}] = \begin{pmatrix} 2 & 3 & 1 \\ 3 & 6 & 5 \\ 1 & 5 & 10 \end{pmatrix}.$$

Apply the method of dimension reduction where the loss of total variance is not allowed to exceed 5%.

Solution. Elementary linear algebra reveals that the eigenvalues and (orthonormal) eigenvectors of $\mathbf{C}$ equal: $\lambda_1 = 7 + 4\sqrt{3}$, $\lambda_2 = 4$, $\lambda_3 = 7 - 4\sqrt{3}$,

$$\mathbf{v}_1 = \frac{1}{\sqrt{6(2-\sqrt{3})}}(2 - \sqrt{3}, \sqrt{3} - 1, 1), \quad \mathbf{v}_2 = \tfrac{1}{3}\sqrt{3}\,(1, 1, -1),$$

and

$$\mathbf{v}_3 = \frac{1}{\sqrt{6(2+\sqrt{3})}}(2 + \sqrt{3}, -\sqrt{3} - 1, 1).$$

For the principal components we now get $Y_1 = \langle \mathbf{X}, \mathbf{v}_1 \rangle, Y_2 = \langle \mathbf{X}, \mathbf{v}_2 \rangle, Y_3 = \langle \mathbf{X}, \mathbf{v}_3 \rangle$ respectively. As $\mathbf{X} = Y_1\mathbf{v}_1 + Y_2\mathbf{v}_2 + Y_3\mathbf{v}_3$ and $\text{var}(Y_3) = \lambda_3 = 0.072$, we look upon Y_3 as being a constant. This implies a loss of variance of $0.072/(\lambda_1 + \lambda_2 + \lambda_3) = 0.072/18 = 0.0040$, that is of 0.40%.

Of course then Y_3 is considered to be equal to

$$\mathbb{E}(\langle \mathbf{X}, \mathbf{v}_3 \rangle) = \langle \boldsymbol{\mu}, \mathbf{v}_3 \rangle = \frac{6 + 3\sqrt{3} - \sqrt{3} - 1 + 2}{\sqrt{6(2+\sqrt{3})}} = \frac{7 + 2\sqrt{3}}{\sqrt{6(2+\sqrt{3})}} = 2.21.$$

So we identify $\mathbf{X}$ with the stochastic vector $\mathbf{X}' = Y_1\mathbf{v}_1 + Y_2\mathbf{v}_2 + 2.21$. □

Exercise 21

Let there be given a cloud of points $(x_1, y_1), \ldots, (x_n, y_n)$, presenting the outcome of a 2-dimensional sample from a population of which both the expectation vector and covariance operator exist. We assume that both the covariance operator and its estimator $\mathbf{S}^2$ are invertible. Suppose that $\mathbf{v} = (v_x, v_y)$ is a normalized eigenvector of $\mathbf{S}^2$ corresponding to its largest eigenvalue. A confidence region for $\boldsymbol{\mu}$ as constructed in the way of §6 has the form of an ellips of which the largest axis has direction $\mathbf{v}$.

Question. Does the straight line, given by

$$\{(\overline{x}, \overline{y}) + t\,(v_x, v_y) \;:\; t \in \mathbb{R}\}$$

necessarily coincide with the 'least squares fit', as described in Chapter IV, belonging to the sample?

Solution. No. To see this, we return to §IV.6, Exercise 9. Here we have a cloud of points given by

$$(1,1),\ (2,1),\ (2,3),\ (3,2),\ (4,3).$$

We already noted that $(\overline{x},\overline{y}) = (\frac{12}{5},2)$ and that the least squares fit is given by the equation

$$y = \tfrac{8}{13} + \tfrac{15}{26}x.$$

Of course this line visits the point $(\overline{x},\overline{y})$, so we wonder whether its slope is the same as that of the line mentioned in the question. That is, we wonder whether $v_y/v_x = \frac{15}{26}$. For this reason we now determine (v_x, v_y).

Using some of the computational rules we derived in Exercise 2, we see that

$$\begin{aligned}
(5-1)\,\mathbf{S}^2 &= \sum_{i=1}^{5}(\mathbf{x}_i - \overline{\mathbf{x}})\otimes(\mathbf{x}_i - \overline{\mathbf{x}}) \\
&= \sum_{i=1}^{5}\mathbf{x}_i\otimes\mathbf{x}_i \;-\; \left(\sum_{i=1}^{5}\mathbf{x}_i\right)\otimes\overline{\mathbf{x}} \;-\; \overline{\mathbf{x}}\otimes\left(\sum_{i=1}^{5}(\mathbf{x}_i - \overline{\mathbf{x}})\right) \\
&= \sum_{i=1}^{5}\mathbf{x}_i\otimes\mathbf{x}_i \;-\; 5\,\overline{\mathbf{x}}\otimes\overline{\mathbf{x}}
\end{aligned}$$

which leads to

$$\begin{aligned}
4\,\mathbf{S}^2 &= \begin{pmatrix}1 & 1\\ 1 & 1\end{pmatrix} + \begin{pmatrix}4 & 2\\ 2 & 1\end{pmatrix} + \begin{pmatrix}4 & 6\\ 6 & 9\end{pmatrix} + \begin{pmatrix}9 & 6\\ 6 & 4\end{pmatrix} \\
&\quad + \begin{pmatrix}16 & 12\\ 12 & 9\end{pmatrix} - 5\begin{pmatrix}144/25 & 24/5\\ 24/5 & 4\end{pmatrix} \\
&= \begin{pmatrix}26/5 & 3\\ 3 & 4\end{pmatrix}.
\end{aligned}$$

The eigenvalues of $4\,\mathbf{S}^2$ are given by $\lambda_{1,2} = \frac{23}{5} \pm \frac{1}{10}\sqrt{936}$; as a consequence the eigenvalues of $\mathbf{S}^2$ are given by $\mu_{1,2} = \frac{23}{20} \pm \frac{1}{40}\sqrt{936}$. The eigenvector corresponding to $\mu_1 = \frac{23}{20} + \frac{1}{40}\sqrt{936}$ equals $(v_x, v_y) = (0.7733,\ 0.6340)$. This leads to a slope of $0.8199 \neq \frac{15}{26}$. So we have a counterexample. □

Exercise 22

Suppose $\mathbf{X}$ is as in Exercise 20; let $\mathfrak{M}$ be the linear span of $(1,1,1)$. Determine $\mathbb{P}(\|\mathbf{P}_{\mathfrak{M}}\mathbf{X}\| \geq 1)$.

Solution. Set $\mathbf{v} := \frac{1}{3}\sqrt{3}\ (1,1,1)$. Then (see Exercise 1 (iv))

$$\mathbf{P}_{\mathfrak{M}}\mathbf{X} = (\mathbf{v}\otimes\mathbf{v})\mathbf{X} = \langle \mathbf{X}, \mathbf{v}\rangle\,\mathbf{v}$$

and therefore $\|\mathbf{P}_{\mathfrak{M}}\mathbf{X}\| = |\langle \mathbf{X}, \mathbf{v}\rangle|$. Using Theorem VIII.4.5 (or §4, Example 2), we see that $\langle \mathbf{X}, \mathbf{v}\rangle$ enjoys a $N(\langle \boldsymbol{\mu}, \mathbf{v}\rangle, \langle \mathbf{C}\mathbf{v}, \mathbf{v}\rangle)$-distribution. That is, $\langle \mathbf{X}, \mathbf{v}\rangle$ is $N(2\sqrt{3}, 12)$-distributed, so that

$$\begin{aligned} \mathbb{P}(\|\mathbf{P}_{\mathfrak{M}}\mathbf{X}\| \geq 1) &= \mathbb{P}(|\langle \mathbf{X}, \mathbf{v}\rangle| \geq 1) \\ &= 1 - \mathbb{P}\left(\frac{-1-2\sqrt{3}}{\sqrt{12}} < \frac{\langle \mathbf{X}, \mathbf{v}\rangle - 2\sqrt{3}}{\sqrt{12}} < \frac{1-2\sqrt{3}}{\sqrt{12}}\right) \\ &= 1 - \left(\Phi\left(\frac{1-2\sqrt{3}}{\sqrt{12}}\right) - \Phi\left(\frac{-1-2\sqrt{3}}{\sqrt{12}}\right)\right) = 0.86. \end{aligned}$$

□

Exercise 23

Prove that for the stochastic p-vectors $\mathbf{X}, \mathbf{X}_1, \mathbf{X}_2, \ldots$ the following statements are equivalent:

(i) $\mathbf{X}_n \to \mathbf{X}$ almost surely,

(ii) For all $\mathbf{v} \in \mathbb{R}^p$: $\langle \mathbf{X}_n, \mathbf{v}\rangle \to \langle \mathbf{X}, \mathbf{v}\rangle$ almost surely,

(iii) For all $i = 1, \ldots, p$: $\langle \mathbf{X}_n, \mathbf{e}_i\rangle \to \langle \mathbf{X}, \mathbf{e}_i\rangle$ almost surely.

Proof. For any arbitrary $\omega \in \Omega$ we have:

$$\mathbf{X}_n(\omega) \to \mathbf{X}(\omega) \quad \Leftrightarrow \quad \langle \mathbf{X}_n(\omega), \mathbf{e}_i\rangle \to \langle \mathbf{X}(\omega), \mathbf{e}_i\rangle \quad \text{for all } i = 1, \ldots, p,$$

as convergence in $\mathbb{R}^p$ is the same as convergence in Cartesian coordinates. It is easy to check that this is also equivalent to

$$\langle \mathbf{X}_n(\omega), \mathbf{v}\rangle \to \langle \mathbf{X}(\omega), \mathbf{v}\rangle \quad \text{for all } \mathbf{v} \in \mathbb{R}^p$$

(each $\mathbf{v}$ can be decomposed as $\sum_i v_i \mathbf{e}_i$ and each $\mathbf{e}_i$ can be considered as a $\mathbf{v}$ respectively). So

$$\begin{aligned} \{\omega \in \Omega : \mathbf{X}_n(\omega) \to \mathbf{X}(\omega)\} &= \{\omega \in \Omega : \langle \mathbf{X}_n(\omega), \mathbf{v}\rangle \to \langle \mathbf{X}(\omega), \mathbf{v}\rangle \quad \text{for each } \mathbf{v}\} \\ &= \{\omega \in \Omega : \langle \mathbf{X}_n(\omega), \mathbf{e}_i\rangle \to \langle \mathbf{X}(\omega), \mathbf{e}_i\rangle \quad \text{for all } i\} \end{aligned}$$

and the statement follows. □

Exercise 24

Prove that for the stochastic p-vectors $\mathbf{X}, \mathbf{X}_1, \mathbf{X}_2, \ldots$ the following statements are equivalent:

(i) $\mathbf{X}_n \to \mathbf{X}$ in probability,

(ii) For all $\mathbf{v} \in \mathbb{R}^p$: $\langle \mathbf{X}_n, \mathbf{v}\rangle \to \langle \mathbf{X}, \mathbf{v}\rangle$ in probability,

(iii) For all $i = 1, \dots, p$: $\langle \mathbf{X}_n, \mathbf{e}_i \rangle \to \langle \mathbf{X}, \mathbf{e}_i \rangle$ in probability.

Proof.

(i) $\Rightarrow$ (ii)
Suppose that (i) applies. We take any arbitrary $\mathbf{v} \neq \mathbf{0}$ from $\mathbb{R}^p$, $n \in \{1, 2, \dots\}$ and $\varepsilon > 0$. Setting $V := \|\mathbf{v}\|$, by the Cauchy–Schwarz inequality we have $|\langle \mathbf{X}_n - \mathbf{X}, \mathbf{v} \rangle| \leq V \|\mathbf{X}_n - \mathbf{X}\|$, so

$$\begin{aligned} \mathbb{P}(|\langle \mathbf{X}_n, \mathbf{v} \rangle - \langle \mathbf{X}, \mathbf{v} \rangle| \geq \varepsilon) &= \mathbb{P}(|\langle \mathbf{X}_n - \mathbf{X}, \mathbf{v} \rangle| \geq \varepsilon) \\ &\leq \mathbb{P}(\|\mathbf{X}_n - \mathbf{X}\| \geq \varepsilon / V), \end{aligned}$$

which by (i) converges to zero for $n \to \infty$. So $\langle \mathbf{X}_n, \mathbf{v} \rangle \to \langle \mathbf{X}, \mathbf{v} \rangle$ in probability. For $\mathbf{v} = \mathbf{0}$ it is evident that (ii) holds.

(ii) $\Rightarrow$ (iii)
Trivial: (iii) is a special case of (ii).

(iii) $\Rightarrow$ (i)
Suppose (iii) applies. For all n we have

$$\|\mathbf{X}_n - \mathbf{X}\|^2 = \sum_{i=1}^{p} |\langle \mathbf{X}_n - \mathbf{X}, \mathbf{e}_i \rangle|^2 \leq p \max_{i=1,\dots,p} |\langle \mathbf{X}_n - \mathbf{X}, \mathbf{e}_i \rangle|^2,$$

so that for any $\varepsilon > 0$

$$\begin{aligned} \mathbb{P}(\|\mathbf{X}_n - \mathbf{X}\| \geq \varepsilon) &\leq \mathbb{P}\left(p \max_{i=1,\dots,p} |\langle \mathbf{X}_n - \mathbf{X}, \mathbf{e}_i \rangle|^2 \geq \varepsilon^2\right) \\ &\leq \sum_{i=1}^{p} \mathbb{P}(|\langle \mathbf{X}_n, \mathbf{e}_i \rangle - \langle \mathbf{X}, \mathbf{e}_i \rangle| \geq \varepsilon / \sqrt{p}). \end{aligned}$$

As (iii) applies, the latter expression tends to zero for $n \to \infty$. Thus we see that (iii) implies (i). □

Exercise 25

Give an example of two sequences $X_1, X_2, \dots$ and $Y_1, Y_2, \dots$ of variables that converge in distribution to variables X and Y, for which *not* $X_n + Y_n \to X + Y$ in distribution. The same with respect to multiplication.

Solution. We choose a variable Z which enjoys a standard normal distribution and take the sequences $Z, -Z, Z, -Z, \dots$ and $-Z, Z, -Z, Z, \dots$, that is, we set $X_k := (-1)^{k+1} Z$ and $Y_k := (-1)^k Z$ for $k = 1, 2, \dots$.

Of course $X_n \to Z$ and $Y_n \to Z$ in distribution and *not* $0 \equiv X_n + Y_n \to Z + Z = 2Z$ in distribution. Moreover, $X_n Y_n = -Z^2$ for all n which most certainly doesn't converge to $ZZ = Z^2$ either. □

Exercise 26

Prove that if $X_n \to X$ and $Y_n \to a$ in distribution, where $a \in \mathbb{R}$ is a certain constant, then also $X_n Y_n \to aX$ in distribution.

Proof. First we note that if $Z_n \to Z$ in distribution, then for *all* $z \in \mathbb{R}$ one has

$$F_Z(z-) \le \underline{\lim}_{n\to\infty} F_{Z_n}(z) \le \overline{\lim}_{n\to\infty} F_{Z_n}(z) \le F_Z(z). \qquad (*)$$

(For any $\varepsilon > 0$ we have some continuity point $y \in (z-\varepsilon, z)$ of F_Z, so that

$$F_Z(z-\varepsilon) \le F_Z(y) = \underline{\lim}_{n\to\infty} F_{Z_n}(y) \le \underline{\lim}_{n\to\infty} F_{Z_n}(z).$$

Letting $\varepsilon \downarrow 0$ we see that $F_Z(z-) \le \underline{\lim}_{n\to\infty} F_{Z_n}(z)$; etcetera.)

As a first step we consider the case where $X_n \to X$ and $Y_n \to 0$ in distribution. By Theorem VIII.7.4 we then have that $|X_n| \to |X|$ and $|Y_n| \to 0$ in distribution. We set $W_n := |Y_n X_n|$ for all n and we choose $w > 0$ and $\varepsilon > 0$ arbitrarily. Then

$$\begin{aligned} F_{W_n}(w) &= \mathbb{P}(|Y_n X_n| \le w) \ge \mathbb{P}(|X_n| \le \tfrac{w}{\varepsilon} \text{ and } |Y_n| \le \varepsilon) \\ &\ge 1 - \mathbb{P}(|X_n| > \tfrac{w}{\varepsilon}) - \mathbb{P}(|Y_n| > \varepsilon) \\ &= 1 - (1 - F_{|X_n|}(\tfrac{w}{\varepsilon})) - (1 - F_{|Y_n|}(\varepsilon)) \\ &= F_{|X_n|}(\tfrac{w}{\varepsilon}) + F_{|Y_n|}(\varepsilon) - 1. \end{aligned}$$

It follows from this (we set $Y := 0$ and apply $(*)$) that

$$\underline{\lim}_{n\to\infty} F_{W_n}(w) \ge F_{|X|}\left((\tfrac{w}{\varepsilon})-\right) + F_{|Y|}(\varepsilon-) - 1 = F_{|X|}((\tfrac{w}{\varepsilon})-).$$

Letting $\varepsilon \downarrow 0$ we see that $\underline{\lim}_{n\to\infty} F_{W_n}(w) \ge 1$ for all $w > 0$. Of course this implies that

$$\lim_{n\to\infty} F_{W_n}(w) = 1 \quad \text{for all } w > 0.$$

As $F_{W_n}(w) = 0$ for all $w < 0$, of course

$$\lim_{n\to\infty} F_{W_n}(w) = 0 \quad \text{for all } w < 0.$$

It follows that $W_n = |X_n Y_n| \to 0$ in distribution. Of course this is only possible if $X_n Y_n \to 0$ in distribution, which concludes the proof in the case $a = 0$.

The general case, where $Y_n \to a$ and where a is not necessarily zero, can easily be reduced to the above by setting

$$Y_n X_n = (Y_n - a) X_n + a X_n.$$

Here $Y_n - a \to 0$ and therefore (by the foregoing) $(Y_n - a) X_n \to 0$ in distribution. Furthermore we have that $a X_n \to aX$ in distribution. Applying Slutsky's theorem for the additive case we see that $Y_n X_n \to aX$ in distribution. □

Exercise 27

Prove that, starting from a population with distribution function F,

$$\sqrt{n}\ \{S_n^2 - \sigma_F^2\} \to N(0, \sigma_F^2)$$

in distribution, where σ_F^2 is the variance of the population.

Proof. In Example 1 of §7 it was proved that

$$\sqrt{n}\ \{\tfrac{n-1}{n} S_n^2 - \sigma_F^2\} \to N(0, \sigma_F^2)$$

in distribution. Multiplying this with a factor $\frac{n}{n-1}$, by Slutsky's theorem (the multiplicative case) we see that

$$\sqrt{n}\ \left\{S_n^2 - \tfrac{n}{n-1}\sigma_F^2\right\} \to N(0, \sigma_F^2)$$

in distribution. Note that

$$\sqrt{n}\ \{S_n^2 - \sigma_F^2\} = \sqrt{n}\ \left\{S_n^2 - \tfrac{n}{n-1}\sigma_F^2\right\} + \tfrac{\sqrt{n}}{n-1}\ \sigma_F^2,$$

where $\frac{\sqrt{n}}{n-1}\,\sigma_F^2 \to 0$ in distribution. The statement now follows from the above, together with Slutsky's theorem (additive case). □

Exercise 28

Using the technique of partial differentiation, find the maximum likelihood estimators for $\mu_X, \mu_Y, \sigma_X, \sigma_Y$ and ρ when sampling from a population that enjoys a 2-dimensional normal distribution.

Solution. We look at the function $\boldsymbol{\theta} \mapsto \log L_{\boldsymbol{\theta}}$ which, given a sample outcome $(x_1, y_1), \ldots, (x_n, y_n)$, converts $\boldsymbol{\theta} = (\mu_X, \mu_Y, \sigma_X, \sigma_Y, \rho)$ into

$$-n\log 2\pi - n\log\sigma_X - n\log\sigma_Y - \tfrac{1}{2}n\log(1-\rho^2)$$
$$-\frac{1}{2(1-\rho^2)}\sum_{i=1}^{n}\left\{\left(\frac{x_i-\mu_X}{\sigma_X}\right)^2 - 2\rho\left(\frac{x_i-\mu_X}{\sigma_X}\right)\left(\frac{y_i-\mu_Y}{\sigma_Y}\right) + \left(\frac{y_i-\mu_Y}{\sigma_Y}\right)^2\right\}$$

and we are looking for the point $\boldsymbol{\theta}$ in which the maximum value is attained. Of course such a $\boldsymbol{\theta}$ necessarily satisfies

$$\frac{\partial \log L_{\boldsymbol{\theta}}}{\partial \mu_X} = \frac{\partial \log L_{\boldsymbol{\theta}}}{\partial \mu_Y} = \frac{\partial \log L_{\boldsymbol{\theta}}}{\partial \sigma_X} = \frac{\partial \log L_{\boldsymbol{\theta}}}{\partial \sigma_Y} = \frac{\partial \log L_{\boldsymbol{\theta}}}{\partial \rho} = 0. \tag{1}$$

For μ_X this gives us

$$-\frac{1}{2(1-\rho^2)}\sum_{i=1}^{n}\left\{2\left(\frac{x_i-\mu_X}{\sigma_X}\right)\cdot\frac{-1}{\sigma_X} - 2\rho\left(\frac{y_i-\mu_Y}{\sigma_Y}\right)\cdot\frac{-1}{\sigma_X}\right\} = 0,$$

or

$$\sum_{i=1}^{n} \frac{2}{\sigma_X^2}(x_i - \mu_X) = \sum_{i=1}^{n} \frac{2\rho}{\sigma_X \sigma_Y}(y_i - \mu_Y).$$

In turn we get

$$n\overline{x} - n\mu_X = \sum_{i=1}^{n} \frac{\rho\sigma_X}{\sigma_Y}(y_i - \mu_Y) = \frac{\rho\sigma_X}{\sigma_Y} n(\overline{y} - \mu_Y)$$

so that

$$\mu_X = \overline{x} - \frac{\rho\sigma_X}{\sigma_Y}(\overline{y} - \mu_Y). \tag{2}$$

By symmetry

$$\mu_Y = \overline{y} - \frac{\rho\sigma_Y}{\sigma_X}(\overline{x} - \mu_X).$$

Substitution of this form into (2) gives $\mu_X = (1 - \rho^2)\overline{x} + \rho^2\mu_X$ or $\mu_X = \overline{x}$; by symmetry then also $\mu_Y = \overline{y}$.

Concerning σ_X, differentiation leads to

$$-\frac{n}{\sigma_X} - \frac{1}{2(1-\rho^2)} \sum_{i=1}^{n} \left\{ -\frac{2(x_i - \mu_X)^2}{\sigma_X^3} + \frac{2\rho(x_i - \mu_X)(y_i - \mu_Y)}{\sigma_Y \sigma_X^2} \right\} = 0$$

which can also be read as

$$\sum_{i=1}^{n} \{(x_i - \overline{x})^2 - \rho(x_i - \overline{x})(y_i - \overline{y})\sigma_Y^{-1}\sigma_X\} = n(1-\rho^2)\sigma_X^2$$

or (see §IV.1 for the notation)

$$\frac{S_{\mathbf{xx}}}{\sigma_X^2} - \rho\frac{S_{\mathbf{xy}}}{\sigma_X \sigma_Y} = n(1 - \rho^2). \tag{3}$$

In the same way

$$\frac{S_{\mathbf{yy}}}{\sigma_Y^2} - \rho\frac{S_{\mathbf{xy}}}{\sigma_X \sigma_Y} = n(1 - \rho^2). \tag{4}$$

From (1), for ρ we get

$$-\tfrac{1}{2}n \cdot \frac{1}{1-\rho^2} \cdot -2\rho + \frac{-2\rho}{2(1-\rho^2)^2} \sum_{i=1}^{n} \{\ldots\}$$
$$- \frac{1}{2(1-\rho^2)} \sum_{i=1}^{n} -2\left(\frac{x_i - \mu_X}{\sigma_X}\right)\left(\frac{y_i - \mu_Y}{\sigma_Y}\right) = 0,$$

or:

$$n\rho - \frac{\rho}{1-\rho^2}\left\{\frac{S_{\mathbf{xx}}}{\sigma_X^2} - 2\rho\frac{S_{\mathbf{xy}}}{\sigma_X\sigma_Y} + \frac{S_{\mathbf{yy}}}{\sigma_Y^2}\right\} + \frac{S_{\mathbf{xy}}}{\sigma_X\sigma_Y} = 0.$$

Addition of (3) and (4) gives us

$$\frac{S_{\mathbf{xx}}}{\sigma_X^2} - 2\rho\frac{S_{\mathbf{xy}}}{\sigma_X\sigma_Y} + \frac{S_{\mathbf{yy}}}{\sigma_Y^2} = 2n(1-\rho^2)\ ;$$

this reduces the last equality to

$$n\rho - 2n\rho + \frac{S_{\mathbf{xy}}}{\sigma_X\sigma_Y} = 0, \quad \text{or:} \quad \rho = \frac{S_{\mathbf{xy}}}{n\sigma_X\sigma_Y}.$$

Substituting this into (3), we get

$$\frac{S_{\mathbf{xx}}}{\sigma_X^2} - \frac{S_{\mathbf{xy}}^2}{n\sigma_X^2\sigma_Y^2} = n - \frac{S_{\mathbf{xy}}^2}{n\sigma_X^2\sigma_Y^2}$$

or $\sigma_X^2 = \frac{1}{n}S_{\mathbf{xx}}$. In the same way $\sigma_Y^2 = \frac{1}{n}S_{\mathbf{yy}}$, and finally

$$\rho = \frac{S_{\mathbf{xy}}}{n(\frac{1}{n}S_{\mathbf{xx}})^{\frac{1}{2}}(\frac{1}{n}S_{\mathbf{yy}})^{\frac{1}{2}}} = \frac{S_{\mathbf{xy}}}{\sqrt{S_{\mathbf{xx}}}\sqrt{S_{\mathbf{yy}}}}.$$

By now we know the optimal $\boldsymbol{\theta}$ completely. □

Exercise 29

Let X be a stochastic variable that is t-distributed with n degrees of freedom. Prove that the variable

$$Y := \frac{X}{\sqrt{X^2+n}}$$

has a density function given by

$$f_Y(y) = \frac{\Gamma(\frac{n+1}{2})}{\sqrt{\pi}\,\Gamma(\frac{n}{2})}\,(1-y^2)^{\frac{1}{2}n-1}$$

if $|y| < 1$ and $f_Y(y) = 0$ elsewhere.

Solution. We use §I.11, Exercise 33. We can write Y as $\varphi(X)$ where for all $x \in \mathbb{R}$ we set

$$\varphi : x \;\mapsto\; \frac{x}{\sqrt{x^2+n}}.$$

Here we have

$$\varphi'(x) = \frac{\sqrt{x^2+n} - x^2/\sqrt{x^2+n}}{x^2+n} = \frac{x^2+n-x^2}{(x^2+n)^{\frac{3}{2}}} = \frac{n}{(x^2+n)^{\frac{3}{2}}},$$

so φ is certainly continuously differentiable, as well as strictly increasing.

We note that in the notation of the mentioned Exercise 33 we have $\varphi(-\infty) = -1$ and $\varphi(+\infty) = 1$. The inverse function φ^{-1} is given by

$$\varphi^{-1}(x) = \frac{\sqrt{n}\, x}{\sqrt{1-x^2}}.$$

Putting all pieces together we obtain:

$$\begin{aligned}
f_Y(y) &= \frac{\frac{\Gamma(\frac{n+1}{2})}{\sqrt{\pi n}\,\Gamma(\frac{n}{2})}\left(1+\frac{\varphi^{-1}(y)^2}{n}\right)^{-\frac{1}{2}(n+1)}}{\varphi'(\varphi^{-1}(y))} \\
&= \frac{(\varphi^{-1}(y)^2+n)^{\frac{3}{2}}}{n}\;\frac{\Gamma(\frac{n+1}{2})}{\sqrt{\pi n}\,\Gamma(\frac{n}{2})}\left(1+\frac{\varphi^{-1}(y)^2}{n}\right)^{-\frac{1}{2}(n+1)} \\
&= \frac{n^{\frac{3}{2}}\,(\frac{y^2}{1-y^2}+1)^{\frac{3}{2}}}{n}\;\frac{\Gamma(\frac{n+1}{2})}{\sqrt{\pi n}\,\Gamma(\frac{n}{2})}\left(1+\frac{y^2}{1-y^2}\right)^{-\frac{1}{2}(n+1)} \\
&= \frac{\Gamma(\frac{n+1}{2})}{\sqrt{\pi}\,\Gamma(\frac{n}{2})}\,(1-y^2)^{-\frac{3}{2}+\frac{1}{2}(n+1)} = \frac{\Gamma(\frac{n+1}{2})}{\sqrt{\pi}\,\Gamma(\frac{n}{2})}\,(1-y^2)^{\frac{1}{2}n-1}
\end{aligned}$$

for $y \in (-1,1)$ and $f_Y(y) = 0$ elsewhere; which is what we had to prove. □

Exercise 30

A sample of size 25 from a 2-dimensional normally distributed population provides $R_P = -0.16$. Test

$$H_0 : \rho = 0 \quad \text{versus} \quad H_1 : \rho \neq 0$$

at a level of significance of 5%.

Solution. By Proposition VIII.8.4 the statistic

$$T = \frac{\sqrt{23}\, R_P}{\sqrt{1-R_P^2}}$$

enjoys a t-distribution with 23 degrees of freedom. The test is two-sided, so we want to know whether $\mathbb{P}(|R_P| \geq 0.16) \geq 0.05$. As the function

$$f : x \mapsto \frac{\sqrt{23}\, x}{\sqrt{1-x^2}}$$

is strictly increasing on $(-1,1)$ (see Exercise 29), we may write

$$|R_P| \geq 0.16 \iff \left|\frac{\sqrt{23}\, R_P}{\sqrt{1-R_P^2}}\right| \geq \frac{\sqrt{23}\cdot 0.16}{\sqrt{1-0.16^2}} = 0.777.$$

Therefore

$$P\text{-value} = \mathbb{P}(|R_P| \geq 0.16) = \mathbb{P}(|T| \geq 0.777) \approx 0.4$$

where T enjoys a t-distribution with 23 degrees of freedom. We accept H_0. □

Exercise 31

A sample of size 50 from a 2-dimensional normally distributed population provides $R_P = 0.70$. Test

$$H_0 : \rho = 0.90 \quad \text{versus} \quad H_1 : \rho < 0.90$$

at a level of significance of 5%.

Solution. As $n = 50$ is rather large, we may use Fisher's Z. That is to say, we use

$$Z = \tfrac{1}{2}\log\left(\frac{1+R_P}{1-R_P}\right),$$

as our test statistic (see also Proposition VIII.8.7). In this case an outcome $Z = 0.867$ is observed. A priori Z enjoyed a $N(\mu, \sigma^2)$-distribution with

$$\mu = \tfrac{1}{2}\log\left(\frac{1+0.90}{1-0.90}\right) = 1.472 \quad \text{and} \quad \sigma^2 = \tfrac{1}{47}.$$

Defining $h : (-1, 1) \to \mathbb{R}$ by

$$h(x) \ := \ \tfrac{1}{2}\log\left(\frac{1+x}{1-x}\right),$$

we have $h'(x) = (1-x^2)^{-1}$, so h is strictly increasing. Consequently the P-value corresponding to the given outcome is

$$\mathbb{P}^{H_0}(R_P \leq 0.70) = \mathbb{P}^{H_0}(Z \leq 0.867) = \mathbb{P}^{H_0}\left(\frac{Z-1.472}{1/\sqrt{47}} \leq -4.148\right) = 0.000.$$

Therefore we reject H_0. □

Exercise 32

Suppose $(X_1, Y_1), \ldots, (X_n, Y_n)$ is a sample from a 2-dimensional normally distributed population with $\rho = 0$. For any non-constant sequence $x_1, \ldots, x_n \in \mathbb{R}$, simply by setting $\beta := 0$ and $\alpha := \mu_Y$, we can say that for all i the variable Y_i is $N(\alpha + \beta x_i, \sigma_Y^2)$-distributed. In that way the sequence $Y_1, \ldots, Y_n$ satisfies the conditions of normal analysis of regression as treated in §IV.3.

(i) Prove that

$$\frac{\sqrt{n-2}\,(R_P)_{\mathbf{x}}}{\sqrt{1-(R_P)_{\mathbf{x}}^2}} = \frac{(\hat{\beta}-\beta)\sqrt{(n-2)S_{\mathbf{xx}}}}{\sqrt{\text{SSE}}}.$$

Proof. Suppose the outcome of $Y_1, \ldots, Y_n$ is given by $y_1, \ldots, y_n$. Now $(x_1, y_1), \ldots, (x_n, y_n)$ can be looked upon as a cloud of points, to which there is the associated Pearson correlation coefficient $(R_P)_{\mathbf{x}}$. By Proposition IV.5.2 we have SSE $= (1-(R_P)^2_{\mathbf{x}})\, S_{\mathbf{yy}}$; by Proposition IV.1.2 $\hat{\beta} = S_{\mathbf{xy}}/S_{\mathbf{xx}}$. The equality to be verified now appears to be equivalent to

$$(R_P)_{\mathbf{x}} = \frac{S_{\mathbf{xy}}}{\sqrt{S_{\mathbf{xx}} S_{\mathbf{yy}}}},$$

which clearly holds. □

(ii) Proposition IV.3.3 states that under the stated conditions the variable

$$\frac{(\hat{\beta}-\beta)\sqrt{(n-2)S_{\mathbf{xx}}}}{\sqrt{\text{SSE}}}$$

enjoys a t-distribution with $n-2$ degrees of freedom. Deduce from this that the variable

$$\frac{\sqrt{n-2}\, R_P}{\sqrt{1-R_P^2}}$$

shares this distribution (Proposition VIII.8.4).

Solution. Just apply (i) together with Lemma VIII.8.3 and Proposition IV.3.3. □

Exercise 33

Again we start from a 2-dimensional normally distributed population, for which $\rho = 0$. A sample $(X_1, Y_1), \ldots, (X_n, Y_n)$ is drawn. The outcome of R_P^2 is surely an element of $[0,1]$. Is R_P^2 beta distributed?

Solution. We determine the probability density of R_P^2. For $z \in [0,1]$ (any other z is out of the question)

$$\begin{aligned}
\mathbb{P}\left(R_P^2 \le z\right) &= \mathbb{P}\left(-\sqrt{z} \le R_P \le \sqrt{z}\right) \\
&= \int_{-\sqrt{z}}^{\sqrt{z}} \frac{\Gamma(\frac{1}{2}(n-1))}{\sqrt{\pi}\,\Gamma(\frac{1}{2}(n-2))} (1-s^2)^{\frac{1}{2}n-2}\, ds \\
&= \int_0^{\sqrt{z}} \frac{2\,\Gamma(\frac{1}{2}(n-1))}{\Gamma(\frac{1}{2})\Gamma(\frac{1}{2}(n-2))} (1-s^2)^{\frac{1}{2}n-2}\, ds \\
(t := s^2) \quad &= \int_0^{z} \frac{\Gamma(\frac{1}{2}n-\frac{1}{2})}{\Gamma(\frac{1}{2})\Gamma(\frac{1}{2}n-1)}\, t^{-\frac{1}{2}}(1-t)^{\frac{1}{2}n-2}\, dt.
\end{aligned}$$

Hence

$$f_{R_P^2}(z) = \frac{\Gamma(\frac{1}{2}n-\frac{1}{2})}{\Gamma(\frac{1}{2})\Gamma(\frac{1}{2}n-1)}\, z^{-\frac{1}{2}}(1-z)^{\frac{1}{2}n-2} \quad \text{for } z \in [0,1]$$

and $f_{R_P^2}(z) = 0$ elsewhere, so that (Definition II.6.2) indeed R_P^2 enjoys a beta distribution, with parameters $p = \frac{1}{2}$ and $q = \frac{1}{2}n - 1$. □

Exercise 34

Prove that in the case of normal multiple analysis of regression the $(m+1)$-vector $(\hat{\alpha}, \hat{\boldsymbol{\beta}})$ and the variable SSE are statistically independent.

Proof. Let $\mathfrak{M}$ be the linear subspace of $\mathbb{R}^n$ spanned by $\mathbf{e}, \mathbf{x}_1, \dots, \mathbf{x}_m$.

For the linear operator corresponding to the matrix $\mathbf{M}$ (as defined in §9) we write $\mathbf{S}$; $\mathbf{S}$ is injective and therefore invertible. Define φ by $\varphi : \mathbf{y} \mapsto \|\mathbf{y}\|^2$ $(\mathbf{y} \in \mathbb{R}^n)$ and let $\psi : \mathfrak{M} \to \mathbb{R}^{m+1}$ be the map defined by

$$\psi(a\mathbf{e} + b_1\mathbf{x}_1 + \dots + b_m\mathbf{x}_m) := (a, b_1, \dots, b_m).$$

Since

$$\mathrm{SSE} = \varphi((\mathbf{I} - \mathbf{P}_{\mathfrak{M}})\mathbf{Y}) \quad \text{and} \quad (\hat{\alpha}, \hat{\boldsymbol{\beta}}) = \psi^{-1}(\mathbf{P}_{\mathfrak{M}}\mathbf{Y}),$$

it will suffice to show that $(\mathbf{I}-\mathbf{P}_{\mathfrak{M}})\mathbf{Y}$ and $\mathbf{P}_{\mathfrak{M}}\mathbf{Y}$ are independent. To this, we define $\tilde{\mathbf{Y}} := (\mathbf{Y} - \mathbb{E}(\mathbf{Y}))/\sigma$, which clearly is $N(\mathbf{0}, \mathbf{I}_n)$-distributed. By Cochran's theorem $\mathbf{P}_{\mathfrak{M}}\tilde{\mathbf{Y}}$ and $(\mathbf{I} - \mathbf{P}_{\mathfrak{M}})\tilde{\mathbf{Y}}$ are statistically independent. Therefore, by Proposition I.4.2, so are $(\mathbf{I} - \mathbf{P}_{\mathfrak{M}})\mathbf{Y}$ and $\mathbf{P}_{\mathfrak{M}}\mathbf{Y}$. □

Exercise 35

(i) Prove that in the case of normal regression analysis the statistic

$$\frac{(\hat{\beta}_i - \beta_i)\sqrt{n-m-1}}{\sqrt{\mathrm{var}(\hat{\beta}_i)}\,\sqrt{\mathrm{SSE}/\sigma^2}}$$

enjoys a t-distribution with $n-m-1$ degrees of freedom.

Proof. Of course $(\hat{\beta}_i - \beta_i)/\sqrt{\mathrm{var}(\hat{\beta}_i)}$ is $N(0,1)$-distributed (by Proposition VIII.9.2 we have $\mathbb{E}(\hat{\beta}_i) = \beta_i$). Moreover, SSE/σ^2 is χ^2-distributed with $n-m-1$ degrees of freedom (Theorem VIII.9.9). Furthermore, by Exercise 34 the statistics $(\hat{\alpha}, \hat{\boldsymbol{\beta}})$ and SSE are statistically independent. For this reason the variables

$$(\hat{\beta}_i - \beta_i)/\sqrt{\mathrm{var}(\hat{\beta}_i)} \quad \text{and} \quad \sqrt{\mathrm{SSE}/\sigma^2}/\sqrt{n-m-1}$$

are also independent. The quotient of these variables is therefore (Proposition II.3.1) t-distributed with $n-m-1$ degrees of freedom. □

(ii) Prove that

$$\mathrm{var}(\hat{\beta}_i) = [(i+1)^{\text{th}} \text{ diagonal element of } (\mathbf{M}^t\mathbf{M})^{-1}]\,\sigma^2.$$

Proof. This immediately follows from the fact that (see Theorem VIII.9.9 (i)) $[\mathcal{C}(\hat{\alpha}, \hat{\boldsymbol{\beta}})] = \sigma^2(\mathbf{M}^t\mathbf{M})^{-1}$. □

(iii) Indicate how (i) and (ii) can be used to perform hypothesis tests on β_i and to construct confidence intervals for β_i.

Solution. By (i) and (ii) the statistic

$$T = \frac{(\hat{\beta}_i - \beta_i)\,\sqrt{n-m-1}}{\sqrt{[(\mathbf{M}^t\mathbf{M})^{-1}]_{i+1,i+1}}\,\sqrt{\mathrm{SSE}}}$$

enjoys a t-distribution with $n-m-1$ degrees of freedom. After the outcomes of $Y_1, \ldots, Y_n$ have become clear, all terms in this expression can be filled in with the exception of β_i. Of course, using the methods of Chapter II and Chapter III, we can now perform hypothesis tests on β_i. □

Exercise 36

In the case of normal regression we start from an independent system $Y_1, \ldots, Y_n$ in which the variable Y_i for each i enjoys an $N(\alpha + \langle \boldsymbol{\beta}, \mathfrak{x}_i \rangle, \sigma^2)$-distribution. Prove that when reducing to the model where each Y_i is $N(\alpha, \sigma^2)$-distributed we have

$$\mathrm{SSE}_{\mathfrak{R}} = \sum_{i=1}^{n} (Y_i - \overline{Y})^2 = S_{\mathbf{YY}}.$$

Proof. Here $\mathfrak{R} = \mathrm{span}\{\mathbf{e}\}$, with $\mathbf{e} = (1, \ldots, 1) \in \mathbb{R}^n$. For $\mathbf{v} := \frac{1}{\sqrt{n}}\,\mathbf{e}$ we have $\mathbf{P}_{\mathfrak{R}} = \mathbf{v} \otimes \mathbf{v}$, so

$$\mathbf{P}_{\mathfrak{R}}\mathbf{Y} = \langle \mathbf{v}, \mathbf{Y} \rangle\,\mathbf{v} = \left(\frac{1}{n} \sum_{i=1}^{n} Y_i \right) \mathbf{e} = \overline{Y}\mathbf{e}$$

and

$$\mathrm{SSE}_{\mathfrak{R}} := \|\mathbf{Y} - \mathbf{P}_{\mathfrak{R}}\mathbf{Y}\|^2 = \|\mathbf{Y} - \overline{Y}\mathbf{e}\|^2 = \sum_{i=1}^{n} (Y_i - \overline{Y})^2 = S_{\mathbf{YY}}.$$

□

Exercise 37

In the case of multiple normal regression, determine the maximum likelihood estimators of $\alpha, \boldsymbol{\beta}$ and σ^2.

Solution. Here, for fixed $\mathbf{y}$, we have to maximize the likelihood function

$$(a, \mathbf{b}, c) \;\mapsto\; f_{\mathbf{Y}}(\mathbf{y}) = f_{Y_1}(y_1) \cdots f_{Y_n}(y_n)$$

on the domain $\mathbb{R} \times \mathbb{R}^m \times (0, +\infty)$. It is easily deduced that

$$f_{\mathbf{Y}}(\mathbf{y}) = \frac{1}{(2\pi)^{n/2}\, c^n}\; \exp\left[-\frac{1}{2c^2}\, \|\mathbf{y} - (a\mathbf{e} + b_1\mathbf{x}_1 + \cdots + b_m\mathbf{x}_m)\|^2 \right].$$

It follows that in a maximizing point $(a, \mathbf{b}, c)$ the expression

$$\|\mathbf{y} - (a\mathbf{e} + b_1\mathbf{x}_1 + \cdots + b_m\mathbf{x}_m)\|^2$$

must be minimal. This is, however, exactly the sum of squares of errors and we know that this sum is minimal if and only if $(a, \mathbf{b})$ is equal to the least squares estimate $(\hat{\alpha}, \hat{\boldsymbol{\beta}})$ for $(\alpha, \boldsymbol{\beta})$. The corresponding minimum is equal to SSE. Thus we arrive at the maximization of the map

$$c \mapsto \frac{1}{(2\pi)^{n/2}\, c^n} \exp\left[-\frac{1}{2c^2}\,\mathrm{SSE}\right]$$

on the domain $(0, +\infty)$. Equivalently, we maximize its logarithm, that is

$$-\tfrac{1}{2}n \log 2\pi - \tfrac{1}{2}n \log c^2 - \frac{1}{2c^2}\,\mathrm{SSE}.$$

Regarding this as a function of c^2, its derivative equals

$$\frac{-\frac{1}{2}n}{c^2} + \frac{\frac{1}{2}\mathrm{SSE}}{(c^2)^2}$$

which vanishes for $c^2 = \mathrm{SSE}/n$. Consequently this expression arises as the maximum likelihood estimate for σ^2. □

Exercise 38

Indicate in what way the theory of multiple linear regression can be used in the case of a model

$$y = \alpha + \beta_1 x_1 + \beta_2 x_1^2 + \beta_3 e^{x_2}.$$

Solution. If we simply define

$$z_1 := x_1,\ z_2 := x_1^2 \quad \text{and} \quad z_3 := e^{x_2},$$

the model reduces to

$$y = \alpha + \beta_1 z_1 + \beta_2 z_2 + \beta_3 z_3.$$

Now the linear model can be applied. □

Exercise 39

Suppose the stochastic p-vector $\mathbf{X}$ has a density function given by $f_{\mathbf{X}}$. We set $\mathbf{Y} := \mathbf{A}\mathbf{X} + \mathbf{b}$, where $\mathbf{A} : \mathbb{R}^p \to \mathbb{R}^p$ is an invertible linear operator and $\mathbf{b} \in \mathbb{R}^p$. Prove that $\mathbf{Y}$ enjoys a density function $f_{\mathbf{Y}}$, given by

$$f_{\mathbf{Y}}(\mathbf{y}) = \frac{1}{|\det \mathbf{A}|}\, f_{\mathbf{X}}(\mathbf{A}^{-1}(\mathbf{y} - \mathbf{b})).$$

Proof. First we define $\varphi : \mathbb{R}^p \to \mathbb{R}^p$ by

$$\varphi \ : \ \mathbf{y} \ \mapsto \ \mathbf{A}^{-1}(\mathbf{y} - \mathbf{b}).$$

We have $\varphi^{-1}(\mathbf{x}) = \mathbf{A}\mathbf{x}+\mathbf{b}$. For each $B = (a_1, b_1) \times \cdots \times (a_p, b_p)$ by the substitution rule for multiple integrals

$$\begin{aligned}\mathbb{P}(\mathbf{Y} \in B) &= \mathbb{P}(\mathbf{X} \in \varphi(B)) = \int_{\varphi(B)} f_{\mathbf{X}}(\mathbf{y})\, d\mathbf{y} \\ &= \int_B f_{\mathbf{X}}(\varphi(\mathbf{y}))\, |\det(D\varphi)_{\mathbf{y}}|\, d\mathbf{y}.\end{aligned}$$

From this it follows that

$$f_{\mathbf{Y}}(\mathbf{y}) = |\det(D\varphi)_{\mathbf{y}}|\; f_{\mathbf{X}}(\mathbf{A}^{-1}(\mathbf{y} - \mathbf{b})).$$

By the chain rule $(D\varphi)_{\mathbf{y}} = (D\mathbf{A}^{-1})_{h(\mathbf{y})} \circ (Dh)_{\mathbf{y}}$, where $h : \mathbf{y} \to \mathbf{y} - \mathbf{b}$; therefore $(D\varphi)_{\mathbf{y}} = \mathbf{A}^{-1}\mathbf{I}_p = \mathbf{A}^{-1}$ and $|\det(D\varphi)_{\mathbf{y}}| = |\det \mathbf{A}^{-1}| = |(\det \mathbf{A})^{-1}|$. □

Statistical tables

Tables II up to V taken over with kind permission of Boom-uitgeverij bv, Meppel, from W. van den Brink and P. Koele, *Statistiek, Deel 3* (Boom Meppel, Amsterdam, 1987).

Table I and Tables VI up to XI computed by Alessandro di Bucchianico and Mark van de Wiel, Eindhoven University of Technology, using Mathematica.

Table I: The binomial distribution

		p								
n	k	0.05	0.10	0.15	0.20	0.25	0.30	0.35	0.40	0.50
1	0	0.9500	0.9000	0.8500	0.8000	0.7500	0.7000	0.6500	0.6000	0.5000
2	0	0.9025	0.8100	0.7225	0.6400	0.5625	0.4900	0.4225	0.3600	0.2500
	1	0.9975	0.9900	0.9775	0.9600	0.9375	0.9100	0.8775	0.8400	0.7500
3	0	0.8574	0.7290	0.6141	0.5120	0.4219	0.3430	0.2746	0.2160	0.1250
	1	0.9928	0.9720	0.9393	0.8960	0.8438	0.7840	0.7183	0.6480	0.5000
	2	0.9999	0.9990	0.9966	0.9920	0.9844	0.9730	0.9571	0.9360	0.8750
4	0	0.8145	0.6561	0.5220	0.4096	0.3164	0.2401	0.1785	0.1296	0.0625
	1	0.9860	0.9477	0.8905	0.8192	0.7383	0.6517	0.5630	0.4752	0.3125
	2	0.9995	0.9963	0.9880	0.9728	0.9492	0.9163	0.8735	0.8208	0.6875
	3	1.0000	0.9999	0.9995	0.9984	0.9961	0.9919	0.9850	0.9744	0.9375
5	0	0.7738	0.5905	0.4437	0.3277	0.2373	0.1681	0.1160	0.0778	0.0313
	1	0.9774	0.9185	0.8352	0.7373	0.6328	0.5282	0.4284	0.3370	0.1875
	2	0.9988	0.9914	0.9734	0.9421	0.8965	0.8369	0.7648	0.6826	0.5000
	3	1.0000	0.9995	0.9978	0.9933	0.9844	0.9692	0.9460	0.9130	0.8125
	4	1.0000	1.0000	0.9999	0.9997	0.9990	0.9976	0.9947	0.9898	0.9688
6	0	0.7351	0.5314	0.3771	0.2621	0.1780	0.1176	0.0754	0.0467	0.0156
	1	0.9672	0.8857	0.7765	0.6554	0.5339	0.4202	0.3191	0.2333	0.1094
	2	0.9978	0.9842	0.9527	0.9011	0.8306	0.7443	0.6471	0.5443	0.3438
	3	0.9999	0.9987	0.9941	0.9830	0.9624	0.9295	0.8826	0.8208	0.6562
	4	1.0000	0.9999	0.9996	0.9984	0.9954	0.9891	0.9777	0.9590	0.8906
	5	1.0000	1.0000	1.0000	0.9999	0.9998	0.9993	0.9982	0.9959	0.9844
7	0	0.6983	0.4783	0.3206	0.2097	0.1335	0.0824	0.0490	0.0280	0.0078
	1	0.9556	0.8503	0.7166	0.5767	0.4449	0.3294	0.2338	0.1586	0.0625
	2	0.9962	0.9743	0.9262	0.8520	0.7564	0.6471	0.5323	0.4199	0.2266
	3	0.9998	0.9973	0.9879	0.9667	0.9294	0.8740	0.8002	0.7102	0.5000
	4	1.0000	0.9998	0.9988	0.9953	0.9871	0.9712	0.9444	0.9037	0.7734
	5	1.0000	1.0000	0.9999	0.9996	0.9987	0.9962	0.9910	0.9812	0.9375
	6	1.0000	1.0000	1.0000	1.0000	0.9999	0.9998	0.9994	0.9984	0.9922
8	0	0.6634	0.4305	0.2725	0.1678	0.1001	0.0576	0.0319	0.0168	0.0039
	1	0.9428	0.8131	0.6572	0.5033	0.3671	0.2553	0.1691	0.1064	0.0352

The distribution function belonging to the binomial distribution. For example: if X is binomially distributed with parameters $n = 5$ and $p = 0.15$, then $\mathbb{P}(X \leq 2) = 0.9734$.

		p								
n	k	0.05	0.10	0.15	0.20	0.25	0.30	0.35	0.40	0.50
8	2	0.9942	0.9619	0.8948	0.7969	0.6785	0.5518	0.4278	0.3154	0.1445
	3	0.9996	0.9950	0.9786	0.9437	0.8862	0.8059	0.7064	0.5941	0.3633
	4	1.0000	0.9996	0.9971	0.9896	0.9727	0.9420	0.8939	0.8263	0.6367
	5	1.0000	1.0000	0.9998	0.9988	0.9958	0.9887	0.9747	0.9502	0.8555
	6	1.0000	1.0000	1.0000	0.9999	0.9996	0.9987	0.9964	0.9915	0.9648
	7	1.0000	1.0000	1.0000	1.0000	1.0000	0.9999	0.9998	0.9993	0.9961
9	0	0.6302	0.3874	0.2316	0.1342	0.0751	0.0404	0.0207	0.0101	0.0020
	1	0.9288	0.7748	0.5995	0.4362	0.3003	0.1960	0.1211	0.0705	0.0195
	2	0.9916	0.9470	0.8591	0.7382	0.6007	0.4628	0.3373	0.2318	0.0898
	3	0.9994	0.9917	0.9661	0.9144	0.8343	0.7297	0.6089	0.4826	0.2539
	4	1.0000	0.9991	0.9944	0.9804	0.9511	0.9012	0.8283	0.7334	0.5000
	5	1.0000	0.9999	0.9994	0.9969	0.9900	0.9747	0.9464	0.9006	0.7461
	6	1.0000	1.0000	1.0000	0.9997	0.9987	0.9957	0.9888	0.9750	0.9102
	7	1.0000	1.0000	1.0000	1.0000	0.9999	0.9996	0.9986	0.9962	0.9805
	8	1.0000	1.0000	1.0000	1.0000	1.0000	1.0000	0.9999	0.9997	0.9980
10	0	0.5987	0.3487	0.1969	0.1074	0.0563	0.0282	0.0135	0.0060	0.0010
	1	0.9139	0.7361	0.5443	0.3758	0.2440	0.1493	0.0860	0.0464	0.0107
	2	0.9885	0.9298	0.8202	0.6778	0.5256	0.3828	0.2616	0.1673	0.0547
	3	0.9990	0.9872	0.9500	0.8791	0.7759	0.6496	0.5138	0.3823	0.1719
	4	0.9999	0.9984	0.9901	0.9672	0.9219	0.8497	0.7515	0.6331	0.3770
	5	1.0000	0.9999	0.9986	0.9936	0.9803	0.9527	0.9051	0.8338	0.6230
	6	1.0000	1.0000	0.9999	0.9991	0.9965	0.9894	0.9740	0.9452	0.8281
	7	1.0000	1.0000	1.0000	0.9999	0.9996	0.9984	0.9952	0.9877	0.9453
	8	1.0000	1.0000	1.0000	1.0000	1.0000	0.9999	0.9995	0.9983	0.9893
	9	1.0000	1.0000	1.0000	1.0000	1.0000	1.0000	1.0000	0.9999	0.9990
11	0	0.5688	0.3138	0.1673	0.0859	0.0422	0.0198	0.0088	0.0036	0.0005
	1	0.8981	0.6974	0.4922	0.3221	0.1971	0.1130	0.0606	0.0302	0.0059
	2	0.9848	0.9104	0.7788	0.6174	0.4552	0.3127	0.2001	0.1189	0.0327
	3	0.9984	0.9815	0.9306	0.8389	0.7133	0.5696	0.4256	0.2963	0.1133
	4	0.9999	0.9972	0.9841	0.9496	0.8854	0.7897	0.6683	0.5328	0.2744
	5	1.0000	0.9997	0.9973	0.9883	0.9657	0.9218	0.8513	0.7535	0.5000
	6	1.0000	1.0000	0.9997	0.9980	0.9924	0.9784	0.9499	0.9006	0.7256
	7	1.0000	1.0000	1.0000	0.9998	0.9988	0.9957	0.9878	0.9707	0.8867
	8	1.0000	1.0000	1.0000	1.0000	0.9999	0.9994	0.9980	0.9941	0.9673
	9	1.0000	1.0000	1.0000	1.0000	1.0000	1.0000	0.9998	0.9993	0.9941
	10	1.0000	1.0000	1.0000	1.0000	1.0000	1.0000	1.0000	1.0000	0.9995
12	0	0.5404	0.2824	0.1422	0.0687	0.0317	0.0138	0.0057	0.0022	0.0002
	1	0.8816	0.6590	0.4435	0.2749	0.1584	0.0850	0.0424	0.0196	0.0032

n	k	p								
		0.05	0.10	0.15	0.20	0.25	0.30	0.35	0.40	0.50
12	2	0.9804	0.8891	0.7358	0.5583	0.3907	0.2528	0.1513	0.0834	0.0193
	3	0.9978	0.9744	0.9078	0.7946	0.6488	0.4925	0.3467	0.2253	0.0730
	4	0.9998	0.9957	0.9761	0.9274	0.8424	0.7237	0.5833	0.4382	0.1938
	5	1.0000	0.9995	0.9954	0.9806	0.9456	0.8822	0.7873	0.6652	0.3872
	6	1.0000	0.9999	0.9993	0.9961	0.9857	0.9614	0.9154	0.8418	0.6128
	7	1.0000	1.0000	0.9999	0.9994	0.9972	0.9905	0.9745	0.9427	0.8062
	8	1.0000	1.0000	1.0000	0.9999	0.9996	0.9983	0.9944	0.9847	0.9270
	9	1.0000	1.0000	1.0000	1.0000	1.0000	0.9998	0.9992	0.9972	0.9807
	10	1.0000	1.0000	1.0000	1.0000	1.0000	1.0000	0.9999	0.9997	0.9968
	11	1.0000	1.0000	1.0000	1.0000	1.0000	1.0000	1.0000	1.0000	0.9998
13	0	0.5133	0.2542	0.1209	0.0550	0.0238	0.0097	0.0037	0.0013	0.0001
	1	0.8646	0.6213	0.3983	0.2336	0.1267	0.0637	0.0296	0.0126	0.0017
	2	0.9755	0.8661	0.6920	0.5017	0.3326	0.2025	0.1132	0.0579	0.0112
	3	0.9969	0.9658	0.8820	0.7473	0.5843	0.4206	0.2783	0.1686	0.0461
	4	0.9997	0.9935	0.9658	0.9009	0.7940	0.6543	0.5005	0.3530	0.1334
	5	1.0000	0.9991	0.9925	0.9700	0.9198	0.8346	0.7159	0.5744	0.2905
	6	1.0000	0.9999	0.9987	0.9930	0.9757	0.9376	0.8705	0.7712	0.5000
	7	1.0000	1.0000	0.9998	0.9988	0.9944	0.9818	0.9538	0.9023	0.7095
	8	1.0000	1.0000	1.0000	0.9998	0.9990	0.9960	0.9874	0.9679	0.8666
	9	1.0000	1.0000	1.0000	1.0000	0.9999	0.9993	0.9975	0.9922	0.9539
	10	1.0000	1.0000	1.0000	1.0000	1.0000	0.9999	0.9997	0.9987	0.9888
	11	1.0000	1.0000	1.0000	1.0000	1.0000	1.0000	1.0000	0.9999	0.9983
	12	1.0000	1.0000	1.0000	1.0000	1.0000	1.0000	1.0000	1.0000	0.9999
14	0	0.4877	0.2288	0.1028	0.0440	0.0178	0.0068	0.0024	0.0008	0.0001
	1	0.8470	0.5846	0.3567	0.1979	0.1010	0.0475	0.0205	0.0081	0.0009
	2	0.9699	0.8416	0.6479	0.4481	0.2811	0.1608	0.0839	0.0398	0.0065
	3	0.9958	0.9559	0.8535	0.6982	0.5213	0.3552	0.2205	0.1243	0.0287
	4	0.9996	0.9908	0.9533	0.8702	0.7415	0.5842	0.4227	0.2793	0.0898
	5	1.0000	0.9985	0.9885	0.9561	0.8883	0.7805	0.6405	0.4859	0.2120
	6	1.0000	0.9998	0.9978	0.9884	0.9617	0.9067	0.8164	0.6925	0.3953
	7	1.0000	1.0000	0.9997	0.9976	0.9897	0.9685	0.9247	0.8499	0.6047
	8	1.0000	1.0000	1.0000	0.9996	0.9978	0.9917	0.9757	0.9417	0.7880
	9	1.0000	1.0000	1.0000	1.0000	0.9997	0.9983	0.9940	0.9825	0.9102
	10	1.0000	1.0000	1.0000	1.0000	1.0000	0.9998	0.9989	0.9961	0.9713
	11	1.0000	1.0000	1.0000	1.0000	1.0000	1.0000	0.9999	0.9994	0.9935
	12	1.0000	1.0000	1.0000	1.0000	1.0000	1.0000	1.0000	0.9999	0.9991
	13	1.0000	1.0000	1.0000	1.0000	1.0000	1.0000	1.0000	1.0000	0.9999

n	k	p								
		0.05	0.10	0.15	0.20	0.25	0.30	0.35	0.40	0.50
15	0	0.4633	0.2059	0.0874	0.0352	0.0134	0.0047	0.0016	0.0005	0.0000
	1	0.8290	0.5490	0.3186	0.1671	0.0802	0.0353	0.0142	0.0052	0.0005
	2	0.9638	0.8159	0.6042	0.3980	0.2361	0.1268	0.0617	0.0271	0.0037
	3	0.9945	0.9444	0.8227	0.6482	0.4613	0.2969	0.1727	0.0905	0.0176
	4	0.9994	0.9873	0.9383	0.8358	0.6865	0.5155	0.3519	0.2173	0.0592
	5	0.9999	0.9978	0.9832	0.9389	0.8516	0.7216	0.5643	0.4032	0.1509
	6	1.0000	0.9997	0.9964	0.9819	0.9434	0.8689	0.7548	0.6098	0.3036
	7	1.0000	1.0000	0.9994	0.9958	0.9827	0.9500	0.8868	0.7869	0.5000
	8	1.0000	1.0000	0.9999	0.9992	0.9958	0.9848	0.9578	0.9050	0.6964
	9	1.0000	1.0000	1.0000	0.9999	0.9992	0.9963	0.9876	0.9662	0.8491
	10	1.0000	1.0000	1.0000	1.0000	0.9999	0.9993	0.9972	0.9907	0.9408
	11	1.0000	1.0000	1.0000	1.0000	1.0000	0.9999	0.9995	0.9981	0.9824
	12	1.0000	1.0000	1.0000	1.0000	1.0000	1.0000	0.9999	0.9997	0.9963
	13	1.0000	1.0000	1.0000	1.0000	1.0000	1.0000	1.0000	1.0000	0.9995
	14	1.0000	1.0000	1.0000	1.0000	1.0000	1.0000	1.0000	1.0000	1.0000
20	0	0.3585	0.1216	0.0388	0.0115	0.0032	0.0008	0.0002	0.0000	0.0000
	1	0.7358	0.3917	0.1756	0.0692	0.0243	0.0076	0.0021	0.0005	0.0000
	2	0.9245	0.6769	0.4049	0.2061	0.0913	0.0355	0.0121	0.0036	0.0002
	3	0.9841	0.8670	0.6477	0.4114	0.2252	0.1071	0.0444	0.0160	0.0013
	4	0.9974	0.9568	0.8298	0.6296	0.4148	0.2375	0.1182	0.0510	0.0059
	5	0.9997	0.9887	0.9327	0.8042	0.6172	0.4164	0.2454	0.1256	0.0207
	6	1.0000	0.9976	0.9781	0.9133	0.7858	0.6080	0.4166	0.2500	0.0577
	7	1.0000	0.9996	0.9941	0.9679	0.8982	0.7723	0.6010	0.4159	0.1316
	8	1.0000	0.9999	0.9987	0.9900	0.9591	0.8867	0.7624	0.5956	0.2517
	9	1.0000	1.0000	0.9998	0.9974	0.9861	0.9520	0.8782	0.7553	0.4119
	10	1.0000	1.0000	1.0000	0.9994	0.9961	0.9829	0.9468	0.8725	0.5881
	11	1.0000	1.0000	1.0000	0.9999	0.9991	0.9949	0.9804	0.9435	0.7483
	12	1.0000	1.0000	1.0000	1.0000	0.9998	0.9987	0.9940	0.9790	0.8684
	13	1.0000	1.0000	1.0000	1.0000	1.0000	0.9997	0.9985	0.9935	0.9423
	14	1.0000	1.0000	1.0000	1.0000	1.0000	1.0000	0.9997	0.9984	0.9793
	15	1.0000	1.0000	1.0000	1.0000	1.0000	1.0000	1.0000	0.9997	0.9941
	16	1.0000	1.0000	1.0000	1.0000	1.0000	1.0000	1.0000	1.0000	0.9987
	17	1.0000	1.0000	1.0000	1.0000	1.0000	1.0000	1.0000	1.0000	0.9998
	18	1.0000	1.0000	1.0000	1.0000	1.0000	1.0000	1.0000	1.0000	1.0000
	19	1.0000	1.0000	1.0000	1.0000	1.0000	1.0000	1.0000	1.0000	1.0000

Table II: The standard normal distribution

	last decimal of z									
z	0	1	2	3	4	5	6	7	8	9
0.0	0.500	0.504	0.508	0.512	0.516	0.520	0.524	0.528	0.532	0.536
0.1	0.540	0.544	0.548	0.552	0.556	0.560	0.564	0.568	0.571	0.575
0.2	0.579	0.583	0.587	0.591	0.595	0.599	0.603	0.606	0.610	0.614
0.3	0.618	0.622	0.626	0.629	0.633	0.637	0.641	0.644	0.648	0.652
0.4	0.655	0.659	0.663	0.666	0.670	0.674	0.677	0.681	0.684	0.688
0.5	0.692	0.695	0.699	0.702	0.705	0.709	0.712	0.716	0.719	0.722
0.6	0.726	0.729	0.732	0.736	0.739	0.742	0.745	0.749	0.752	0.755
0.7	0.758	0.761	0.764	0.767	0.770	0.773	0.776	0.779	0.782	0.785
0.8	0.788	0.791	0.794	0.797	0.800	0.802	0.805	0.808	0.811	0.813
0.9	0.816	0.819	0.821	0.824	0.826	0.829	0.832	0.834	0.837	0.839
1.0	0.841	0.844	0.846	0.849	0.851	0.853	0.855	0.858	0.860	0.862
1.1	0.864	0.867	0.869	0.871	0.873	0.875	0.877	0.879	0.881	0.883
1.2	0.885	0.887	0.889	0.891	0.893	0.894	0.896	0.898	0.900	0.902
1.3	0.903	0.905	0.907	0.908	0.910	0.912	0.913	0.915	0.916	0.918
1.4	0.919	0.921	0.922	0.924	0.925	0.927	0.928	0.929	0.931	0.932
1.5	0.933	0.935	0.936	0.937	0.938	0.939	0.941	0.942	0.943	0.944
1.6	0.945	0.946	0.947	0.948	0.950	0.951	0.952	0.953	0.954	0.955
1.7	0.955	0.956	0.957	0.958	0.959	0.960	0.961	0.962	0.963	0.963
1.8	0.964	0.965	0.966	0.966	0.967	0.968	0.969	0.969	0.970	0.971
1.9	0.971	0.972	0.973	0.973	0.974	0.974	0.975	0.976	0.976	0.977
2.0	0.977	0.978	0.978	0.979	0.979	0.980	0.980	0.981	0.981	0.982
2.1	0.982	0.983	0.983	0.983	0.984	0.984	0.985	0.985	0.985	0.986
2.2	0.986	0.986	0.987	0.987	0.988	0.988	0.988	0.988	0.989	0.989
2.3	0.989	0.990	0.990	0.990	0.990	0.991	0.991	0.991	0.991	0.992
2.4	0.992	0.992	0.992	0.993	0.993	0.993	0.993	0.993	0.993	0.994
2.5	0.994	0.994	0.994	0.994	0.995	0.995	0.995	0.995	0.995	0.995
2.6	0.995	0.995	0.996	0.996	0.996	0.996	0.996	0.996	0.996	0.996
2.7	0.997	0.997	0.997	0.997	0.997	0.997	0.997	0.997	0.997	0.997
2.8	0.997	0.998	0.998	0.998	0.998	0.998	0.998	0.998	0.998	0.998
2.9	0.998	0.998	0.998	0.998	0.998	0.998	0.999	0.999	0.999	0.999
3.0	0.999	0.999	0.999	0.999	0.999	0.999	0.999	0.999	0.999	0.999

The distribution function corresponding to the $N(0,1)$-distribution. For example: if Z enjoys this distribution then $\mathbb{P}(Z \leq 1.84) = 0.967$ and $\mathbb{P}(|Z| \leq 1.84) = 0.967 - (1 - 0.967) = 0.934$. In general we have: $\mathbb{P}(Z \leq -z) = \mathbb{P}(Z \geq z) = 1 - \mathbb{P}(Z \leq z)$.

Table III: The t-distribution

n	$1-\alpha$ 0.90	0.95	0.975	0.99	0.995	n	$1-\alpha$ 0.90	0.95	0.975	0.99	0.995
1	3.078	6.314	12.71	31.82	63.66	22	1.321	1.717	2.074	2.508	2.819
2	1.886	2.920	4.303	6.965	9.925	23	1.319	1.714	2.069	2.500	2.807
3	1.638	2.353	3.182	4.541	5.841	24	1.318	1.711	2.064	2.492	2.797
4	1.533	2.132	2.776	3.747	4.604	25	1.316	1.708	2.060	2.485	2.787
5	1.476	2.015	2.571	3.365	4.032	26	1.315	1.706	2.056	2.479	2.779
6	1.440	1.943	2.447	3.143	3.707	27	1.314	1.703	2.052	2.473	2.771
7	1.415	1.895	2.365	2.998	3.499	28	1.313	1.701	2.048	2.467	2.763
8	1.397	1.860	2.306	2.896	3.355	29	1.311	1.699	2.045	2.462	2.756
9	1.383	1.833	2.262	2.821	3.250	30	1.310	1.697	2.042	2.457	2.750
10	1.372	1.812	2.228	2.764	3.169	35	1.306	1.690	2.030	2.438	2.724
11	1.363	1.796	2.201	2.718	3.106	40	1.303	1.684	2.021	2.423	2.704
12	1.356	1.782	2.179	2.681	3.055	45	1.301	1.679	2.014	2.412	2.690
13	1.350	1.771	2.160	2.650	3.012	50	1.299	1.676	2.009	2.403	2.678
14	1.345	1.761	2.145	2.624	2.977	60	1.296	1.671	2.000	2.390	2.660
15	1.341	1.753	2.131	2.602	2.947	70	1.294	1.667	1.994	2.381	2.648
16	1.337	1.746	2.120	2.583	2.921	80	1.292	1.664	1.990	2.374	2.639
17	1.333	1.740	2.110	2.567	2.898	90	1.291	1.662	1.987	2.368	2.632
18	1.330	1.734	2.101	2.552	2.878	100	1.290	1.660	1.984	2.364	2.626
19	1.328	1.729	2.093	2.539	2.861	200	1.286	1.652	1.972	2.345	2.601
20	1.325	1.725	2.086	2.528	2.845	500	1.283	1.648	1.965	2.334	2.586
21	1.323	1.721	2.080	2.518	2.831	1000	1.282	1.646	1.962	2.330	2.581

Quantiles with the t-distribution. For example: if X enjoys the t-distribution with 20 degrees of freedom, then $\mathbb{P}(X \leq 2.086) = 0.975$. Or: if Y enjoys the t-distribution with 30 degrees of freedom, then $\mathbb{P}(|Y| \leq 2.457) = 0.98$.

Table IV: The χ^2-distribution

	α					$1-\alpha$				
n	0.005	0.01	0.025	0.05	0.10	0.90	0.95	0.975	0.99	0.995
1	.000	.000	.001	.004	.016	2.71	3.84	5.02	6.63	7.88
2	.010	.020	.051	.103	.211	4.61	5.99	7.38	9.21	10.6
3	.072	.115	.216	.352	.584	6.25	7.81	9.35	11.3	12.8
4	.207	.297	.484	.711	1.06	7.78	9.49	11.1	13.3	14.9
5	.412	.554	.831	1.15	1.61	9.24	11.1	12.8	15.1	16.7
6	.676	.872	1.24	1.64	2.20	10.6	12.6	14.4	16.8	18.5
7	.989	1.24	1.69	2.17	2.83	12.0	14.1	16.0	18.5	20.3
8	1.34	1.65	2.18	2.73	3.49	13.4	15.5	17.5	20.1	22.0
9	1.73	2.09	2.70	3.33	4.17	14.7	16.9	19.0	21.7	23.6
10	2.16	2.56	3.25	3.94	4.87	16.0	18.3	20.5	23.2	25.2
11	2.60	3.05	3.82	4.57	5.58	17.3	19.7	21.9	24.7	26.8
12	3.07	3.57	4.40	5.23	6.30	18.5	21.0	23.3	26.2	28.3
13	3.57	4.11	5.01	5.89	7.04	19.8	22.4	24.7	27.7	29.8
14	4.07	4.66	5.63	6.57	7.79	21.1	23.7	26.1	29.1	31.3
15	4.60	5.23	6.26	7.26	8.55	22.3	25.0	27.5	30.6	32.8
16	5.14	5.81	6.91	7.96	9.31	23.5	26.3	28.8	32.0	34.3
17	5.70	6.41	7.56	8.67	10.1	24.8	27.6	30.2	33.4	35.7
18	6.26	7.01	8.23	9.39	10.9	26.0	28.9	31.5	34.8	37.2
19	6.84	7.63	8.91	10.1	11.7	27.2	30.1	32.9	36.2	38.6
20	7.43	8.26	9.59	10.9	12.4	28.4	31.4	34.2	37.6	40.0
21	8.03	8.90	10.3	11.6	13.2	29.6	32.7	35.5	38.9	41.4
22	8.64	9.54	11.0	12.3	14.0	30.8	33.9	36.8	40.3	42.8
23	9.26	10.2	11.7	13.1	14.8	32.0	35.2	38.1	41.6	44.2
24	9.89	10.9	12.4	13.8	15.7	33.2	36.4	39.4	43.0	45.6
25	10.5	11.5	13.1	14.6	16.5	34.4	37.7	40.6	44.3	46.9
30	13.8	15.0	16.8	18.5	20.6	40.3	43.8	47.0	50.9	53.7
40	20.7	22.2	24.4	26.5	29.1	51.8	55.8	59.3	63.7	66.8
50	28.0	29.7	32.4	34.8	37.7	63.2	67.5	71.4	76.2	79.5
60	35.5	37.5	40.5	43.2	46.5	74.4	79.1	83.3	88.4	92.0
70	43.3	45.4	48.8	51.7	55.3	85.5	90.5	95.0	100.4	104.2
80	51.2	53.5	57.2	60.4	64.3	96.6	101.9	106.6	112.3	116.3
90	59.2	61.8	65.6	69.1	73.3	107.6	113.1	118.1	124.1	128.3
100	67.3	70.1	74.2	77.9	82.4	118.5	124.3	129.6	135.8	140.2

Quantiles with the χ^2-distribution. For example: if X is χ^2-distributed with 18 degrees of freedom, then $\mathbb{P}(X \leq 31.5) = 0.975$, as well as $\mathbb{P}(X \geq 10.9) = 0.90$.

Table V: The F-distribution

n	$1-\alpha$	m : degrees of freedom in numerator																		
		1	2	3	4	5	6	7	8	9	10	15	20	30	40	50	60	100	500	∞
1	0.90	39.9	49.5	53.6	55.8	57.2	58.2	58.9	59.4	59.9	60.2	61.2	61.7	62.3	62.5	62.7	62.8	63.0	63.3	63.3
	0.95	161	200	216	225	230	234	237	239	241	242	246	248	250	251	252	252	253	254	254
2	0.90	8.53	9.00	9.16	9.24	9.29	9.33	9.35	9.37	9.38	9.39	9.42	9.44	9.46	9.47	9.47	9.47	9.48	9.49	9.49
	0.95	18.5	19.0	19.2	19.2	19.3	19.3	19.4	19.4	19.4	19.4	19.4	19.4	19.5	19.5	19.5	19.5	19.5	19.5	19.5
	0.975	38.5	39.0	39.2	39.3	39.3	39.3	39.4	39.4	39.4	39.4	39.4	39.5	39.5	39.5	39.5	39.5	39.5	39.5	39.5
	0.99	98.5	99.0	99.2	99.2	99.3	99.3	99.4	99.4	99.4	99.4	99.4	99.4	99.5	99.5	99.5	99.5	99.5	99.5	99.5
3	0.90	5.54	5.46	5.39	5.34	5.31	5.28	5.27	5.25	5.24	5.23	5.20	5.18	5.17	5.16	5.15	5.15	5.14	5.14	5.13
	0.95	10.1	9.55	9.28	9.12	9.01	8.94	8.89	8.85	8.81	8.79	8.70	8.66	8.62	8.59	8.58	8.57	8.55	8.53	8.53
	0.975	17.4	16.0	15.4	15.1	14.9	14.7	14.6	14.5	14.5	14.4	14.3	14.2	14.1	14.0	14.0	14.0	14.0	13.9	13.9
	0.99	34.1	30.8	29.5	28.7	28.2	27.9	27.7	27.5	27.3	27.2	26.9	26.7	26.5	26.4	26.4	26.3	26.2	26.1	26.1
4	0.90	4.54	4.32	4.19	4.11	4.05	4.01	3.98	3.95	3.94	3.92	3.87	3.84	3.82	3.80	3.80	3.79	3.78	3.76	3.76
	0.95	7.71	6.94	6.59	6.39	6.26	6.16	6.09	6.04	6.00	5.96	5.86	5.80	5.75	5.72	5.70	5.69	5.66	5.64	5.63
	0.975	12.2	10.7	10.0	9.60	9.36	9.20	9.07	8.98	8.90	8.84	8.66	8.56	8.46	8.41	8.38	8.36	8.32	8.27	8.26
	0.99	21.2	18.0	16.7	16.0	15.5	15.2	15.0	14.8	14.7	14.5	14.2	14.0	13.8	13.7	13.7	13.7	13.6	13.5	13.5
5	0.90	4.06	3.78	3.62	3.52	3.45	3.40	3.37	3.34	3.32	3.30	3.24	3.21	3.17	3.16	3.15	3.14	3.13	3.11	3.10
	0.95	6.61	5.79	5.41	5.19	5.05	4.95	4.88	4.82	4.77	4.74	4.62	4.56	4.50	4.46	4.44	4.43	4.41	4.37	4.36
	0.975	10.0	8.43	7.76	7.39	7.15	6.98	6.85	6.76	6.68	6.62	6.43	6.33	6.23	6.18	6.14	6.12	6.08	6.03	6.02
	0.99	16.3	13.3	12.1	11.4	11.0	10.7	10.5	10.3	10.2	10.1	9.72	9.55	9.38	9.29	9.24	9.20	9.13	9.04	9.02
6	0.90	3.78	3.46	3.29	3.18	3.11	3.05	3.01	2.98	2.96	2.94	2.87	2.84	2.80	2.78	2.77	2.76	2.75	2.73	2.72
	0.95	5.99	5.14	4.76	4.53	4.39	4.28	4.21	4.15	4.10	4.06	3.94	3.87	3.81	3.77	3.75	3.74	3.71	3.68	3.67
	0.975	8.81	7.26	6.60	6.23	5.99	5.82	5.70	5.60	5.52	5.46	5.27	5.17	5.07	5.01	4.98	4.96	4.92	4.86	4.85
	0.99	13.7	10.9	9.78	9.15	8.75	8.47	8.26	8.10	7.98	7.87	7.56	7.40	7.23	7.14	7.09	7.06	6.99	6.90	6.88

n	$1-\alpha$	m : degrees of freedom in numerator																		
		1	2	3	4	5	6	7	8	9	10	15	20	30	40	50	60	100	500	∞
7	0.90	3.59	3.26	3.07	2.96	2.88	2.83	2.78	2.75	2.72	2.70	2.63	2.59	2.56	2.54	2.52	2.51	2.50	2.48	2.47
	0.95	5.59	4.74	4.35	4.12	3.97	3.87	3.79	3.73	3.68	3.64	3.51	3.44	3.38	3.34	3.32	3.30	3.27	3.24	3.23
	0.975	8.07	6.54	5.89	5.52	5.29	5.12	4.99	4.90	4.82	4.76	4.57	4.47	4.36	4.31	4.28	4.25	4.21	4.16	4.14
	0.99	12.2	9.55	8.45	7.85	7.46	7.19	6.99	6.84	6.72	6.62	6.31	6.16	5.99	5.91	5.86	5.82	5.75	5.67	5.65
8	0.90	3.46	3.11	2.92	2.81	2.73	2.67	2.62	2.59	2.56	2.54	2.46	2.42	2.38	2.36	2.35	2.34	2.32	2.30	2.29
	0.95	5.32	4.46	4.07	3.84	3.69	3.58	3.50	3.44	3.39	3.35	3.22	3.15	3.08	3.04	3.02	3.01	2.97	2.94	2.93
	0.975	7.57	6.06	5.42	5.05	4.82	4.65	4.53	4.43	4.36	4.30	4.10	4.00	3.89	3.84	3.81	3.78	3.74	3.68	3.67
	0.99	11.3	8.65	7.59	7.01	6.63	6.37	6.18	6.03	5.91	5.81	5.52	5.36	5.20	5.12	5.07	5.03	4.96	4.88	4.86
9	0.90	3.36	3.01	2.81	2.69	2.61	2.55	2.51	2.47	2.44	2.42	2.34	2.30	2.25	2.23	2.22	2.21	2.19	2.17	2.16
	0.95	5.12	4.26	3.86	3.63	3.48	3.37	3.29	3.23	3.18	3.14	3.01	2.94	2.86	2.83	2.80	2.79	2.76	2.72	2.71
	0.975	7.21	5.71	5.08	4.72	4.48	4.32	4.20	4.10	4.03	3.96	3.77	3.67	3.56	3.51	3.47	3.45	3.40	3.35	3.33
	0.99	10.6	8.02	6.99	6.42	6.06	5.80	5.61	5.47	5.35	5.26	4.96	4.81	4.65	4.57	4.52	4.48	4.41	4.33	4.31
10	0.90	3.29	2.92	2.73	2.61	2.52	2.46	2.41	2.38	2.35	2.32	2.24	2.20	2.16	2.13	2.12	2.11	2.09	2.06	2.06
	0.95	4.96	4.10	3.71	3.48	3.33	3.22	3.14	3.07	3.02	2.98	2.85	2.77	2.70	2.66	2.64	2.62	2.59	2.55	2.54
	0.975	6.94	5.46	4.83	4.47	4.24	4.07	3.95	3.85	3.78	3.72	3.52	3.42	3.31	3.26	3.22	3.20	3.15	3.09	3.08
	0.99	10.0	7.56	6.55	5.99	5.64	5.39	5.20	5.06	4.94	4.85	4.56	4.41	4.25	4.17	4.12	4.08	4.01	3.93	3.91
11	0.90	3.23	2.86	2.66	2.54	2.45	2.39	2.34	2.30	2.27	2.25	2.17	2.12	2.08	2.05	2.04	2.03	2.01	1.98	1.97
	0.95	4.84	3.98	3.59	3.36	3.20	3.09	3.01	2.95	2.90	2.85	2.72	2.65	2.57	2.53	2.51	2.49	2.46	2.42	2.40
	0.975	6.72	5.26	4.63	4.28	4.04	3.88	3.76	3.66	3.59	3.53	3.33	3.23	3.12	3.06	3.03	3.00	2.96	2.90	2.88
	0.99	9.65	7.21	6.22	5.67	5.32	5.07	4.89	4.74	4.63	4.54	4.25	4.10	3.94	3.86	3.81	3.78	3.71	3.62	3.60
12	0.90	3.18	2.81	2.61	2.48	2.39	2.33	2.28	2.24	2.21	2.19	2.10	2.06	2.01	1.99	1.97	1.96	1.94	1.91	1.90
	0.95	4.75	3.89	3.49	3.26	3.11	3.00	2.91	2.85	2.80	2.75	2.62	2.54	2.47	2.43	2.40	2.38	2.35	2.31	2.30
	0.975	6.55	5.10	4.47	4.12	3.89	3.73	3.61	3.51	3.44	3.37	3.18	3.07	2.96	2.91	2.87	2.85	2.80	2.74	2.72
	0.99	9.33	6.93	5.95	5.41	5.06	4.82	4.64	4.50	4.39	4.30	4.01	3.86	3.70	3.62	3.57	3.54	3.47	3.38	3.36

n	$1-\alpha$	m : degrees of freedom in numerator																		
		1	2	3	4	5	6	7	8	9	10	15	20	30	40	50	60	100	500	∞
	0.90	3.14	2.76	2.56	2.43	2.35	2.28	2.23	2.20	2.16	2.14	2.05	2.01	1.96	1.93	1.92	1.90	1.88	1.85	1.85
13	0.95	4.67	3.81	3.41	3.18	3.03	2.92	2.83	2.77	2.71	2.67	2.53	2.46	2.38	2.34	2.31	2.30	2.26	2.22	2.21
	0.975	6.41	4.97	4.35	4.00	3.77	3.60	3.48	3.39	3.31	3.25	3.05	2.95	2.84	2.78	2.74	2.72	2.67	2.61	2.60
	0.99	9.07	6.70	5.74	5.21	4.86	4.62	4.44	4.30	4.19	4.10	3.82	3.66	3.51	3.43	3.38	3.34	3.27	3.19	3.17
	0.90	3.10	2.73	2.52	2.39	2.31	2.24	2.19	2.15	2.12	2.10	2.01	1.96	1.91	1.89	1.87	1.86	1.83	1.80	1.80
14	0.95	4.60	3.74	3.34	3.11	2.96	2.85	2.76	2.70	2.65	2.60	2.46	2.39	2.31	2.27	2.24	2.22	2.19	2.14	2.13
	0.975	6.30	4.86	4.24	3.89	3.66	3.50	3.38	3.29	3.21	3.15	2.95	2.84	2.73	2.67	2.64	2.61	2.56	2.50	2.49
	0.99	8.86	6.51	5.56	5.04	4.69	4.46	4.28	4.14	4.03	3.94	3.66	3.51	3.35	3.27	3.22	3.18	3.11	3.03	3.00
	0.90	3.07	2.70	2.49	2.36	2.27	2.21	2.16	2.12	2.09	2.06	1.97	1.92	1.87	1.85	1.83	1.82	1.79	1.76	1.76
15	0.95	4.54	3.68	3.29	3.06	2.90	2.79	2.71	2.64	2.59	2.54	2.40	2.33	2.25	2.20	2.18	2.16	2.12	2.08	2.07
	0.975	6.20	4.77	4.15	3.80	3.58	3.41	3.29	3.20	3.12	3.06	2.86	2.76	2.64	2.59	2.55	2.52	2.47	2.41	2.40
	0.99	8.68	6.36	5.42	4.89	4.56	4.32	4.14	4.00	3.89	3.80	3.52	3.37	3.21	3.13	3.08	3.05	2.98	2.89	2.87
	0.90	3.05	2.67	2.46	2.33	2.24	2.18	2.13	2.09	2.06	2.03	1.94	1.89	1.84	1.81	1.79	1.78	1.76	1.73	1.72
16	0.95	4.49	3.63	3.24	3.01	2.85	2.74	2.66	2.59	2.54	2.49	2.35	2.28	2.19	2.15	2.12	2.11	2.07	2.02	2.01
	0.975	6.12	4.69	4.08	3.73	3.50	3.34	3.22	3.12	3.05	2.99	2.79	2.68	2.57	2.51	2.47	2.45	2.40	2.33	2.32
	0.99	8.53	6.23	5.29	4.77	4.44	4.20	4.03	3.89	3.78	3.69	3.41	3.26	3.10	3.02	2.97	2.93	2.86	2.78	2.75
	0.90	3.03	2.64	2.44	2.31	2.22	2.15	2.10	2.06	2.03	2.00	1.91	1.86	1.81	1.78	1.76	1.75	1.73	1.69	1.69
17	0.95	4.45	3.59	3.20	2.96	2.81	2.70	2.61	2.55	2.49	2.45	2.31	2.23	2.15	2.10	2.08	2.06	2.02	1.97	1.96
	0.975	6.04	4.62	4.01	3.66	3.44	3.28	3.16	3.06	2.98	2.92	2.72	2.62	2.50	2.44	2.41	2.38	2.33	2.26	2.25
	0.99	8.40	6.11	5.18	4.67	4.34	4.10	3.93	3.79	3.68	3.59	3.31	3.16	3.00	2.92	2.87	2.83	2.76	2.68	2.65
	0.90	3.01	2.62	2.42	2.29	2.20	2.13	2.08	2.04	2.00	1.98	1.89	1.84	1.78	1.75	1.74	1.72	1.70	1.67	1.66
18	0.95	4.41	3.55	3.16	2.93	2.77	2.66	2.58	2.51	2.46	2.41	2.27	2.19	2.11	2.06	2.04	2.02	1.98	1.93	1.92
	0.975	5.98	4.56	3.95	3.61	3.38	3.22	3.10	3.01	2.93	2.87	2.67	2.56	2.44	2.38	2.35	2.32	2.27	2.20	2.19
	0.99	8.29	6.01	5.09	4.58	4.25	4.01	3.84	3.71	3.60	3.51	3.23	3.08	2.92	2.84	2.78	2.75	2.68	2.59	2.57

n	$1-\alpha$	m : degrees of freedom in numerator																		
		1	2	3	4	5	6	7	8	9	10	15	20	30	40	50	60	100	500	∞
	0.90	2.99	2.61	2.40	2.27	2.18	2.11	2.06	2.02	1.98	1.96	1.86	1.81	1.76	1.73	1.71	1.70	1.67	1.64	1.63
19	0.95	4.38	3.52	3.13	2.90	2.74	2.63	2.54	2.48	2.42	2.38	2.23	2.16	2.07	2.03	2.00	1.98	1.94	1.89	1.88
	0.975	5.92	4.51	3.90	3.56	3.33	3.17	3.05	2.96	2.88	2.82	2.62	2.51	2.39	2.33	2.30	2.27	2.22	2.15	2.13
	0.99	8.18	5.93	5.01	4.50	4.17	3.94	3.77	3.63	3.52	3.43	3.15	3.00	2.84	2.76	2.71	2.67	2.60	2.51	2.49
	0.90	2.97	2.59	2.38	2.25	2.16	2.09	2.04	2.00	1.96	1.94	1.84	1.79	1.74	1.71	1.69	1.68	1.65	1.62	1.61
20	0.95	4.35	3.49	3.10	2.87	2.71	2.60	2.51	2.45	2.39	2.35	2.20	2.12	2.04	1.99	1.97	1.95	1.91	1.86	1.84
	0.975	5.87	4.46	3.86	3.51	3.29	3.13	3.01	2.91	2.84	2.77	2.57	2.46	2.35	2.29	2.25	2.22	2.17	2.10	2.09
	0.99	8.10	5.85	4.94	4.43	4.10	3.87	3.70	3.56	3.46	3.37	3.09	2.94	2.78	2.69	2.64	2.61	2.54	2.44	2.42
	0.90	2.79	2.39	2.18	2.04	1.95	1.87	1.82	1.77	1.74	1.71	1.60	1.54	1.48	1.44	1.41	1.40	1.36	1.31	1.29
60	0.95	4.00	3.15	2.76	2.53	2.37	2.25	2.17	2.10	2.04	1.99	1.84	1.75	1.65	1.59	1.56	1.53	1.48	1.41	1.39
	0.975	5.29	3.93	3.34	3.01	2.79	2.63	2.51	2.41	2.33	2.27	2.06	1.94	1.82	1.74	1.70	1.67	1.60	1.51	1.48
	0.99	7.08	4.98	4.13	3.65	3.34	3.12	2.95	2.82	2.72	2.63	2.35	2.20	2.03	1.94	1.88	1.84	1.75	1.63	1.60
	0.90	2.75	2.35	2.13	1.99	1.90	1.82	1.77	1.72	1.68	1.65	1.55	1.48	1.41	1.37	1.34	1.32	1.28	1.21	1.19
120	0.95	3.92	3.07	2.68	2.45	2.29	2.17	2.09	2.02	1.96	1.91	1.75	1.66	1.55	1.50	1.46	1.43	1.37	1.28	1.25
	0.975	5.15	3.80	3.23	2.89	2.67	2.52	2.39	2.30	2.22	2.16	1.94	1.82	1.69	1.61	1.56	1.53	1.45	1.34	1.31
	0.99	6.85	4.79	3.95	3.48	3.17	2.96	2.79	2.66	2.56	2.47	2.19	2.03	1.86	1.76	1.70	1.66	1.56	1.42	1.38
	0.90	2.71	2.30	2.08	1.94	1.85	1.77	1.72	1.67	1.63	1.60	1.49	1.42	1.34	1.30	1.26	1.24	1.18	1.08	1.00
∞	0.95	3.84	3.00	2.60	2.37	2.21	2.10	2.01	1.94	1.88	1.83	1.67	1.57	1.46	1.39	1.35	1.32	1.24	1.11	1.00
	0.975	5.02	3.69	3.12	2.79	2.57	2.41	2.29	2.19	2.11	2.05	1.83	1.71	1.57	1.48	1.43	1.39	1.30	1.13	1.00
	0.99	6.63	4.61	3.78	3.32	3.02	2.80	2.64	2.51	2.41	2.32	2.04	1.88	1.70	1.59	1.52	1.47	1.36	1.15	1.00

Quantiles with the F-distribution. For example: if X enjoys the F-distribution with 10 and 20 degrees of freedom in the numerator and in the denominator respectively, then $\mathbb{P}(X \leq 2.35) = 0.95$. On the other hand, in such a case $1/X$ enjoys the F-distribution with 20 and 10 degrees of freedom respectively, which leads to $\mathbb{P}(X \geq 1/2.20) = 0.90$.

Table VI: Wilcoxon's signed-rank test

n	0.005	0.01	0.025	0.05	0.10
3	*	*	*	*	*
4	*	*	*	*	0
5	*	*	*	0	2
6	*	*	0	2	3
7	*	0	2	3	5
8	0	1	3	5	8
9	1	3	5	8	10
10	3	5	8	10	14
11	5	7	10	13	17
12	7	9	13	17	21
13	9	12	17	21	26
14	12	15	21	25	31
15	15	19	25	30	36
16	19	23	29	35	42
17	23	27	34	41	48
18	27	32	40	47	55
19	32	37	46	53	62
20	37	43	52	60	69
21	42	49	58	67	77
22	48	55	65	75	86
23	54	62	73	83	94
24	61	69	81	91	104
25	68	76	89	100	113
26	75	84	98	110	124

n	0.005	0.01	0.025	0.05	0.10
27	83	92	107	119	134
28	91	101	116	130	145
29	100	110	126	140	157
30	109	120	137	151	169
31	118	130	147	163	181
32	128	140	159	175	194
33	138	151	170	187	207
34	148	162	182	200	221
35	159	173	195	213	235
36	171	185	208	227	250
37	182	198	221	241	265
38	194	211	235	256	281
39	207	224	249	271	297
40	220	238	264	286	313
41	233	252	279	302	330
42	247	266	294	319	348
43	261	281	310	336	365
44	276	296	327	353	384
45	291	312	343	371	402
46	307	328	361	389	422
47	322	345	378	407	441
48	339	362	396	426	462
49	355	379	415	446	482
50	373	397	434	466	503

Left critical values with Wilcoxon's signed-rank test. For example: for a sample size of $n = 18$, significance level $\alpha = 0.05$ and left-sided alternative (the median of the population is $< m_0$), we accept H_0 ($m = m_0$) precisely if T^+ is > 47. If the alternative is right-sided we accept H_0 if $T^+ < \frac{1}{2} \cdot 18 \cdot 19 - 47$.

Table VII: Wilcoxon's rank-sum test

n	m	0.005	0.01	0.025	0.05	0.10	0.10	0.05	0.025	0.01	0.005
3	2	*	*	*	*	3	9	*	*	*	*
	3	*	*	*	6	7	14	15	*	*	*
4	2	*	*	*	*	3	11	*	*	*	*
	3	*	*	*	6	7	17	18	*	*	*
	4	*	*	10	11	13	23	25	26	*	*
5	2	*	*	*	3	4	12	13	*	*	*
	3	*	*	6	7	8	19	20	21	*	*
	4	*	10	11	12	14	26	28	29	30	*
	5	15	16	17	19	20	35	36	38	39	40
6	2	*	*	*	3	4	14	15	*	*	*
	3	*	*	7	8	9	21	22	23	*	*
	4	10	11	12	13	15	29	31	32	33	34
	5	16	17	18	20	22	38	40	42	43	44
	6	23	24	26	28	30	48	50	52	54	55
7	2	*	*	*	3	4	16	17	*	*	*
	3	*	6	7	8	10	23	25	26	27	*
	4	10	11	13	14	16	32	34	35	37	38
	5	16	18	20	21	23	42	44	45	47	49
	6	24	25	27	29	32	52	55	57	59	60
	7	32	34	36	39	41	64	66	69	71	73
8	2	*	*	3	4	5	17	18	19	*	*
	3	*	6	8	9	11	25	27	28	30	*
	4	11	12	14	15	17	35	37	38	40	41
	5	17	19	21	23	25	45	47	49	51	53
	6	25	27	29	31	34	56	59	61	63	65
	7	34	35	38	41	44	68	71	74	77	78
	8	43	45	49	51	55	81	85	87	91	93
9	1	*	*	*	*	1	10	*	*	*	*
	2	*	*	3	4	5	19	20	21	*	*
	3	6	7	8	10	11	28	29	31	32	33
	4	11	13	14	16	19	37	40	42	43	45
	5	18	20	22	24	27	48	51	53	55	57
	6	26	28	31	33	36	60	63	65	68	70
	7	35	37	40	43	46	73	76	79	82	84
	8	45	47	51	54	58	86	90	93	97	99
	9	56	59	62	66	70	101	105	109	112	115

n	m	0.005	0.01	0.025	0.05	0.10	0.10	0.05	0.025	0.01	0.005
10	1	*	*	*	*	1	11	*	*	*	*
	2	*	*	3	4	6	20	22	23	*	*
	3	6	7	9	10	12	30	32	33	35	36
	4	12	13	15	17	20	40	43	45	47	48
	5	19	21	23	26	28	52	54	57	59	61
	6	27	29	32	35	38	64	67	70	73	75
	7	37	39	42	45	49	77	81	84	87	89
	8	47	49	53	56	60	92	96	99	103	105
	9	58	61	65	69	73	107	111	115	119	122
	10	71	74	78	82	87	123	128	132	136	139
11	1	*	*	*	*	1	12	*	*	*	*
	2	*	*	3	4	6	22	24	25	*	*
	3	6	7	9	11	13	32	34	36	38	39
	4	12	14	16	18	21	43	46	48	50	52
	5	20	22	24	27	30	55	58	61	63	65
	6	28	30	34	37	40	68	71	74	78	80
	7	38	40	44	47	51	82	86	89	93	95
	8	49	51	55	59	63	97	101	105	109	111
	9	61	63	68	72	76	113	117	121	126	128
	10	73	77	81	86	91	129	134	139	143	147
	11	87	91	96	100	106	147	153	157	162	166
12	1	*	*	*	*	1	13	*	*	*	*
	2	*	*	4	5	7	23	25	26	*	*
	3	7	8	10	11	14	34	37	38	40	41
	4	13	15	17	19	22	46	49	51	53	55
	5	21	23	26	28	32	58	62	64	67	69
	6	30	32	35	38	42	72	76	79	82	84
	7	40	42	46	49	54	86	91	94	98	100
	8	51	53	58	62	66	102	106	110	115	117
	9	63	66	71	75	80	118	123	127	132	135
	10	76	79	84	89	94	136	141	146	151	154
	11	90	94	99	104	110	154	160	165	170	174
	12	105	109	115	120	127	173	180	185	191	195

Left and right critical values with Wilcoxon's rank-sum test. The critical values belong to the sum of the ranks of the *smallest* sample, say T_Y. For example: given sample sizes $n = 11$ and $m = 5$ and given $\alpha = 0.10$, in a two-sided test we accept H_0 (the medians of the two populations are the same) precisely if $27 < T_Y < 58$.

Table VIII: The runs test

		$1-\alpha$									
n	m	0.005	0.01	0.025	0.05	0.10	0.10	0.05	0.025	0.01	0.005
4	1	*	*	*	*	*	*	*	*	*	*
	2	*	*	*	*	*	*	*	*	*	*
	3	*	*	*	*	2	7	7	*	*	*
	4	*	*	*	2	2	8	8	*	*	*
5	1	*	*	*	*	*	*	*	*	*	*
	2	*	*	*	*	2	*	*	*	*	*
	3	*	*	*	2	2	7	*	*	*	*
	4	*	*	2	2	3	8	9	9	9	*
	5	*	2	2	3	3	9	9	10	10	*
6	1	*	*	*	*	*	*	*	*	*	*
	2	*	*	*	*	2	*	*	*	*	*
	3	*	*	2	2	2	*	*	*	*	*
	4	*	2	2	3	3	9	9	9	*	*
	5	2	2	3	3	3	9	10	10	11	11
	6	2	2	3	3	4	10	11	11	12	12
7	1	*	*	*	*	*	*	*	*	*	*
	2	*	*	*	*	2	*	*	*	*	*
	3	*	*	2	2	3	*	*	*	*	*
	4	*	2	2	3	3	9	9	*	*	*
	5	2	2	3	3	4	10	10	11	11	*
	6	2	3	3	4	4	11	11	12	12	13
	7	3	3	3	4	5	11	12	13	13	13
8	1	*	*	*	*	*	*	*	*	*	*
	2	*	*	*	2	2	*	*	*	*	*
	3	*	*	2	2	3	*	*	*	*	*
	4	2	2	3	3	3	9	*	*	*	*
	5	2	2	3	3	4	10	11	11	*	*
	6	3	3	3	4	5	11	12	12	13	13
	7	3	3	4	4	5	12	13	13	14	14
	8	3	4	4	5	5	13	13	14	14	15
9	1	*	*	*	*	*	*	*	*	*	*
	2	*	*	*	2	2	*	*	*	*	*
	3	*	2	2	2	3	*	*	*	*	*
	4	2	2	3	3	4	9	*	*	*	*
	5	2	3	3	4	4	10	11	*	*	*
	6	3	3	4	4	5	11	12	13	13	*

		$1-\alpha$									
n	m	0.005	0.01	0.025	0.05	0.10	0.10	0.05	0.025	0.01	0.005
9	7	3	4	4	5	5	12	13	14	14	15
	8	3	4	5	5	6	13	14	14	15	15
	9	4	4	5	6	6	14	14	15	16	16
10	2	*	*	*	2	2	*	*	*	*	*
	3	*	2	2	3	3	*	*	*	*	*
	4	2	2	3	3	4	*	*	*	*	*
	5	3	3	3	4	5	11	11	*	*	*
	6	3	3	4	5	5	12	12	13	*	*
	7	3	4	5	5	6	13	13	14	15	15
	8	4	4	5	6	6	13	14	15	15	16
	9	4	5	5	6	7	14	15	16	16	17
	10	5	5	6	6	7	15	16	16	17	17
11	2	*	*	*	2	2	*	*	*	*	*
	3	*	2	2	3	3	*	*	*	*	*
	4	2	2	3	3	4	*	*	*	*	*
	5	3	3	4	4	5	11	*	*	*	*
	6	3	4	4	5	5	12	13	13	*	*
	7	4	4	5	5	6	13	14	14	15	15
	8	4	5	5	6	7	14	15	15	16	16
	9	5	5	6	6	7	15	15	16	17	17
	10	5	5	6	7	8	15	16	17	18	18
	11	5	6	7	7	8	16	17	17	18	19
12	2	*	*	2	2	2	*	*	*	*	*
	3	2	2	2	3	3	*	*	*	*	*
	4	2	3	3	4	4	*	*	*	*	*
	5	3	3	4	4	5	11	*	*	*	*
	6	3	4	4	5	6	12	13	13	*	*
	7	4	4	5	6	6	13	14	14	15	*
	8	4	5	6	6	7	14	15	16	16	17
	9	5	5	6	7	7	15	16	16	17	18
	10	5	6	7	7	8	16	17	17	18	19
	11	6	6	7	8	9	16	17	18	19	19
	12	6	7	7	8	9	17	18	19	19	20

Critical values belonging to the runs test. For example: if we have two groups, consisting of 12 and 10 members respectively, then under H_0 the total number of chains T satisfies the inequality $\mathbb{P}(T \leq 6) \leq 0.01$, and 6 is the largest number for which this inequality is still correct.

Table IX: Spearman's rank correlation test

n	α 0.20	0.10	0.05	0.02	0.01
5	0.800	0.900	1.000	1.000	*
6	0.657	0.829	0.886	0.943	1.000
7	0.571	0.714	0.786	0.893	0.929
8	0.524	0.643	0.738	0.833	0.881
9	0.483	0.600	0.700	0.783	0.833
10	0.455	0.564	0.648	0.745	0.794
11	0.427	0.536	0.618	0.709	0.755
12	0.406	0.503	0.587	0.678	0.727
13	0.385	0.484	0.560	0.648	0.703
14	0.367	0.464	0.538	0.626	0.679
15	0.354	0.446	0.521	0.604	0.654

Critical values for Spearman's rank correlation coefficient. For example: for a sample size of $n = 13$ and $\alpha = 0.10$ we accept H_0 (no correlation) if and only if $|R_S| < 0.484$. The values are for two-sided tests.

Table X: Kendall's tau

n	α 0.20	0.10	0.05	0.02	0.01
4	6	6	*	*	*
5	8	8	10	10	*
6	9	11	13	13	15
7	11	13	15	17	19
8	12	16	18	20	22
9	14	18	20	24	26
10	17	21	23	27	29
11	19	23	27	31	33
12	20	26	30	36	38
13	24	28	34	40	44
14	25	33	37	43	47
15	29	35	41	49	53
16	30	38	46	52	58
17	34	42	50	58	64
18	37	45	53	63	69
19	39	49	57	67	75
20	42	52	62	72	80

n	α 0.20	0.10	0.05	0.02	0.01
21	44	56	66	78	86
22	47	61	71	83	91
23	51	65	75	89	99
24	54	68	80	94	104
25	58	72	86	100	110
26	61	77	91	107	117
27	63	81	95	113	125
28	68	86	100	118	130
29	70	90	106	126	138
30	75	95	111	131	145
31	77	99	117	137	151
32	82	104	122	144	160
33	86	108	128	152	166
34	89	113	133	157	175
35	93	117	139	165	181
40	112	144	170	200	222

Critical values for the sum of concordances with Kendall's tau. Two-sided: for $n = 19$ and $\alpha = 0.10$ we accept H_0 (no correlation) if $(\frac{1}{2} \cdot 19 \cdot 18)|R_K| < 49$.

Table XI: The Kolmogorov–Smirnov test

	α			
n	0.10	0.05	0.02	0.01
1	0.9500	0.9750	0.9900	0.9950
2	0.7764	0.8419	0.9000	0.9293
3	0.6360	0.7076	0.7846	0.8290
4	0.5652	0.6239	0.6889	0.7342
5	0.5094	0.5633	0.6272	0.6685
6	0.4680	0.5193	0.5774	0.6166
7	0.4361	0.4834	0.5384	0.5758
8	0.4096	0.4543	0.5065	0.5418
9	0.3875	0.4300	0.4796	0.5133
10	0.3687	0.4092	0.4566	0.4889
11	0.3524	0.3912	0.4367	0.4677
12	0.3381	0.3754	0.4192	0.4490
13	0.3255	0.3614	0.4036	0.4325
14	0.3142	0.3489	0.3897	0.4176
15	0.3040	0.3376	0.3771	0.4042
16	0.2947	0.3273	0.3657	0.3920
17	0.2863	0.3180	0.3553	0.3809
18	0.2785	0.3094	0.3457	0.3706
19	0.2714	0.3014	0.3369	0.3612
20	0.2647	0.2941	0.3287	0.3524
21	0.2586	0.2872	0.3210	0.3443
22	0.2528	0.2809	0.3139	0.3367
23	0.2475	0.2749	0.3073	0.3295
24	0.2424	0.2693	0.3010	0.3229
25	0.2377	0.2640	0.2952	0.3166
26	0.2332	0.2591	0.2896	0.3106
27	0.2290	0.2544	0.2844	0.3050
28	0.2250	0.2499	0.2794	0.2997
29	0.2212	0.2457	0.2747	0.2947
30	0.2176	0.2417	0.2702	0.2898
31	0.2141	0.2379	0.2659	0.2868
32	0.2108	0.2342	0.2619	0.2808
33	0.2077	0.2308	0.2580	0.2768
34	0.2047	0.2274	0.2543	0.2728
35	0.2018	0.2242	0.2507	0.2686
36	0.1991	0.2212	0.2474	0.2654
37	0.1965	0.2183	0.2439	0.2621
38	0.1939	0.2154	0.2411	0.2585
39	0.1915	0.2127	0.2379	0.2562
40	0.1891	0.2101	0.2349	0.2522
41	0.1869	0.2076	0.2321	0.2492
42	0.1847	0.2052	0.2296	0.2459
43	0.1826	0.2028	0.2266	0.2440
44	0.1805	0.2006	0.2240	0.2422
45	0.1786	0.1983	0.2421	0.2394
46	0.2359	0.2360	0.2676	0.2450
47	0.1748	0.1946	0.2312	0.2312
48	0.1730	0.2246	0.2045	0.2139
49	0.1718	0.1897	0.1874	0.1990
50	0.1693	0.1908	0.1865	0.1889

Quantiles belonging to the Kolmogorov–Smirnov test statistic. For example: if we denote this test statistic by D_n, we have: $\mathbb{P}(D_{20} \leq 0.2941) = 0.95$.